Statistical Field Theory

GIORGIO PARISI
UNIVERSITY OF ROME "TOR VERGATA"

ADDISON-WESLEY PUBLISHING COMPANY
THE ADVANCED BOOK PROGRAM

Redwood City, California • Menlo Park, California
Reading, Massachusetts • New York • Amsterdam
Don Mills, Ontario • Sydney • Bonn • Madrid
Singapore • Tokyo • San Juan
Wokingham, United Kingdom

Publisher: Allan M. Wylde
Production Administrator: Lorraine Ferrier
Editorial Coordinator: Pearline Randall
Promotions Manager: Celina Gonzales

To the memory of my friend Kurt Symanzik, who so greatly contributed to the subject of this book

Copyright © 1988 by Addison-Wesley Publishing Company, Inc.

All rights reserved. No part of this publication may be reproduced, stored in a retrieval system, or transmitted in any form or by any means, electronic, mechanical, photocopying, recording, or otherwise, without the prior written permission of the publisher. Printed in the United States of America. Published simultaneously in Canada.

Library of Congress Cataloging-in-Publication Data

Parisi, Giorgio.
  Statistical field theory.

  Includes index.
  1. Quantum field theory. 2. Statistical mechanics.
  I. Title.
  QC174.45.P36 1987   530.1′4   86-18839
  ISBN 0-201-05985-1

ABCDEFGHIJK-AL-898

Frontiers in Physics

DAVID PINES / EDITOR

Volumes of the Series published from 1961 to 1973 are not officially numbered. The parenthetical numbers shown are designed to aid librarians and bibliographers to check the completeness of their holdings.

Titles published in this series prior to 1987 appear under either the W. A. Benjamin or the Benjamin/Cummings imprint; titles published since 1986 appear under the Addison-Wesley imprint.

(1)	N. Bloembergen	Nuclear Magnetic Relaxation: A Reprint Volume, 1961
(2)	G. F. Chew	S-Matrix Theory of Strong Interactions: A Lecture Note and Reprint Volume, 1961
(3)	R. P. Feynman	Quantum Electrodynamics: A Lecture Note and Reprint Volume, 1961
(4)	R. P. Feynman	The Theory of Fundamental Processes: A Lecture Note Volume, 1961
(5)	L. Van Hove, N. M. Hugenholtz, and L. P. Howland	Problems in Quantum Theory of Many-Particle Systems: A Lecture Note and Reprint Volume, 1961
(6)	D. Pines	The Many-Body Problem: A Lecture Note and Reprint Volume, 1961
(7)	H. Frauenfelder	The Mössbauer Effect: A Review—with a Collection of Reprints, 1962
(8)	L. P. Kadanoff G. Baym	Quantum Statistical Mechanics: Green's Function Methods in Equilibrium and Nonequilibrium Problems, 1962
(9)	G. E. Pake	Paramagnetic Resonance: An Introductory Monograph, 1962 [cr. (42)—2nd edition]
(10)	P. W. Anderson	Concepts in Solids: Lectures on the Theory of Solids, 1963
(11)	S. C. Frautschi	Regge Poles and S-Matrix Theory, 1963
(12)	R. Hofstadter	Electron Scattering and Nuclear and Nucleon Structure: A Collection of Reprints with an Introduction, 1963
(13)	A. M. Lane	Nuclear Theory: Pairing Force Correlations to Collective Motion, 1964
(14)	R. Omnès M. Froissart	Mandelstam Theory and Regge Poles: An Introduction for Experimentalists, 1963
(15)	E. J. Squires	Complex Angular Momenta and Particle Physics: A Lecture Note and Reprint Volume, 1963
(16)	H. L. Frisch J. L. Lebowitz	The Equilibrium Theory of Classical Fluids: A Lecture Note and Reprint Volume, 1964

(17)	M. Gell-Mann Y. Ne'eman	The Eightfold Way: (A Review—with a Collection of Reprints), 1964
(18)	M. Jacob G. F. Chew	Strong-Interaction Physics: A Lecture Note Volume, 1964
(19)	P. Nozières	Theory of Interacting Fermi Systems, 1964
(20)	J. R. Schrieffer	Theory of Superconductivity, 1964 (revised 3rd printing, 1983)
(21)	N. Bloembergen	Nonlinear Optics: A Lecture Note and Reprint Volume, 1965
(22)	R. Brout	Phase Transitions, 1965
(23)	I. M. Khalatnikov	An Introduction to the Theory of Superfluidity, 1965
(24)	P. G. deGennes	Superconductivity of Metals and Alloys, 1966
(25)	W. A. Harrison	Pseudopotentials in the Theory of Metals, 1966
(26)	V. Barger D. Cline	Phenomenological Theories of High Energy Scattering: An Experimental Evaluation, 1967
(27)	P. Choquàrd	The Anharmonic Crystal, 1967
(28)	T. Loucks	Augmented Plane Wave Method: A Guide to Performing Electronic Structure Calculations—A Lecture Note and Reprint Volume, 1967
(29)	Y. Ne'eman	Algebraic Theory of Particle Physics: Hadron Dynamics in Terms of Unitary Spin Currents, 1967
(30)	S. L. Adler R. F. Dashen	Current Algebras and Applications to Particle Physics, 1968
(31)	A. B. Migdal	Nuclear Theory: The Quasiparticle Method, 1968
(32)	J. J. J. Kokkedee	The Quark Model, 1969
(33)	A. B. Migdal V. Krainov	Approximation Methods in Quantum Mechanics, 1969
(34)	R. Z. Sagdeev and A. A. Galeev	Nonlinear Plasma Theory, 1969
(35)	J. Schwinger	Quantum Kinematics and Dynamics, 1970
(36)	R. P. Feynman	Statistical Mechanics: A Set of Lectures, 1972
(37)	R. P. Feynman	Photo-Hadron Interactions, 1972
(38)	E. R. Caianiello	Combinatorics and Renormalization in Quantum Field Theory, 1973
(39)	G. B. Field, H. Arp, and J. N. Bahcall	The Redshift Controversy, 1973
(40)	D. Horn F. Zachariasen	Hadron Physics at Very High Energies, 1973
(41)	S. Ichimaru	Basic Principles of Plasma Physics: A Statistical Approach, 1973 (2nd printing, with revisions, 1980)
(42)	G. E. Pake T. L. Estle	The Physical Principles of Electron Paramagnetic Resonance, 2nd Edition, completely revised, enlarged, and reset, 1973 [cf. (9)—1st edition]

Volumes published from 1974 onward are being numbered as an integral part of the bibliography.

43	R. C. Davidson	Theory of Nonneutral Plasmas, 1974
44	S. Doniach E. H. Sondheimer	Green's Functions for Solid State Physicists, 1974
45	P. H. Frampton	Dual Resonance Models, 1974
46	S. K. Ma	Modern Theory of Critical Phenomena, 1976
47	D. Forster	Hydrodynamic Fluctuations, Broken Symmetry, and Correlation Functions, 1975
48	A. B. Migdal	Qualitative Methods in Quantum Theory, 1977
49	S. W. Lovesey	Condensed Matter Physics: Dynamic Correlations, 1980
50	L. D. Faddeev A. A. Slavnov	Gauge Fields: Introduction to Quantum Theory, 1980
51	P. Ramond	Field Theory: A Modern Primer, 1981 [cf. 74—2nd edition]
52	R. A. Broglia A. Winther	Heavy Ion Reactions: Lecture Notes Vol. I: Elastic and Inelastic Reactions, 1981
53	R. A. Broglia A. Winther	Heavy Ion Reactions: Lecture Notes Vol. II, *in preparation*
54	H. Georgi	Lie Algebras in Particle Physics: From Isospin to Unified Theories, 1982
55	P. W. Anderson	Basic Notions of Condensed Matter Physics, 1983
56	C. Quigg	Gauge Theories of the Strong, Weak, and Electromagnetic Interactions, 1983
57	S. I. Pekar	Crystal Optics and Additional Light Waves, 1983
58	S. J. Gates, M. T. Grisaru, M. Roček, and W. Siegel	Superspace *or* One Thousand and One Lessons in Supersymmetry, 1983
59	R. N. Cahn	Semi-Simple Lie Algebras and Their Representations, 1984
60	G. G. Ross	Grand Unified Theories, 1984
61	S. W. Lovesey	Condensed Matter Physics: Dynamic Correlations, 2nd Edition, 1986
62	P. H. Frampton	Gauge Field Theories, 1986
63	J. I. Katz	High Energy Astrophysics, 1987
64	T. J. Ferbel	Experimental Techniques in High Energy Physics, 1987
65	T. Appelquist, A. Chodos, and P. G. O. Freund	Modern Kaluza-Klein Theories, 1987
66	G. Parisi	Statistical Field Theory, 1988
67	R. C. Richardson E. N. Smith	Techniques in Low-Temperature Condensed Matter Physics, 1988

68	J. W. Negele H. Orland	Quantum Many-Particle Systems, 1987
69	E. Kolb M. Turner	The Early Universe, 1988
70	E. Kolb M. Turner	The Early Universe: Reprints, 1988
71	V. Barger R. J. N. Phillips	Collider Physics, 1987
72	T. Tajima	Computational Plasma Physics, 1988
73	W. Kruer	The Physics of Laser Plasma Interactions, 1987

Editor's Foreword

The problem of communicating in a coherent fashion recent developments in the most exciting and active fields of physics continues to be with us. The enormous growth in the number of physicists has tended to make the familiar channels of communication considerably less effective. It has become increasingly difficult for experts in a given field to keep up with the current literature; the novice can only be confused. What is needed is both a consistent account of a field and the presentation of a definite "point of view" concerning it. Formal monographs cannot meet such a need in a rapidly developing field, while the review article seems to have fallen into disfavor. Indeed, it would seem that the people most actively engaged in developing a given field are the people least likely to write at length about it.

FRONTIERS IN PHYSICS was conceived in 1961 in an effort to improve the situation in several ways. Leading physicists frequently give a series of lectures, a graduate seminar, or a graduate course in their special fields of interest. Such lectures serve to summarize the present status of a rapidly developing field and may well constitute the only coherent account available at the time. Often, notes on lectures exist (prepared by the lecturer himself, by graduate students, or by postdoctoral fellows) and are distributed in mimeographed form on a limited basis. One of the principal purposes of the FRONTIERS IN PHYSICS Series is to make such notes available to a wider audience of physicists.

It should be emphasized that lecture notes are necessarily rough and informal, both in style and content; and those in the series will prove no exception. This is as it should be. One point of the series is to offer new, rapid, more informal, and, it is hoped, more effective ways for physicists to teach one another. The point is lost if only elegant notes qualify.

As FRONTIERS IN PHYSICS has evolved, a third category of book, the informal text/monograph, an intermediate step between lecture notes and formal texts or monographs, has played an increasingly important role in the series. In an informal text or monograph an author has reworked his or her lecture notes to the point at which the manuscript represents a coherent summation of a newly-developed field, complete with references and problems, suitable for either classroom teaching or individual study.

The past decade has witnessed the emergence of certain powerful unifying ideas in physics, ideas that have lead theorists working on problems in particle physics and those working in condensed matter physics to the realization that they share a significant common intellectual ground. That common ground, which goes far beyond the utilization of field theoretic techniques, is the subject matter of *Statistical Field Theory*. As a theoretical physicist who has made significant seminal contributions in both condensed matter physics and particle physics, Giorgio Parisi is unusually well qualified to write on this topic. He provides the reader with both a lucid introduction and a current overview of our present understanding of problems in statistical field theory; his book can be read with profit by physicists at every stage of development, from the beginning graduate student to the experienced researcher.

David Pines

Urbana, Illinois
November, 1987

Preface

In the academic year 1979–1980 I taught a course in theoretical physics for the fourth-year undergraduate students at Rome University. My aim was to convey the functional integral approach to quantum field theory. I knew, however, from previous experience, that functional methods look very abstract, especially to a beginner.

After some reflection I decided to start from traditional statistical physics and to arrive at the functional representation of Euclidean field theory via the Ising model, the Ginsburg-Landau model, and the physics of second-order phase transitions. At this point the functional integral tool would be in our hands and we could proceed easily; I derived the Feynman path-integral representation for quantum mechanics first for imaginary and later for real time: it was now straightforward to obtain the functional integral representation for relativistic quantum field theory.

In my original project I meant to continue by discussing those properties peculiar to a real-time quantum field (e.g., second quantization, Fock space, scattering theory). However, the academic year came to an end and (fortunately for the students) this last material had to be omitted from the course.

When the students came to me asking what they should read to prepare for the examination, I realized that, while there are plenty of books on relativistic quantum field theory (many of which are excellent), the material corresponding to my course was scattered throughout many books and review articles; I was finally compelled by the students to write down some lecture notes (in Italian) which formed the skeleton of this book.

When I started to expand these lecture notes into a book, I saw that my original presentation, although self-contained, was merely an introduction to the subject: the possible connections with work in related fields and the most recent developments were missing. I was thus faced with a dilemma: either do without any additional materials or discuss these other subjects. In the first case the book would have too narrow a point of view, while in the second case it could easily expand to one thousand pages. The final choice was a compromise: I have presented the various new subjects only briefly, trying to give the reader a feeling for the achievements and for the unresolved problems. The book in its present version is part introduction and part review.

I am grateful to the students in my course at Rome University and in particular to F. Fucito and E. Marinari, who took beautiful notes that were quite useful to me in reconstructing what I had said. It is a pleasure for me to thank all the people who had the patience to read the manuscript and/or to offer suggestions at various stages of this work. In particular I should like to thank G. Ciccotti, M. d'Eramo, H. Hamber, J. Kogut, G. Martinelli, A. Sagnotti, N. Sourlas, and Zhang Yi-cheng. I have also benefited from the stimulating atmosphere of the IHES, where part of this book was written.

Rome G.P.

Contents

Preface ix

CHAPTER 1
CLASSICAL EQUILIBRIUM STATISTICAL MECHANICS 1

1.1 The basic axiom 1
1.2 The entropy 3
1.3 The absolute temperature 7
Notes for Chapter 1 13

CHAPTER 2
MAGNETIC SYSTEMS 10

2.1 General properties 10
2.2 Spontaneous symmetry breaking 13
Appendix to Chapter 2 18
Notes for Chapter 2 19

CHAPTER 3
THE ISING MODEL 22

3.1 The model 22
3.2 The mean-field approximation 24
3.3 A soluble model: weak long-range forces 31
3.4 The correlation functions 33
3.5 Second-order phase transitions 39

xi

3.6 The infinite-volume limit 41

Notes for Chapter 3 43

CHAPTER 4
THE LOW-TEMPERATURE AND HIGH-TEMPERATURE EXPANSIONS 46

4.1 The low-temperature expansion 46

4.2 The Gaussian model 49

4.3 The random walk 54

4.4 The high-temperature expansion of the Gaussian model 56

4.5 The high-temperature expansion of the Ising model 59

4.6 The free energy of the two-dimensional Ising problem 61

4.7 Self-duality 65

Notes for Chapter 4 66

CHAPTER 5
THE LANDAU-GINSBURG MODEL 68

5.1 The model 68

5.2 The Gaussian model again 70

5.3 On the zero lattice spacing limit 73

5.4 The perturbative expansion 75

5.5 The removal of ultraviolet divergences 83

5.6 A constructive approach 85

5.7 The Borel transform 87

5.8 The singularities of the Borel transform 90

Appendix to Chapter 5 92

Notes for Chapter 5 101

CHAPTER 6
NEAR THE TRANSITION 105

6.1 Infrared divergences 105

6.2 The Hartree-Fock approximation 106

6.3 The large-N expansion 108

Notes for Chapter 6 111

CHAPTER 7
THE RENORMALIZATION GROUP 112

7.1 Block variables 112
7.2 The recursion relation 117
7.3 Some mathematical preliminaries 118
7.4 The scaling laws 121
7.5 Approximations 126
7.6 The Migdal-Kadanoff approximation 131
7.7 The hierarchical model 133
Notes for Chapter 7 136

CHAPTER 8
PERTURBATIVE EVALUATION OF THE CRITICAL EXPONENTS 140

8.1 The basic definitions 140
8.2 A simple example 144
8.3 A more sophisticated computation 146
8.4 On the scaling laws 150
8.5 The fully renormalized expansion 154
Notes for Chapter 8 160

CHAPTER 9
NEAR FOUR DIMENSIONS 163

9.1 Noninteger dimensions 163
9.2 At four dimensions 165
9.3 The massless theory 169
9.4 The ϵ expansion 174
9.5 The range of the renormalized coupling 176
Notes for Chapter 9 179

CHAPTER 10
ON SPONTANEOUS SYMMETRY BREAKING 183

10.1 Perturbation theory 183
10.2 No-go theorems 187

10.3 Goldstone bosons 192

10.4 Near two dimensions 195

10.5 On metastability 201

Notes for Chapter 10 205

CHAPTER 11
OTHER MODELS 209

11.1 The Ising model again 209

11.2 The real gas 213

11.3 On universality 215

Notes for Chapter 11 219

CHAPTER 12
THE TRANSFER MATRIX 221

12.1 One-dimensional models 221

12.2 A few examples 225

12.3 The zero lattice spacing limit 229

12.4 Higher dimensions 232

Notes for Chapter 12 232

CHAPTER 13
PATH INTEGRALS FOR QUANTUM MECHANICS 234

13.1 Quantum mechanics 234

13.2 Path integrals 236

13.3 Equation of motion 241

13.4 An example: the harmonic oscillator 243

13.5 Path integrals in phase space 248

Appendix to Chapter 13 249

Notes for Chapter 13 254

CHAPTER 14
SEMICLASSICAL METHODS 257

14.1 The method of the stationary phase 257

14.2 The classical limit 258

14.3 The density of eigenvalues 260

14.4 The WKB method and its generalization 264

14.5 Trace identities 268

14.6 Tunneling effects 271

14.7 Fermions 275

Notes for Chapter 14 280

CHAPTER 15
RELATIVISTIC QUANTUM FIELD THEORY 283

15.1 General properties 283

15.2 The Osterwalder-Schrader condition 286

15.3 The particle interpretation 290

Notes for Chapter 15 291

CHAPTER 16
PARTICLE-FIELD DUALITY 292

16.1 The free case 292

16.2 The φ^4 interaction 294

16.3 Rigorous considerations 297

Notes for Chapter 16 298

CHAPTER 17
TIME-DEPENDENT CORRELATIONS 300

17.1 Quantum-statistical mechanics 300

17.2 The quantum fluctuation-dissipation theorem 303

17.3 The classical KMS condition 306

17.4 A classical example 309

17.5 A quantum example 311

Notes for Chapter 17 314

CHAPTER 18
THE APPROACH TO EQUILIBRIUM 316

18.1 The microcanonical ensemble 316

18.2 The KAM theorem 321

Notes for Chapter 18 324

CHAPTER 19
THE STOCHASTIC APPROACH 325

19.1 Brownian motion 325

19.2 The Langevin and Fokker-Planck equations 329

19.3 An example 332

19.4 Nelson's quantum mechanics 334

Notes for Chapter 19 336

CHAPTER 20
COMPUTER SIMULATION 338

20.1 Molecular dynamics 338

20.2 The Monte Carlo method 340

Notes for Chapter 20 347

Index 351

CHAPTER 1

Classical Equilibrium Statistical Mechanics

1.1. The basic axiom

The main aim of statistical mechanics is to derive the thermodynamic properties of macroscopic bodies starting from a description of the motion of the microscopic components (e.g., atoms, electrons, etc.). This would be an impossible and hopeless task if one took the normal approach of mechanics (given the Hamiltonian and the initial conditions, to compute the trajectories), since the number of degrees of freedom is so huge (see, however, Chapter 20): probabilistic methods are mandatory.

The problem can be divided into two parts: (a) Find the probability distribution of the microscopic components in thermal equilibrium, (i.e., after a sufficiently long time). (b) Compute the macroscopic properties of the system from the microscopic probability distribution.

We perform step (a) simply by assuming that the equilibrium probability distribution is given by the Boltzmann formula [see Eq. (1.5)]. While we shall try to justify this hypothesis in the last part of the book, in this chapter we discuss only some general consequences of this hypothesis, most notably with reference to elementary thermodynamics. Step (b) will be the main subject of this book. The reader already familiar with the principles of equilibrium statistical mechanics may prefer to skip this chapter.

Let us consider a classical Hamiltonian system with $2N$ degrees of freedom in a box of volume V: the coordinates q_i and the momenta p_i satisfy the classical equations of motion.

$$\dot{q}_i = \frac{\partial H}{\partial p_i}, \qquad \dot{p}_i = -\frac{\partial H}{\partial q_i}, \qquad (1.1)$$

where H is the Hamiltonian. Let us consider an observable $A(q, p)$ (we use q and p without labels to indicate the set of all q's and p's), i.e., an arbitrary measurable quantity. The equilibrium average of A ($\bar{A}$) is defined by

$$\bar{A} = \lim_{t \to \infty} \frac{1}{t} \int_0^t A[q(t'), p(t')] \, dt'$$

$$= \int dq \, dp \, A(q, p) P_{eq}(q, p) \,. \tag{1.2}$$

Equation (1.2) defines the equilibrium probability P_{eq}. It is generally believed that for macroscopic systems ($N \to \infty$ at fixed density $\rho \equiv N/V$) the expectation values that have a good limit when $N \to \infty$, i.e., intensive quantitives in the thermodynamic language, such as the kinetic energy per particle, satisfy the following relation in most cases:

$$A[q(t'), p(t')] \simeq \bar{A} + 0(N^{-1/2}) \,, \tag{1.3}$$

if the observation time t' is large enough (so that, for example, thermal diffusion will have eliminated all the initial differences of temperature between different parts of the system).[1]

The function $P_{eq}(q, p)$ is never negative and satisfies the normalization condition

$$\int dq \, dp \, P_{eq}(q, p) = 1 \,. \tag{1.4}$$

The fundamental hypothesis of equilibrium statistical mechanics is that P_{eq} follows the canonical distribution, i.e.

$$P_{eq}(q, p) = \frac{\exp[-\beta H(q, p)]}{Z} \,, \tag{1.5}$$

where Z (the so-called partition function) is fixed by the condition (1.4),

$$Z = \int dq \, dp \, \exp[-\beta H(q, p)] \,. \tag{1.6}$$

As we shall see later, β is related to the absolute temperature T by

$$\beta = \frac{1}{kT} \,. \tag{1.7}$$

The Boltzmann coefficient k relates the empirical scales of temperature and energy; experimentally it is about 1.38×10^{-16} erg/degree. If we use the same units for both temperature and energy, k can be set equal to 1, as will be done in most of this book.

It is evident that Eq. (1.5) cannot be true for a finite system: energy is conserved by the equations of motion (1.1). For a given initial condition the energy can take only one value $[(P_{eq}(q, p) = \delta(H - E)\tilde{P}_{eq}(q, p)]$, while in Eq. (1.5) configurations with different energies have a nonzero probability. Equation (1.5) can be correct only in the infinite-volume limit $(N \to \infty)$; indeed we shall argue later that the microcanonical distribution $(\tilde{P}_{eq}(q, p) = \text{const.})$ coincides in the infinite-volume limit with the canonical distribution. The reasons for this hope are presented later in the book. However, it is possible to give a physical meaning to Eq. (1.5) for a finite-volume system as well. Let us divide a very large system into two parts and call C_1 and C_2 the generalized coordinates of the first and second parts, respectively. We suppose that the total Hamiltonian can be decomposed as

$$H(C_1, C_2) = H_1(C_1) + H_2(C_2) + \varepsilon H_{12}(C_1, C_2), \qquad (1.8)$$

where ε is a small number.

We now consider the case in which the size of system 2 goes to infinity while the size of system 1 remains constant. In this limit, as long as ε remains different from zero, the large-time probability distribution of the whole system is

$$P_{eq}(C_1, C_2) \propto \exp[-\beta H_1(C_1) - \beta H_2(C_2) - \varepsilon \beta H_{12}(C_1, C_2)]. \qquad (1.9)$$

If we now send ε to zero but keep the observation time very large (in some cases an explicit computation shows that the time to reach equilibrium for the small system is of order ε^{-2}; see Chapter 18), the probability distribution (1.9) factors into the product of two independent probability distributions for each of the two subsystems.

Equation (1.5) can therefore be justified for finite systems by considering system 1 to be a part (weakly coupled to the rest) of a nearly infinite system. In other words, the small system is in thermal contact with a thermal reservoir. It is rather important to note that two systems in thermal contact have a distribution probability characterized by the same β; from elementary thermodynamics these two systems must have the same temperature. It is thus consistent to assume that there is a universal (system-independent) relation between β and the temperature. We shall see in section 1.3 that β^{-1} is proportional to the absolute temperature.

1.2. The entropy

We argued earlier that the probability distribution (1.5) describes a Hamiltonian system at large times; we can now generalize Eq. (1.5). We consider a set of configurations C, with a measure $d\mu(C)$ and a "Hamiltonian" $H(C)$. The "equilibrium" probability distribution is defined by

$$dP_\beta(C) = \frac{d\mu(C)\exp[-\beta H(C)]}{Z(\beta)} = \int d\mu(C) P_\beta(C)$$

$$Z(\beta) = \int d\mu(C) \exp[-\beta H(C)] \qquad (1.10)$$

$$\langle A \rangle_\beta = \int d\mu(C) A(C) P_\beta(C) .$$

In this case the configuration space is arbitrary: We are not restricted as in Eq. (1.5) to spaces having a Hamiltonian structure, i.e., N coordinates and N momenta. In many of the applications the configuration space will be discrete.

We should like to find a characterization of the canonical distribution (1.10) that will distinguish it from all other possible probability distributions. It is convenient to introduce the entropy of a distribution $(S[P])$, defined as follows:

$$S[P] = -\int d\mu(C) P(C) \ln P(C) = -\langle \ln P(C) \rangle . \qquad (1.11)$$

The more ordered the system (i.e., the more concentrated the probability distribution in a restricted region of phase space), the smaller is the entropy; the more disordered the system (i.e., the more uniform the probability distribution), the larger is the entropy.

In order to better understand the definition of the entropy of a probability distribution, let us consider the case of a discrete configuration space, whose different configurations are labeled by an index j ($j = 1, \ldots, M$ where M can also be infinite). Equation (1.10) now becomes

$$P_j = \frac{\exp[-\beta H_j]}{Z} \qquad Z = \sum_j^M \exp[-\beta H_j]$$

$$\langle A \rangle = \sum_j^M P_j A_j \qquad S[P] = -\sum_j^M P_j \ln P_j . \qquad (1.12)$$

A true Hamiltonian system can be reasonably well approximated by Eq. (1.12) if we divide the phase space into cells of size ε^{2N} and set H equal to the average Hamiltonian in the jth cell. In the limit $\varepsilon \to 0$ we recover Eq. (1.5).

Notice that the entropy defined by Eq. (1.12) is always nonnegative ($P_j \leq 1 \,\forall j$), while that defined in Eq. (1.11) may have any sign.[2] The entropy computed for a continuous d-dimensional system by using Eq.

1.2. The Entropy

(1.11) (S_{con}) and the entropy computed by dividing the phase space into cells of size ε^d (S_{dis}) are related by a trivial normalization factor when ε is very small:[3]

$$S_{dis} \simeq S_{con} - d \ln \varepsilon . \qquad (1.13)$$

We shall see later that the quantum entropy for a system of $2N$ degrees of freedom is well approximated by the entropy of the classical system with ε equal to the Planck constant.

Let us compute the entropy for a very simple example. The system has M possible configurations, and only L configurations (let us say the first L) have a nonzero equal probability:

$$\begin{aligned} P_j &= \frac{1}{L} \quad j \leq L \\ P_j &= 0 \quad j > L . \end{aligned} \qquad (1.14)$$

The corresponding entropy is $\ln L$. From this example we can draw the following general conclusions: If $S[P] = 0$, only one configuration has nonzero probability; if $S[P] = \ln M$, all configurations have the same probability; roughly speaking, we could say that $\exp[S[P]]$ is the number of configurations in which the system may stay with a not-too-small probability. This last statement can be sharpened by using some concepts from information theory.[4]

Let us consider a system that can stay in 2^k configurations. If we want to communicate which configuration the system is in, we need to transmit a number between 1 and 2^k, which can be coded as a sequence of k digits that may be 0 or 1: by definition the message contains k bits of information. Now if we have N copies of the same system, we need Nk bits to transmit the information as to which configuration these N systems occupy.

Let us suppose that the probability distribution of these N systems is known; in this case we already have information on the systems, and it is reasonable to assume that we need fewer than Nk bits to know the configurations of the N systems. Unfortunately this last statement is not so clear. Our information is probabilistic; if the N systems stay in a rather unlikely configuration and the message has been coded using fewer than Nk bits, the message may be interpreted in the wrong way.

A precise way to formulate the problem is the following: If we transmit a message describing the configurations of the N systems, what is the minimum number of bits (B_N) that we must communicate to ensure that the probability of misinterpretation of the message (after decoding) goes to zero when N goes to infinity? The answer is given by Shannon's theorem, and it is very simple:

$$\frac{B_N}{N} \xrightarrow[N\to\infty]{} \frac{S[P]}{\ln 2} . \tag{1.15}$$

We see that the way in which the message is coded may depend on the distribution probability P. This theorem tells us that in order to know the configurations of N systems (for large N) we need the same number of bits as in the case in which only $\exp[S[P]]$ configurations have equal nonzero probability.

In other words, the amount of information[5] (about the system) contained in the probability distribution is (in bits)

$$I[P] = \frac{\ln M - S[P]}{\ln 2} . \tag{1.16}$$

We can now gain some insight into the meaning of the probability distribution (1.5). We consider the evolution of an ensemble of real Hamiltonian systems with a large number of degrees of freedom, and we suppose that the only information we have on the microscopic state of the system is the average value of the energy U. It is reasonable to assume that the probability distribution will minimize the information (maximize $S[P]$) at fixed $E[P] = \langle H \rangle_P$; let us try to maximize $S[P]$ as defined in Eq. (1.12) with the constraints

$$U = E[P] \equiv \sum_{j=1}^{M} P_j H_j ; \quad \sum_{j=1}^{M} P_j = 1 . \tag{1.17}$$

The method of Lagrange multipliers can be used to reproduce Eq. (1.5), where β is fixed by the condition $\langle H \rangle = U$. Equation (1.5) can therefore be derived from a variational principle: The maximum of the entropy at fixed expectation value of the energy.

Another variational formulation (which plays a very important role) leading to Eq. (1.5) is the principle of minimal free energy $\Phi[P]$:

$$\Phi[P] = E[P] - \frac{S[P]}{\beta} , \tag{1.18}$$

as can be seen using the method of the Lagrangian multipliers.[6]

We also have the following relations:

$$\Phi[P_{eq}] \equiv F(\beta) = -\frac{1}{\beta} \ln Z = U - \frac{S_{eq}}{\beta} ,$$

$$S_{eq} \equiv S[P_{eq}] = \beta^2 \frac{\partial F}{\partial \beta} , \tag{1.19}$$

$$U \equiv E[P_{eq}] = -\frac{\partial}{\partial \beta} \ln Z = F(\beta) + \beta \frac{\partial F}{\partial \beta} ,$$

as can be easily checked.

1.3. The absolute temperature

It is very important to find the connection between β and the absolute temperature and between the microscopic entropy and the thermodynamic entropy. Let us first recall some results of elementary thermodynamics.

Consider a system that can receive energy from outside in the form of work and that is also in thermal contact with a heat reservoir. The heat received by the system during an infinitesimal transformation can be computed from the first law of the thermodynamics:

$$\delta Q = dU - \delta L . \tag{1.20}$$

From the second law we know that if T is the absolute temperature, then $\delta Q/T$ is an exact differential (this is the definition of the absolute temperature, apart from a multiplicative scale factor):

$$\frac{\delta Q}{T} = dS_M , \tag{1.21}$$

where S_M is the macroscopic entropy, i.e., the entropy defined in the thermodynamical approach. In other words, $1/T$ is the integrating factor of the heat. If we prove that

$$\beta \delta Q = \beta(dU - \delta L) = dS_{eq} , \tag{1.22}$$

we find that β is proportional to $1/T$ and that $S_M = S_{eq}$, apart from an additive constant.

Let us prove Eq. (1.22). In order to do work on the system, the Hamiltonian must depend on a control parameter λ.[7] We can give energy to the system by changing λ. We now consider an infinitesimal variation $d\lambda$ and $d\beta$ of λ and β. Neglecting higher-order infinitesimals, we have

$$\delta L = \left\langle \frac{dH}{d\lambda} \right\rangle d\lambda . \tag{1.23}$$

It is evident that

$$d[-\ln Z] = U \, d\beta + \beta \left\langle \frac{dH}{d\lambda} \right\rangle d\lambda = U \, d\beta + \beta \delta L , \tag{1.24}$$

and therefore

$$dS_{eq} = d(\beta U - \beta F) = \beta(dU - \delta L) , \tag{1.25}$$

which is the desired result; β is therefore $1/kT$ (k can be set equal to 1 using the appropriate units). Equation (1.19) reduces to

$$\frac{\partial F}{\partial T} = S_{eq} \qquad F = U_{eq} - TS_{eq}. \qquad (1.26)$$

F is thus the usual thermodynamic free energy. This result is very nice. Indeed, if we consider a mechanically isolated system that may exchange heat with a thermal reservoir at fixed temperature T, the second principle tells us that the thermodynamic free energy ($F_M \equiv U - TS_M$) cannot increase; if the different components of the system are at equilibrium (e.g., if they have the same temperature, pressure, etc.) the free energy cannot decrease further. It is satisfying to have arrived at the same conclusion (that the equilibrium probability distribution minimizes the free energy) by using both the microscopic distribution (1.5) and the standard macroscopic thermodynamic arguments.

Notes for Chapter 1

1. The usual argument runs as follows: after the system has reached equilibrium, if we measure the value of an intensive quantity inside a small region, the output of the measurement will not be the equilibrium value, but thermal fluctuations will be present; when the volume goes to infinity the number of regions in which we can measure A will be proportional to the volume (i.e., to N), and fluctuations from different regions will cancel out in the average over the regions. In other words, we divide the volume into V/V_0 parts, V_0 being a fixed volume. We have $A = (V/V_0) \sum_{k=1}^{V/V_0} A_k$, where A_k is the observable in the kth subvolume. If V_0 is not too small, we can assume that the A_k's are independent random variables with a common average $\bar{A}$. (As we shall see in Chapter 7, the assumption of independent fluctuations fails at the phase transition point.) Equation (1.13) follows from the central limit theorem, which states that the probability distribution of the sum of L independent random variables ($A = (1/L) \sum_{k=1}^{L} A_k$) becomes Gaussian when L goes to infinity:

$$P(A) \simeq \left(\frac{L}{2\pi\delta}\right)^{1/2} \exp\left[-(A - \bar{A})^2 \cdot \frac{L}{(2\delta)}\right],$$

Notes for Chapter 1

where

$$\bar{A} = \lim_{L\to\infty} \frac{1}{L} \sum_{k=1}^{L} A_k \quad \text{and} \quad \delta = \lim_{L\to\infty} \frac{1}{L} \sum_{k=1}^{L} (A_k - \bar{A})^2.$$

The theorem obviously fails if δ is not finite.

For more information on this very important theorem see, for example, B. Gnedenko, *The Theory of Probability*, Mir, Moscow (1973).

2. As an example we consider the probability distribution

$$P(x) = \frac{1}{(2\pi A)^{1/2}} \exp\left(-\frac{x^2}{2A}\right)$$

for x in the range ∞, $+\infty$. The corresponding entropy is $\frac{1}{2}[1 - \ln(2\pi A)]$, which goes to $-\infty$ when A goes to zero [$P(x)$ becomes a δ function].

3. Let us check Eq. (1.13) in the case of a one-dimensional distribution: The probability of the system staying in the ith cell $x_i < x < x_{i+1}$, $x_i = \varepsilon i$, is given by $\int_{x_i}^{x_{i+1}} dx\, P(x) \simeq \varepsilon P(x_i) \equiv \varepsilon P_i$. The entropy is thus $-\sum_{-\infty i}^{+\infty} \ln(\varepsilon P_i)(\varepsilon P_i) = -\varepsilon \sum_{-\infty i}^{+\infty} P_i(\ln P_i + \ln \varepsilon) \simeq -\int_{-\infty}^{+\infty} dx\, P(x) \ln P(x) - \ln \varepsilon$.

4. There are many books on information theory. The one with which I am personally most familiar is A. Kintchin, *Mathematical Foundations of Information Theory*, Dover, London (1963).

5. The average number of bits we can gain in coding the state of the system is by definition the amount of information contained in the probability distribution.

6. If we want to maximize a function $\Phi(P)$ with the constraint $g(P) = 0$, we can maximize the function $\Phi_\lambda(P) \equiv \Phi(P) + \lambda g(P)$, where the so-called Lagrange multiplier λ is chosen in such a way that the maximum of $\Phi_\lambda(P)$ satisfies the constraint. In our case, $g(P) = \sum_{j=1}^{N} P_j - 1$, we get

$$\frac{\partial}{\partial P_j}[\Phi_\lambda(P)] = H_j + \left(\frac{1}{\beta} + \lambda\right) + \frac{1}{\beta} \ln P_j = 0$$

$$P_j = \frac{\exp(-\beta H_j)}{Z}, \quad Z = \sum_{j=1}^{N} \exp(-\beta H_j).$$

7. This point is well discussed by E. Schrödinger, in his book *Statistical Thermodynamics*, Cambridge University Press, Cambridge (1962).

CHAPTER 2

Magnetic Systems

2.1. General properties

The formalism of the previous chapter can be applied to a very large variety of systems; however, in the next chapters we shall deal mainly with magnetic systems. We have made this choice because magnetic systems offer a very wide spectrum of problems, ranging from very simple to very difficult. Moreover, the physics of magnetic systems has strong ties with problems that concern other materials and with relativistic field theory. It is, therefore, an ideal starting point.

Let us consider a generic magnetic system; for each configuration of the system we can define the corresponding magnetization $\vec{S}(x)$, x being a point of the physical space. Microscopically $\vec{S}(x)$ receives contributions both from the motion of the electrons and from the spins of the electrons (and in a very small amount from the nuclei); if the electrons that mainly contribute to $\vec{S}(x)$ are localized near the atoms of a lattice, it may be convenient to define $\vec{S}$ only on the points of the lattice.

From the experimental and theoretical points of view it is interesting to study the response of the system to the action of an external magnetic field $\vec{h}(x)$. Let us therefore consider the following Hamiltonian:

$$H(C) = H_0(C) - \int_V d^3x \, \vec{h}(x) \cdot \vec{S}(x) = H_0(C) - \int_V d^3x \sum_{\nu=1}^{3} h_\nu(x) S_\nu(x),$$
(2.1)

where $H_0(C)$ is the Hamiltonian in absence of magnetic field, V is the volume occupied by the system, and, in order to lighten the notation, we have not indicated explicitly the dependence of $\vec{S}$ on C. As in Sec. 1.2, C denotes a generic configuration of the system.

Most of the information on the magnetic properties of the system is contained in the free energy $F[h]$ as a functional of $\vec{h}(x)$. If the magnetic field does not depend on x, we can write the free energy as a function of h

2.1. General Properties

$(F(h) \equiv F[h]|_{\tilde{h}(x)=\tilde{h}})$ and define a free-energy density $f(h) = F(h)/V$. (We shall follow as far as possible the rule of indicating functionals by square brackets and functions by curved brackets.) The expectation values of the local magnetization $(m_\nu(x))$ and of the total magnetization (M_ν) are given as

$$m_\nu(x) = \langle S_\nu(x) \rangle \, ; \quad M_\nu = \langle S_\nu^T \rangle : \quad S_\nu^T \equiv \int_V d^3x \, S_\nu(x) \, . \quad (2.2)$$

If the Hamiltonian system is spatially homogeneous (i.e., invariant under continuous or discrete translations), and the magnetic field is consequently constant, the magnetization will be x independent (let us neglect the case of antiferromagnets): $m_\nu(x) = m_\nu$.

It is easy to see that

$$m_\nu(x) = -\frac{\delta F[h]}{\delta h_\nu(x)} \, , \quad m_\nu = -\frac{\partial f(h)}{\partial h_\nu} \quad (2.3)$$

if we use the definitions

$$F[h] = -\frac{1}{\beta} \ln(Z[h]) \, ,$$

$$Z[h] = \int dC \exp[-\beta H(C)] \, . \quad (2.4)$$

There are already nontrivial experimental consequences that follow from Eqs. (2.2), (2.3), and (2.4). Let us start from the relation

$$u(h) = f(h) - T \frac{\partial f}{\partial T} \, , \quad (2.5)$$

where $u(h)$ is the internal energy density.

If we differentiate Eq. (2.5) with respect to h_ν we find

$$\frac{\partial u(h)}{\partial h_\nu} = \frac{\partial f}{\partial h_\nu} - T \frac{\partial^2 f}{\partial T \partial h_\nu} = -m_\nu + T \frac{\partial m_\nu}{\partial T} \, . \quad (2.6)$$

In the same way, if we define nonmagnetic internal energy $u_0 = \langle H_0 \rangle = u + \sum_\nu^3 h_\nu m_\nu$, we get

$$\frac{\partial u_0}{\partial h_\nu} = \sum_\mu^3 h_\mu \frac{\partial m_\nu}{\partial h_\mu} + T \frac{\partial m_\nu}{\partial T} \, . \quad (2.7)$$

Equations (2.6) and (2.7) are consequences of the existence of the free energy (which in a pure thermodynamic treatment follows from the second law of thermodynamics) and of the equality of cross derivatives; these kinds of relations are normally referred to as Maxwell equations.

A very important quantity is the magnetic susceptibility: $\chi_{\mu\nu} = \partial m_\mu / \partial h_\nu$; in the isotropic case $\chi_{\mu\nu} = \delta_{\mu\nu} \chi$. We can also consider a point-dependent magnetic susceptibility:

$$\chi_{\mu\nu}(x, y) = \frac{\delta m_\mu(x)}{\delta h_\nu(y)}.$$

In a translationally invariant system we have

$$\chi_{\mu\nu}(x, y) = \chi_{\mu\nu}(x - y), \qquad \chi_{\mu\nu} = \frac{1}{V} \int d^3x \, \chi_{\mu\nu}(x). \tag{2.8}$$

Starting from the relations

$$\chi_{\mu\nu} = -\frac{\partial^2 f}{\partial h_\mu \partial h_\nu}, \qquad \chi_{\mu\nu}(x, y) = -\frac{\delta^2 F}{\delta h_\mu(x) \delta h_\nu(y)}, \tag{2.9}$$

we get the following symmetry properties of the susceptibility:

$$\chi_{\mu\nu} = \chi_{\nu\mu}, \qquad \chi_{\mu\nu}(x, y) = \chi_{\nu\mu}(y, x). \tag{2.10}$$

If we perform explicitly the derivatives of Eq. (2.4) we find[1]

$$V\chi_{\mu\nu} = \beta \langle S_\mu^T S_\nu^T \rangle_C \equiv \beta [\langle S_\mu^T S_\nu^T \rangle - \langle S_\mu^T \rangle \langle S_\nu^T \rangle]$$

$$= \beta [\langle (S_\mu^T - \langle S_\mu^T \rangle)(S_\nu^T - \langle S_\nu^T \rangle) \rangle].$$

$$\chi_{\mu\nu}(x, y) = \beta [\langle S_\mu(x) S_\nu(y) \rangle - \langle S_\mu(x) \rangle \langle S_\nu(y) \rangle] \tag{2.11}$$

$$\equiv \beta \langle S_\mu(x) S_\nu(y) \rangle_C.$$

If the probability distribution $P(C)$ is concentrated on a single configuration (i.e., if it is a delta function), no thermodynamical fluctuations are present and the right-hand sides of Eqs. (2.10) and (2.11) are zero. If we consider the diagonal elements of the susceptibility, we get $V\chi_{\nu\nu} = \langle (S_\nu^T - \langle S_\nu^T \rangle)^2 \rangle$; in other words, the diagonal elements of the susceptibility tensor are given by the variance of the probability distribution of the quantity S_ν^T.

Equation (2.11) is a particular case of the more general theorem, which states that the linear response of a system is proportional to the random thermodynamical fluctuations. If the Hamiltonian is $H_0 + \lambda H_1$, we have

$$-\frac{1}{\beta} \frac{d}{d\lambda} \langle A \rangle = \langle H_1 A \rangle_C \equiv \langle H_1 A \rangle - \langle H_1 \rangle \langle A \rangle. \tag{2.12}$$

Although the proof of Eq. (2.12) is trivial,[2] there is an extension to the dynamic statistical mechanics that is considerably less trivial. It is often called "the fluctuation-dissipation theorem" and is described in Chapter 17. In this context the word "dissipation" is synonymous with "response." Although this name may sound strange, in many cases the linear response is proportional to the energy dissipated in the presence of an external perturbation; for example, if an electric field E acts on a conductor, the energy dissipated (E_d) is given by

$$E_d = EI = \frac{E^2}{R}, \qquad (2.13)$$

where the electric current I is fixed by Ohm's law, $I = E/R$.

2.2. Spontaneous symmetry breaking

Ferromagnetic systems have spontaneous magnetization at low temperatures[3] i.e., below the Curie temperature. If a system in zero external magnetic field has a magnetization pointing in a given direction, in the absence of an external perturbation (if we stay below the Curie temperature) the magnetization will not change with time; however, when we apply an external magnetic field and then remove it, we can have the same system with a time-independent spontaneous magnetization that is the opposite of the previous one. In other words the systems may stay in two (or more) states characterized by the direction of the magnetization; however, in the presence of an external magnetic field (neglecting possible nonequilibrium effects, e.g., hysteresis) the magnetization will always be oriented in the direction of the external field.

From the microscopic point of view it is clear that for each configuration C there is another configuration C' such that C and C' have the same energy but opposite magnetization, i.e., $S(C) = -S(C')$. The configuration C' can be obtained from the configuration C by inverting the sign of all the velocities and the spins of all the particles of the system.[4]

When spontaneous magnetization is present, the Hamiltonian of the system is invariant under a transformation that changes the sign of the magnetization; however, the state of the system, i.e., its probability distribution, is not invariant under this transformation. When this happens we say that the symmetry is spontaneously broken (it would be explicitly broken in the presence of an external magnetic field).

At first sight spontaneous symmetry breaking seems to be in conflict with the fundamental principles of statistical mechanics, i.e., Eq. (1.5): if H is invariant, P must also be invariant. This argument is absolutely correct for a finite volume, but fails for an infinite system in which Eq. (1.5) is only formal because Z is infinite. If we start from Eq. (1.5) for finite systems and we go to the infinite-volume limit with the magnetic field h exactly equal to zero, the magnetization will always be equal to zero (the integral of an odd function is zero).

However, this last result is true only if h is exactly equal to zero; a better insight on the problem may be obtained if we consider the magnetization density and the free energy as functions of h in the infinite-volume limit. The symmetry argument tells us that $f(h) = f(-h)$ and $m(h) \equiv -\partial f/\partial h = -m(-h)$, i.e., f and m are even and odd functions of h, respectively. We have two possibilities: either $m(h)$ is continuous at $h = 0$ and $m(0)$ is consequently zero, or $m(h)$ is discontinuous at $h = 0$ and $m(0)$ is not so well defined; this second case corresponds to spontaneous symmetry breaking, as we shall see. More precisely, these two possibilities correspond to[5]

a) $\qquad f(h) = f(0) + 0(h^\alpha), \qquad \alpha > 1$

b) $\qquad f(h) = f(0) - m_s|h| + 0(h^\alpha).$ $\qquad$ (2.14)

In case (a), f is differentiable with respect to h and $m(h)$ is continuous. In case (b), f is not differentiable with respect to h at $h = 0$ and m is a discontinuous function;

$$m^+ \equiv \lim_{h \to 0^+} m(h) = m_s \neq m^- \equiv \lim_{h \to 0^-} m(h) = -m_s. \qquad (2.15)$$

The symmetry relation implies only $m^+ = -m^-$.

In case (a) the magnetization at $h = 0$ is really zero, whereas in case (b) an infinitesimal magnetic field is enough to produce a nonzero magnetization (see Fig. 3.2). If that happens, we say that the system has a spontaneous magnetization. In the limit $h \to 0$ the Hamiltonian becomes invariant under the transformation $S \leftrightarrow -S$, but the statistical expectation values are not invariant under the same transformation, so we say that the symmetry is spontaneously broken. For most of the ferromagnets, case (b) is realized at low enough temperature (below the Curie temperature), at least in three dimensions.

The nondifferentiability of the free energy with respect to a parameter and the existence of two or more equilibrium states is the distinctive feature of a first-order transition: according to the Ehrenfest classification, if the free energy is differentiable $(k-1)$ times, but not k times, the singularity point is called a transition of order k (notice that the free energy is always continuous);[6] however, it is customary to call any transition of order greater than the first a second-order transition. If a first-order phase transition is induced by changing the temperature, we are in the presence of a latent heat:

2.2. Spontaneous Symmetry Breaking

$$u_+ \equiv \lim_{T \to T_c^+} \left(\frac{\partial(\beta f)}{\partial \beta}\right) \neq u_- \equiv \lim_{T \to T_c^-} \left(\frac{\partial(\beta f)}{\partial \beta}\right). \tag{2.16}$$

It is crucial to understand that phase transitions (and by consequence spontaneous symmetry breaking) are possible only if the volume of the system is infinite. Indeed, in a finite volume, if the configuration space is compact, Z is an entire function (we assume that the Hamiltonian is not singular): $|Z(\beta)| = |\int dC \exp(-\beta H(C))| \leq V_C \exp|\beta||H|$ even for complex β, where $V_C \equiv \int dC$ is the volume of the configuration space and $|H| = \max[|H(C)|]$; in the extreme case the configuration space is a finite set of points and Z is a linear combination of exponentials. If the configuration space is not compact but $Z(\beta) = \int dC \exp(-\beta H(C))$ is finite (i.e., the integral is convergent) for any real positive β, the same integral must exist for any complex β such that Re $\beta > 0$ ($|Z(a+ib)| \leq Z(a)$); it is clear, however, that the integral defining Z will be divergent at $\beta \leq 0$ when the volume of the configuration space is infinite (if H is positive).

We have proved that the partition function of a reasonable Hamiltonian Z (such that $Z(\beta)$ is defined for positive β) is an analytic function in the positive half of the complex β plane; a similar result may be obtained for the behavior of Z as a function of the magnetic field.

What happens to the free energy (i.e., $(-1/V\beta)\ln Z(\beta)$)? If Z is analytic, its logarithm may be singular only at the points where $Z = 0$. Now for real values of the parameters (temperature and magnetic field) Z is the sum of positive terms, so it cannot have zeros; consequently the free energy is analytic near the positive real axis for a finite-volume system. However, in the infinite-volume limit a nonanalytic free energy may be produced if the complex zeros of the partition function pinch the real parameter axis.[7] A toy model is given by a system of volume V, which can stay in only two states (of equal energy at zero magnetic field) having total magnetization $\pm V$, respectively. We easily obtain for the free energy as a function of the magnetic field h

$$Z(h) = 2\cosh(\beta h V), \quad f(h) = -\frac{1}{\beta V}\ln[2\cosh(\beta h V)],$$
$$m(h) = \tanh(\beta h V), \quad Z\left(\frac{(2n+1)\pi i}{2\beta V}\right) = 0. \tag{2.17}$$

When V goes to infinity, the zeros of Z pinch the real h axis at $h = 0$, and the expectation value of the magnetization (m) becomes $m = \text{sign}(h)$.

In a real macroscopic system of finite extent the internal energy will change a finite amount (in units of k) when the temperature changes by 10^{-23} Kelvin, and for all practical purposes we are in the presence of a discontinuity. In order to have a mathematical discontinuity we must go to the infinite-volume limit. This is one of the main reasons why in this book we shall concentrate on infinite-volume systems: in this limiting case sharper statements can be made.

In order to specify the finite system Hamiltonian, we must fix the Hamiltonian of the bulk and the boundary conditions (i.e., the Hamiltonian at the surface); if we do not stay at a first-order phase transition point, by definition there is only one thermodynamic stable phase, and the distribution probability in the infinite-volume limit will not depend on the boundary condition. We shall see later that the effect of the boundary conditions on the expectation values at the center of a box of side L vanishes as $\exp(-L/\xi)$ in the generic case (as a power at a second-order phase-transition point). At a first-order transition point there are two or more phases, and the final result will depend on the boundary conditions. For example, in the magnetic case, the expectation value of the magnetization will be $\pm m_s$ if we constrain the magnetization on the boundary to be positive or negative, respectively; if we set free our periodic boundary conditions for the magnetization at the surface, the expectation value of the magnetization will be zero by symmetry arguments. In other words the boundary conditions for a nearly infinite system play the same role as an infinitesimal magnetic field for an infinite system: in both cases we produce a difference in energy between the two most likely configurations (characterized by having mostly positive or negative magnetization). This energy difference goes to infinity with the volume or with the surface, so that only one of the two possible configurations dominates the sum in the partition function, when the volume goes to infinity.

If we consider the time evolution[8] of a real system at the first-order phase-transition point inside a box of side L, the large-time behavior strongly depends on the initial conditions: at exactly 0°C we can have either ice or water. Similarly the spontaneous magnetization of a piece of iron may point in any direction. A detailed computation shows that we need to wait for a time proportional to $\exp(L^{D-1})$ (an incredibly large time!) to see the spontaneous reverse of the magnetization.[9] In order to reproduce the experimental situation we must first send the volume to infinity, and only later the observation time to infinity: The two limits do not commute.

If we define

$$\langle \ \rangle_\pm = \lim_{h \to 0^\pm} \langle \ \rangle_h \qquad (2.18)$$

we expect that the large-time behavior of a single large (or infinite-volume) system will be described by $\langle \ \rangle_+$ or $\langle \ \rangle_-$ if the $t=0$ initial conditions are homogeneous in space (i.e., translationally invariant).[10] On the contrary, if we average over the possible initial conditions we get

2.2. Spontaneous Symmetry Breaking

the symmetric $h = 0$ state:

$$\langle \ \rangle_0 \equiv \tfrac{1}{2} \langle \ \rangle_+ + \tfrac{1}{2} \langle \ \rangle_- . \tag{2.19}$$

From the foregoing discussion it is clear that $\langle \ \rangle_0$ describes a statistical mixture, and only $\langle \ \rangle_\pm$ can describe the large-time behavior of a single system. The general situation in mathematical terms is the following: let us consider the set of all possible translationally invariant states. This set is convex[11] (e.g., we could obtain the state $\langle \ \rangle_p \equiv p \langle \ \rangle_+ + (1-p) \langle \ \rangle_-$ by imposing free boundary conditions but adding an external magnetic field equal to $\operatorname{arctanh}[(2p-1)/(\beta V)]$ before sending the volume to infinity). A state that cannot be decomposed as the superposition of two different states (i.e., $\langle \ \rangle_a = \alpha \langle \ \rangle_b + (1-\alpha) \langle \ \rangle_c$ implies $\langle \ \rangle_a = \langle \ \rangle_b = \langle \ \rangle_c$ if $0 < \alpha < 1$) is said to be pure or extremal. It can be proved that each state can be decomposed in a unique way as the convex combination of pure states.[12] In this language, a symmetry group G is spontaneously broken if the symmetric state is not pure and it can be decomposed as the convex combination of pure states that transform as a nontrivial representation of G.

The argument for identifying the pure states with large-time behavior of a single, spatially homogeneous system is mainly the following: If we add to the Hamiltonian a small perturbation localized around the point x ($\delta H = B(x)$), we expect on physical grounds that the expectation value of an observable $A(y)$ localized around the point y does not change at all when $|x - y|$ goes to infinity.[13] From the linear response theorem this may happen if the connected correlation functions $\langle B(x) A(y) \rangle_C$ go to zero at large distance for any B and A. The vanishing of the connected correlation functions at large distances is the clustering property.[14] If it is satisfied, the state is said to be clustering or is said to satisfy the "cluster decomposition property." Now the stability of the large-time evolution of a single physical system implies that the corresponding state is clustering. Up to now our argument may be described as mere handwaving; however, there is a rigorous theorem which states that pure states are clustering and vice versa. Spontaneous symmetry breaking can be related to the failure of the clustering decomposition in the symmetric state. Indeed, coming back to our magnetic case, it is easy to show that

$$\langle S(x) S(y) \rangle_{Cp} \equiv \langle S(x) S(y) \rangle_p - \langle S(x) \rangle_p \langle S(y) \rangle_p$$
$$= 4p(1-p) m^2 \quad \text{for } |x - y| \to \infty . \tag{2.20}$$

According to the theorem, the only two clustering states are the two extremal states at $p = 0$ and $p = 1$ (i.e., $\langle \ \rangle_-$ and $\langle \ \rangle_+$).

Appendix to Chapter 2

Functionals were introduced in the last century by the great Italian mathematician, Vito Volterra.[15] The basic idea behind them is very simple: in the same way that a function $f(\ )$ associates a number x with another number, i.e., $f(x)$, a functional $F[\ ]$ associates a function $g(\ )$ with a number, i.e., $F[g]$.

Very elementary examples of functionals are the definite integral (e.g., $I[g] \equiv \int_{-\infty}^{+\infty} g(x)\,dx$), the value of the function at a given point (e.g., $F_0[g] = g(0)$), and the maximum of a function in a given interval (e.g., $M[g] = \max_{0 \le x \le 1}[g(x)]$). The functional $S[g] \equiv -\int dx\, g(x) \ln|g(x)|$ is the entropy of $g(\ )$ if the function g is positive definite and normalized to one (i.e., $\int dx\, g(x) = 1$).

Generally speaking we must be careful when dealing with functionals; very often they can take an infinite value even when g seems to be "well behaved" (e.g., $I[g] = \infty$ if $g = 1/(1 + x^2)^{1/2}$). If we do not work within well-defined mathematical framework, paradoxical results may be obtained. However, in most cases it is sufficient to consider the functionals as acting only on a restricted space of functions (e.g., $I[g]$ will be well defined when g is a bounded measurable function with fast decrease at infinity, $F_0[g]$ and $M[g]$ when g is a continuous function).

If for simplicity we skip all the mathematical details, we can define the functional derivative of a functional by starting from the Taylor expansion. We suppose that

$$F[g + \varepsilon h] = F[g] + \varepsilon \int dx\, \frac{\delta F}{\delta g(x)} h(x) + O(\varepsilon^2)$$

for any reasonable $h(x)$, where $\delta F/\delta g(x)$ is by definition the functional derivative; we stress that the functional derivative is a function of x and a functional of g. In this book all the functional derivatives we need can be computed by inspection. For example, we easily obtain

$$\frac{\delta I}{\delta g(x)} = 1, \qquad \frac{\delta}{\delta g(x)} \int_{-\infty}^{\infty} f(g(x))\, dx = f'(g(x))$$

$$\frac{\delta F_0}{\delta g(x)} = \delta(x), \qquad \frac{\delta}{\delta g(x)}[f(g(0))] = f'(g(0))\delta(x)$$

$$\frac{\delta}{\delta g(x)} \int_{-\infty}^{+\infty} \left(\frac{dg}{dy}\right)^2 dy = -2\,\frac{d^2 g(x)}{dx^2}.$$

This last result was obtained by integration by parts. A slightly less trivial example is

$$\frac{\delta M}{\delta g(x)} = \delta(x - x_M),$$

where x_M is the value of x at which the function $g(x)$ reaches its maximum in the interval 0-1; if the function $g(x)$ has more than one maximum the functional M is not differentiable at this point.

We recall that the Lagrangian approach to classical mechanics can easily be cast into functional language: The action is a functional of the trajectory $q(t)$ in the time interval from t_0 to t_1: its explicit form is

$$S[q] = \int_{t_0}^{t_1} L(q, \mathring{q}) \, dt,$$

where L is the Lagrangian function. The classical equations of motion[16] are $\delta S/\delta q(t) = 0$; indeed at $t \neq t_0$ and $t \neq t_1$ we have

$$\frac{\delta S}{\delta q(t)} = \frac{\partial}{\partial q} L(q(t), \mathring{q}(t)) - \frac{d}{dt} \left[\frac{\partial}{\partial \mathring{q}} L(q(t), \mathring{q}(t)) \right].$$

Most of the techniques used for functions of many variables can also be applied to functionals; for example, the proof in the last chapter of the variational principle [Eq (1.18)] can easily be extended to the case where the configuration space is continuous and $\Phi[P]$ is defined as

$$\Phi[P] = \int d\mu\,(C) P(C) \left[H(C) + \frac{1}{\beta} \ln P(C) \right].$$

Sometimes it is convenient to approximate the space of functions, on which the functional is defined, by a finite-dimensional space (of dimension L); in this way the functional is reduced to a function of many variables, and formal manipulations are much easier. This may be a reasonable way to prove theorems on functionals, when we are able to keep under control the results in the limit L going to infinity.

Notes for Chapter 2

1. $\langle AB \rangle_C$ stands for the connected average, i.e., $\langle AB \rangle_C = \langle AB \rangle - \langle A \rangle \langle B \rangle$. For the definition of the connected average of many functions, see p. 53.

2. Indeed, we have

$$\frac{d}{d\lambda}\langle A\rangle = \frac{d}{d\lambda}\left\{\frac{\int dC[\exp(-\beta H_0 - \beta\lambda H_1)A]}{\int dC \exp(-\beta H_0 - \beta\lambda H_1)}\right\}$$

$$= -\beta\int dC \frac{[\exp(-\beta H_0 - \beta H_1\lambda)AH_1]}{Z}$$

$$+ \beta\int dC \exp(-\beta H_0 - \lambda\beta H_1)A \cdot \int dC$$

$$\cdot \frac{\exp(-\beta H_0 - \beta\lambda H_1)H_1}{Z^2},$$

$$Z = \int dC[\exp(-\beta H_0 - \beta\lambda H_1)].$$

3. For a nice presentation of what happens in real materials see, for example, C. Kittel, *Introduction to Solid State Theory*, John Wiley and Sons, New York (1966).
4. This transformation is called time reversal. The reader should notice that in this chapter, from here on, we suppress the indices of the magnetization; our general rule will be to suppress indices as much as possible unless this suppression creates ambiguities.
5. A value of α less than 1 is impossible because this leads to an infinite value of the magnetization at $h = 0$. In case (b) there is a first-order transition in the magnetic field; in case (a), a higher-order transition or no transition at all (e.g., when α is even).
6. The free energy is a convex function, i.e., $f(\alpha\beta_1 + (1-\alpha)\beta_2) \geq \alpha f(\beta_1) + (1-\alpha)f(\beta_2)$. The proof of this is rather simple: In a finite volume, where $f(\beta)$ is an analytic function, its second derivative with respect to any parameter is always negative because it can be written as the expectation value of minus a perfect square, e.g., $(d^2/d\beta^2)f(\beta) = -(1/\beta V)\langle(H - \langle H\rangle)^2\rangle \equiv -(1/\beta V)\langle HH\rangle_C$. It is well known that a function whose second derivative is never positive is convex. Now the limit of differentiable functions may not be differentiable, but the limit of convex functions must be convex (convexity is written under the form of an inequality). It is also well known that convexity implies continuity.
7. This observation is mainly due to T. D. Lee and C. N. Yang. An elementary discussion of the subject can be found in K. Huang, *Statistical Mechanics*, John Wiley and Sons, New York (1963).

Notes for Chapter 2

8. From a purely logical point of view the discussion of the time evolution of a system undergoing a first-order phase transition is out of place here. It is useful, however, for understanding from a physical point of view the importance of the pure (clustering) states, to be defined soon.

9. This kind of estimate is based on the hypothesis that the time needed to go from configuration A to configuration B is proportional to $\exp[-\beta(F_M - F_A)]$, F_M being the maximal free energy that the system must assume in its trajectory (which is supposed to be continuous) from A to B. A very nice quantitative application of this idea can be found in J. S. Langer and V. Ambegaokar, *Phys. Rev. 164*, 498 (1967).

   The same approach can be used for obtaining the mean life of a metastable state.

10. We want to exclude a macroscopically inhomogeneous initial condition, e.g., half space with positive magnetization and half space with negative magnetization.

11. A subset of a linear space is convex if, when a and b belong to this subset, $c = \alpha a + (1 - \alpha)b$ belongs to the same subset for $0 \leq \alpha \leq 1$.

12. A presentation of spontaneous symmetry breaking from this point of view can be found in D. Ruelle, *Statistical Mechanics*, Benjamin, New York (1969). See also D. Kastler in the *Proceedings of the 1967 International Conference on Particles and Fields*, ed. by C. R. Hagen, G. Guralnik, and V. A. Mathur, John Wiley and Sons, New York (1968).

13. This condition is very strong. A similar condition under some restrictive technical hypothesis can be shown to imply the canonical distribution.

14. If we define the intensive quantities $\bar{A}$ and $\bar{B}$ as $\bar{A} = 1/V \int_V dx\, A(x)$ and $\bar{B} = 1/V \int_V dx\, B(x)$, in the infinite-volume limit these intensive quantities do not fluctuate (e.g., $\langle \bar{A}\bar{B} \rangle = \langle \bar{A} \rangle \langle \bar{B} \rangle$ only if the connected correlations go to zero at infinity).

15. V. Volterra, *Theory of Functionals and of Integral and Integro-differential Equations*, Dover, New York (1959). An introduction to functionals and functional derivatives can be found in J. Tarski, *Quantum Theory and Statistical Physics*, ed. by A. O. Barut and W. E. Brittin, Gordon and Breach, New York (1968).

16. See any book on classical mechanics; for example, H. Goldstein, *Classical Mechanics*, Addison-Wesley, Reading, MA (1950).

CHAPTER 3

The Ising Model

3.1. The model

In the last forty years, models have played a key role in the development of condensed matter physics; they have been the point of contact between the physical-mathematical studies and the experimental data. It is a very difficult if not impossible task in most cases to compute the properties of a piece of matter starting from the quantum Hamiltonian (which can be derived from first principles) describing the interaction of electrons with nuclei. A numerical computation of the spectrum of monoatomic bismuth is yet to be attained. However, this is not a serious difficulty. Nowadays, condensed matter physics is not a testing ground for quantum mechanics (whose validity in this energy range is beyond any doubt) nor for electromagnetism; the aim of most of the research is to understand both qualitatively and quantitatively the enormous variety of phenomena, which range from magnetism to vulcanization, from superconductivity to metal-insulator transitions and so on. This can be done by picking from among all degrees of freedom of the system a given number of variables that are relevant to the problem one is considering (e.g., the magnetization for the magnetism). An effective reduced Hamiltonian can be written as a function of the relevant variables. For example, in the magnetic case we have

$$\exp[-\beta H_\beta^{ef}(S)] = \int d[C]\delta(S(x, C) - S(x)) \exp[-\beta H(C)] \quad (3.1)$$

or, equivalently,

$$\int d[S] \exp[-\beta H_\beta^{ef}[S]] g[S] = \int d[C] \exp[-\beta H[C]] g[S[C]] \quad (3.2)$$

for any functional $g[S]$. In other words, $\exp(-\beta H_\beta^{ef}(S))$ is proportional to the probability distribution of the S variables.

3.1. The Model

It is clear that $H^{ef}_\beta[S]$ will depend on β in an explicit way. Now if we could exactly perform the integration over the other degrees of freedom in Eq. (3.1), the final result would be a terrible mess; the normal procedure consists in postulating a simple model form for the effective Hamiltonian, depending on a few parameters that can be fitted from the experiments or approximately computed in a microscopic way. The art of model building is rather subtle; the model must be simple enough to be investigated analytically and flexible enough to reproduce the essential properties of the systems we want to study. Although in most cases one tries to construct models that are as realistic as possible, for very well known models this process is inverted: one tries to perform experiments on systems whose effective Hamiltonian is, as close as possible to that of a well-known model.

In many magnetic materials the electrons responsible for magnetic behavior are localized near the atoms of a lattice, and the force, which tends to orient the spins, is the exchange interaction (which is a very short-range force). The most popular models, which describe this situation, are the Ising and Heisenberg models: the magnetization is defined only on the points of the lattice. (For simplicity we shall consider mainly the square lattice in two dimensions, the cubic lattice in three dimensions, and the hypercubic lattice in higher dimensions. D will denote the dimensions of the space.) The magnetization S_i can take only the values ± 1 for the Ising model or belongs to the unit three-dimensional sphere for the Heisenberg model ($\sum_{\nu}^{3}(S_i^\nu)^2 = 1$). One can also consider a generalized n-component Heisenberg model in which $\vec{S}_i$ belongs to the n-dimensional unit sphere; for $n = 1$ and 3 we recover the Ising and the Heisenberg models, respectively.[1] The Hamiltonian in the presence of a site-dependent external magnetic field h_i is, in the Ising and n-component Heisenberg cases, respectively,

$$H_I = -J \sum_{(i,k)} S_i S_k - \sum_i h_i S_i$$

$$H_H = -J \sum_{(i,k)} \sum_{a}^{n} S_i^a S_k^a - \sum_i \sum_{a}^{n} h_i^a S_i^a, \qquad (3.3)$$

where the sum over i and k runs over all possible nearest-neighbor pairs of the lattice. If the exchange constant J is positive, the system is ferromagnetic, and parallel "spins" are energetically favored; if J is negative, the system is antiferromagnetic and nearby spins tend to stay antiparallel. On square or cubic lattices at zero external magnetic fields, ferromagnetic and antiferromagnetic models, having the same value of $|J|$, have the same partition function (i.e., $Z(J)$ is an even function of J): indeed one can switch from the ferromagnetic to the antiferromagnetic case by writing the Hamiltonian terms of new variables,

$$S'_i = S_i p(i),\tag{3.4}$$

where $p(i) = (-1)^{i_x+i_y+i_z}$ is the "parity" of the site, and i_x, i_y, and i_z are the integer coordinates of the site i (nearby sites have opposite parity). This equivalence does not hold for antiferromagnets on other lattices (e.g., the triangular or the *fcc* lattices); in these cases it is impossible to find a transformation like (3.4) that has a similar effect, i.e., to convert an antiferromagnetic Hamiltonian into a ferromagnetic one. The system is said to be frustrated; new phenomena, beyond the scope of this book, are present in this case.[2]

3.2. The mean-field approximation

In spite of the apparent simplicity of the Ising model and of the efforts of more than one generation of physicists, an exact solution (i.e., evaluation of the free energy and of the correlation functions) is available only in one dimension or in two dimensions at zero magnetic field. In the general case, approximate solutions must be considered. The simplest of these is the mean-field approximation, which we now describe.

The starting point of mean-field theory is the variational principle (1.18). However, we do not look for the true minimum of the free-energy functional, but restrict ourselves to simple forms of the probability distribution;[3] in the simplest approximation, the probability distribution is assumed to be factorized. The most general form is

$$P[S] = \prod_i P_i(S_i)$$

$$P_i(S) = \frac{1+m_i}{2} \cdot \delta_{S,1} + \frac{1-m_i}{2} \cdot \delta_{S,-1},\tag{3.5}$$

where $\delta_{a,b}$ is the Kronecker delta.

The probability P is already normalized to one:

$$\sum_{\{S\}} P[S] = \prod_i \left[\sum_{S_i=\pm 1} P_i(S_i)\right] = 1.\tag{3.6}$$

We denote by $\Sigma_{\{S\}}$ the sum over all configurations of the system (in this case their number is 2^N). The factorization of the probability implies the

3.2. The Mean-Field Approximation

following rules for the expectation values:

$$\langle g_1(S_1)g_2(S_2)\rangle_P = \langle g_1(S_1)\rangle_P \langle g_2(S_2)\rangle_P,$$

$$\langle g(S_i)\rangle_P = \frac{1+m_i}{2} g(1) + \frac{1-m_i}{2} g(-1), \quad (3.7)$$

where the g's are arbitrary functions. We finally find

$$\langle S_i \rangle = m_i,$$

$$\langle H \rangle = -J \sum_{(i,k)} m_i m_k - \sum_i h_i m_i,$$

$$S[P] = -\langle \ln P(S) \rangle_P = -\sum_i \langle \ln P_i(S_i) \rangle \quad (3.8)$$

$$= \sum_i s(m_i)$$

$$s(m) \equiv \frac{1+m}{2} \ln\left(\frac{1+m}{2}\right) + \frac{1-m}{2} \ln\left(\frac{1-m}{2}\right)$$

$$\Phi[P] = \langle H \rangle_P - \frac{S[P]}{\beta}.$$

The necessary condition for a minimum, $\partial\Phi/\partial m_i = 0$, becomes here

$$-\sum_k J_{ik} m_k - h_i + \beta \operatorname{arctanh} m_i = 0,$$

$$m_i = \tanh\left[\beta\left(\sum_k J_{ik} m_k + h_i\right)\right], \quad (3.9)$$

where we have introduced the matrix J_{ik}, this matrix is equal to J if i and k are nearest neighbors and it is zero elsewhere: $H = -\frac{1}{2}\sum_{i,k} J_{ik} S_i S_k - \sum_i h_i S_i$.

Equation (3.9) is the starting point of the so-called mean-field approximation, which states that each spin S_i feels on the average an external effective magnetic field equal to $h_i + J_{ik} S_k$. Here we have derived Eq. (3.9) starting from the hypothesis of factorized probabilities in order to stress its variational origin.

An alternative derivation of the mean-field equation starts from the equality[4]

$$\langle S_i \rangle \equiv \frac{\sum_{\{S\}} S_i \exp(-\beta H[S])}{\sum_{\{S\}} \exp(-\beta H[S])}$$

$$= \frac{\sum_{\{S\}} \exp(-\beta H[S]) \cdot \frac{\sum_{\mu_i = \pm 1} \exp(-\beta H[\mu_i, \hat{S}]) \mu_i}{\sum_{\mu_i = \pm 1} \exp(-\beta H[\mu_i, \hat{S}])}}{\sum_{\{S\}} \exp(-\beta H[S])}$$

$$= \frac{\sum_{\{S\}} \exp(-\beta H[S]) \tanh\left(\beta h_i + \beta \sum_k J_{ik} S_k\right)}{\sum_{\{S\}} \exp(-\beta H[S])}$$

$$= \langle \tanh(\beta h_i + \beta J_{ik} S_k) \rangle , \qquad (3.10)$$

where $\hat{S}$ denotes the set of all S's but S_i. The proof is very simple: If we sum over S_i in the numerator of the third expression in Eq. (3.10), we get $\Sigma_{S_i = \pm 1} \exp[-\beta H(S_i, \hat{S})]$, which cancels exactly with $\Sigma_{\mu_i = \pm 1} \times \exp(-\beta H(\mu_i, \hat{S}))$. The remaining term at the numerator, $\Sigma_{\{\hat{S}\}, \mu_i = \pm 1} \mu_i \times \exp[-\beta H(\mu_i, \hat{S})]$, becomes $\Sigma_{\{S\}} S_i \exp[-\beta H(S)]$ when μ_i is renamed S_i. On the other hand, if we sum over μ_i we get the hyperbolic tangent, indeed

$$\sum_{\mu = \pm 1} [\mu \exp(\beta h_{ef} \mu)] \bigg/ \sum_{\mu \pm 1} [\exp(\beta h_{ef} \mu)] = \tanh(\beta h_{ef}).$$

Equation (3.10) states that the expectation value of the spin S_i is exactly equal to the expectation value of the hyperbolic tangent of the effective field acting on it (i.e., $h_{ef} = h_i + J_{ik} S_k$).

If we approximate S_k with $m = \langle S_k \rangle$ (i.e., we set $\langle \tanh(h_i + J_{ik} S_k) \rangle = \tanh(h_i + J_{ik} m_k)$, we get the mean-field equations (3.9).

There are more refined approximations, which have as their starting points Eq. (3.10). Let us discuss for simplicity only the two-dimensional case, at zero external magnetic field for $J = 1$. If we call S_0, S_1, S_2, S_3, S_4 the spin S_i and its four neighbors, we get

3.2. The Mean-Field Approximation

$$m = \langle S_0 \rangle = \langle \tanh(\beta(S_1 + S_2 + S_3 + S_4)) \rangle$$

$$= \langle A(S_1 + S_2 + S_3 + S_4)$$

$$+ B(S_1 S_2 S_3 + S_2 S_3 S_4 + S_3 S_4 S_1 + S_4 S_1 S_2) \rangle \quad (3.11)$$

$$= 4mA + 4B \langle S_1 S_2 S_3 \rangle$$

$$A = \frac{\tanh(4\beta) + 2\tanh(2\beta)}{8} \qquad B = \frac{\tanh(4\beta) - 2\tanh(2\beta)}{8}$$

where we have used the fact that the spins can take only values ± 1. If now we approximate $\langle S_1 S_2 S_3 \rangle$ with m^3, we get $m = 4Am + 4Bm^3$. When we compare the result for the critical temperature in this approximation with the mean-field result (which will be computed in subsection 3.4), we see that we find $\beta_c = 0.357$ (i.e., $A = \frac{1}{4}$), which agrees better with the exact result $\beta_c \simeq 0.4404$ (see next chapter) than does the mean-field theory result $\beta_c = \frac{1}{4}$.

Many equations similar to Eqs. (3.11) and (3.12) can be obtained for the correlation functions by summing over not one spin, but many. Equations of this kind are normally called Dobrushin-Lanford-Ruelle equations[5] and are often used to control the probability distribution in the infinite-volume limit in a rigorous way. These equations imply that a given region (in the previous case one spin) is in thermal equilibrium with its surroundings.

Let us come back to the study of the usual mean-field theory equations, i.e., Eq. (3.9). We consider a ferromagnetic ($J > 0$) and set $J = 1$ for simplicity; if the h_i's do not depend on i (constant magnetic field), it is reasonable to suppose (and it is indeed true) that the minimum of Φ corresponds to constant magnetization (the m_i's do not depend on i). In this case all the terms of the sum over k in Eq. (3.9) give the same contribution and their number is equal to $2D$. We finally find

$$\langle S_i \rangle \equiv m_i = m$$

$$\langle H \rangle / N \equiv u = -Dm^2 - hm$$

$$\varphi(m) = \frac{\Phi(m)}{N} = -Dm^2 - hm + \frac{1}{\beta} s(m) \quad (3.12)$$

$$m = \tanh((2Dm + h)\beta).$$

Of course the condition $d\varphi/dm = 0$ is not sufficient. We must also impose the condition $d^2\varphi/dm^2 > 0$ in order to be sure that we have found a local minimum and not a local maximum of φ. Finally we must compare all the local minima to find the global one.

Let us first discuss the case $h = 0$. The point $m = 0$ is always a solution of the equation $d\varphi/dm = 0$; an elementary computation shows that $d^2\varphi/dm^2 = \chi^{-1} = -2D + 1/\beta(1 - m^2)$.

In the region $2D\beta < 1$, $d^2\varphi/dm^2$ is always positive in the physical interval $-1 \le m \le 1$. Therefore there is only one solution of the equation $d\varphi/dm = 0$, and the solution is a (global) minimum. However, in the region $2D\beta > 1$, $d^2\varphi/dm^2$ is negative for m close to zero (i.e., $m^2 < 1 - \frac{1}{2}D\beta$) and positive elsewhere: the point at $m = 0$ is a local maximum. There are two symmetric minima at $m = \pm m_s$, where

$$m_s \approx 3^{1/2}(2D\beta - 1)^{1/2} \qquad 2D\beta \sim 1$$
$$m_s \approx 1 - 2\exp(-4D\beta) \qquad 2D\beta \to \infty. \qquad (3.13)$$

If $|h|$ is small, two minima and one maximum will be present. The minimum at which m has the same sign as h will be the global one; if $|h|$ is large enough, the higher minimum and the maximum coalesce, and the function $\varphi(m)$ has only one minimum. The functions corresponding to the various situations are sketched in Fig. 3.1.

The corresponding behavior of $m(h)$ is shown in Fig. 3.2 in the two regions $\beta \gtrless \beta_c = \frac{1}{2}D$. For $\beta > \beta_c$ spontaneous magnetization is present,

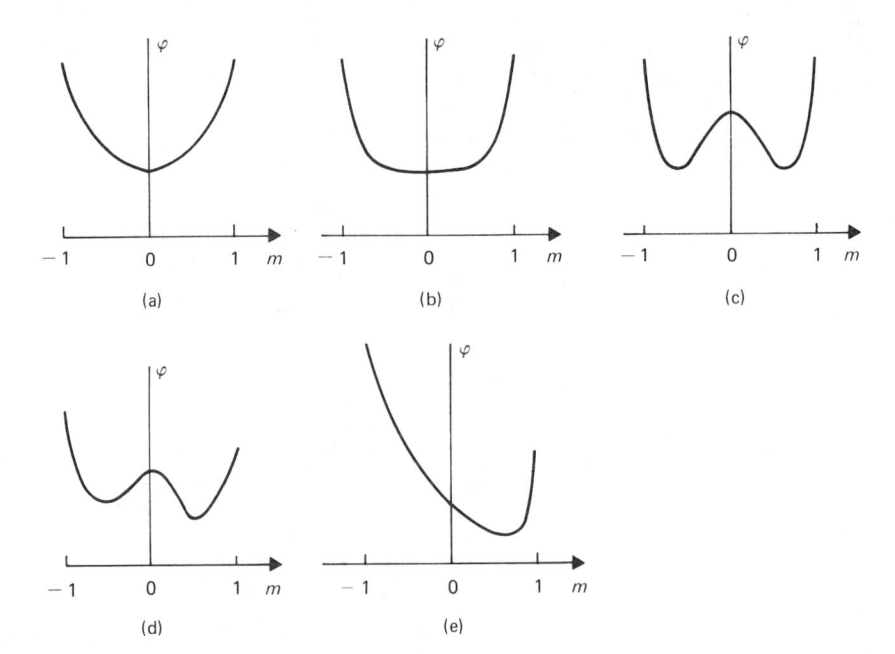

Figure 3.1. A qualitative sketch of the free energy φ as a function of the magnetization m in various situations: (a) $h = 0$, $T > T_c$; (b) $h = 0$, $T = T_c$; (c) $h = 0$, $T < T_c$; (d) positive small h, $T < T_c$; (e) positive large h, $T < T_c$ or positive small h, $T > T_c$.

3.2. The Mean-Field Approximation

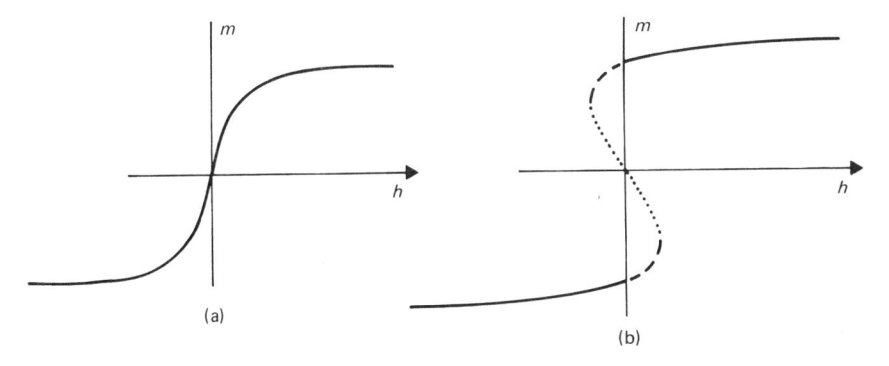

Figure 3.2. The magnetization m as a function of the magnetic field h: (a) $T > T_c$; (b) $T < T_c$. In Fig. 3.2b the solid, dashed, and dotted lines denote the stable, metastable, and unstable states, respectively.

and $T_c = 1/\beta_c$, the Curie temperature. The corresponding curve for the free energy in the low-temperature region is shown in Fig. 3.3. Both in Fig. 3.2b and in Fig. 3.3 we can distinguish three regions: the physical one (solid line), which corresponds to the global minimum, the unphysical one (dotted line), which corresponds to the local maximum (here the susceptibility is negative!), and a third intermediate region (dashed line), which corresponds to a local minimum.

Handwaving arguments suggest that if we prepare the state at $t = 0$ in the metastable region, i.e., $m > 0$ but $h < 0$ (this can be done by decreasing the magnetic field adiabatically from positive to negative), the system will remain there (local stability) for a very long time, up to the moment when a thermodynamical fluctuation will allow the system to go

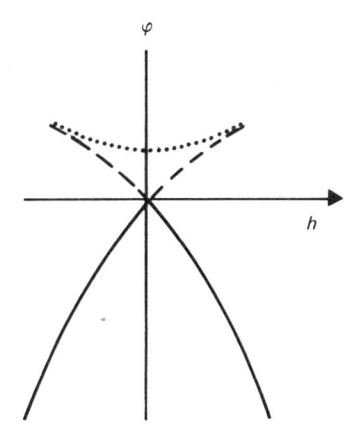

Figure 3.3. The free energy φ as a function of the magnetic field h: The solid, dashed, and dotted lines denote the stable, metastable, and unstable states, respectively.

to a more stable state $m < 0$. This region is called metastable; the mean life of a metastable state may range from seconds to more than a century.[6] Intuitively we expect that we cannot define with infinite precision the expectation values for a metastable state, the observation time being large but finite; if t is finite in Eq. (1.2), the expectation values will have a small but nontrivial dependence on the initial condition. That is precisely what happens in a purely static approach, as we shall see in Chapter 10.

To summarize, one finds the following result near T_c at $h = 0$:

$$u = 0, \quad m = 0 \quad \text{for } T > T_c;$$
$$m_s \propto (T_c - T)^\beta, \quad u \propto T - T_c \quad \text{for } T < T_c; \quad (3.14)$$
$$\chi \propto |T - T_c|^{-\gamma};$$
$$\beta = \tfrac{1}{2} \quad \gamma = 1.$$

At $T = T_c$ but $h \neq 0$, one finds

$$m \propto h^{1/\delta}, \quad \delta = 3. \quad (3.15)$$

The specific heat is discontinuous, but there is no latent heat at $T = T_c$; we are in presence of a second-order transition; however, no singularity is present if h is different from zero.

In this way we have derived a rigorous upper bound to the free energy. We can, however, approximate the free energy with the upper bound; in this case the spins S_i are assumed not to be correlated.

We shall soon see how one can compute the correlation functions using the linear response theorem: we can thus obtain the first correction to the mean-field approximation and therefore we can decide in which cases the mean-field approximation is reliable. Before doing this, let us sketch how we can reproduce the results of the mean-field approximation in a slightly different way. The key ingredient is the identity

$$\int d\mu\,(C) \exp[-\beta H] \geq \int d\mu\,(C) \exp[-\beta H_0 - \beta \langle H - H_0 \rangle_{H_0}], \quad (3.16)$$

where

$$\langle A \rangle_{H_0} = \frac{\int d\mu\,(C) \exp(-\beta H_0) A}{\int d\mu\,(C) \exp(-\beta H_0)}. \quad (3.17)$$

3.3. A Soluble Model: Weak Long-Range Forces

Equation (3.16) is essentially a convexity inequality that can easily be proved by considering the function

$$g(\lambda) = \ln \int d[C] \exp[-\beta H_0 - \beta\lambda(H_1 - H_0)]. \tag{3.18}$$

Obviously $d^2g/d\lambda^2 \geq 0$; therefore

$$g(\lambda) \geq g(0) + \lambda \left.\frac{dg}{d\lambda}\right|_{\lambda=0}. \tag{3.19}$$

Specializing Eq. (3.19) to the case $\lambda = 1$, we recover inequality (3.16). Let us look at the case $h_i = 0$ and consider a trial Hamiltonian $H_0 = r \sum_i S_i$. One finds for the free-energy density

$$f \leq -\frac{1}{\beta} \ln(2 \cosh(r\beta)) + D \tanh^2(r\beta) - r \tanh(\beta r), \tag{3.20}$$

$$\langle H - H_0 \rangle_{H_0} = Dm^2 - rm, \quad m = \tanh(\beta r).$$

Now if we substitute $(1/\beta) \text{arctanh}(m)$ for r, we recover Eq. (3.11). This should not be a surprise: indeed, we define

$$P_{H_0}(C) = \frac{\exp(-\beta H_0(C))}{\int d\mu(C) \exp(-\beta H_0(C))}. \tag{3.21}$$

The convexity inequality reduces to the variational principle, and for the H_0 we have considered, P_{H_0} has exactly the form given by Eq. (3.5).

3.3. A soluble model: weak long-range forces

In this subsection, we present an Ising model that is exactly soluble by the mean-field approximation.

The model has the same Hamiltonian as Eq. (3.3), where now the sum on i and k runs over all possible pairs of a lattice of N sites, the variable J being equal to $1/N$. The force among spins is weak and long range; the thermodynamic (i.e., infinite-volume) limit is obtained by sending N to infinity. The model can be solved using the key identity

$$\exp(-\beta H) \equiv \exp\left(+\frac{\beta}{2N} \sum_{i,k} S_i S_k + \beta h \sum_i S_i\right)$$

$$= \left(\frac{N\beta}{2\pi}\right)^{1/2} \int_{-\infty}^{+\infty} d\lambda \exp\left[-\frac{N\beta\lambda^2}{2} + \sum_i (\beta\lambda + \beta h) S_i\right]. \tag{3.22}$$

Indeed, the partition function can be written as

$$Z = \sum_{\{S\}} \exp(-\beta H) = \left(\frac{N\beta}{2\pi}\right)^{1/2}$$
$$\times \int_{-\infty}^{\infty} d\lambda \exp\left(-\frac{N\beta\lambda^2}{2}\right) \cdot [2\cosh(\beta\lambda + \beta h)]^N \qquad (3.23)$$
$$= \left(\frac{N\beta}{2\pi}\right)^{1/2} \int_{-\infty}^{+\infty} d\lambda \exp(-N\beta A(\lambda))$$

$$A(\lambda) = \frac{\lambda^2}{2} - \beta^{-1} \ln(2\cosh(\beta\lambda + \beta h)),$$

where $\Sigma_{\{S\}}$ stands for the sum over the 2^N possible configurations of the N spins. The introduction of the variable λ has allowed us to disentangle the contribution of the various spins and to study independent site problems. In the limit $N \to \infty$ the integral may be evaluated using the method of steepest descent.[7] One easily finds that

$$f = \min_\lambda A(\lambda) + O\left(\frac{1}{N}\right). \qquad (3.24)$$

The value $\bar{\lambda}$ for which $A(\lambda)$ is minimized has the physical meaning of the magnetization density; the minimum condition implies

$$\left.\frac{dA}{d\lambda}\right|_{\lambda=\bar\lambda} = 0 \Rightarrow \lambda = \tanh[\beta(\lambda + h)]$$
$$m = -\frac{\partial f}{\partial h} = -\frac{\partial A}{\partial h} - \frac{\partial \lambda}{\partial h}\frac{\partial A}{\partial \lambda} = \tanh[\beta(\lambda + h)] = \lambda. \qquad (3.25)$$

Equation (3.25) can be obtained in the mean-field approximation, as we have seen. In other words, the mean-field approximation becomes exact in the infinite-range model. A simple solution of the model is possible owing to the fact that the connected correlation function $\langle S_i S_k \rangle_C$ (which is i and k independent) is of order $1/N$ and vanishes when N goes to infinity. Indeed, from the linear response theorem, it is given by

$$\langle S_i S_k \rangle_C = \frac{\chi}{N\beta} = \frac{1}{N\beta} \frac{d\bar\lambda}{dh}. \qquad (3.26)$$

As we shall see later, the mean-field approximation is also exact for finite-range forces in the limit in which the spatial dimensions D go to infinity.[8] In this case the connected correlation functions also go to zero.

The existence of this soluble model allows us to rederive the mean-field

3.4. The Correlation Functions

approximation in the following way. We consider the Hamiltonian

$$H_\alpha = (1-\alpha)\frac{D}{N-1}\sum_{i,k}{}' S_i S_k + \frac{\alpha}{2}\sum_{i,k} J_{i,k} S_i S_k, \qquad (3.27)$$

where the sum Σ' runs over all i and k with the condition $i \neq k$.

At $\alpha = 0$ and $\alpha = 1$ we recover, respectively, the infinite-range and short-range models. If we call f_α the free energy density corresponding to the Hamiltonian H_α, it is evident that

$$\frac{d}{d\alpha} f_\alpha\bigg|_{\alpha=0} = \frac{\left\langle \frac{\partial H_\alpha}{\partial \alpha}\right\rangle_{\alpha=0}}{N}$$

$$= \sum_{i,k}\left(\frac{D}{N-1} - \frac{J_{i,k}}{2}\right)\frac{\langle S_i S_k\rangle_{\alpha=0}}{N}$$

$$= \left(m^2 + \frac{\chi}{\beta N}\right)\sum_{i,k}{}'\left[\frac{D}{N-1} - \frac{J_{i,k}}{2}\right] = 0. \qquad (3.28)$$

Indeed, for fixed i there are $N-1$ and $2D$ contributions proportional to $D/(N-1)$ and to $-|J|/2 \equiv -\frac{1}{2}$, respectively.

The convexity inequality (3.19) tells us that $f_0 \geq f_1$. We have thus recovered the standard inequality of mean-field theory. We shall see later how a similar procedure can be used to derive another soluble model of the mean-field approximation, the hierarchical model, which, however, has a much richer structure.

3.4. The correlation functions

We now compute the correlation functions in the framework of the mean-field approximation. At first sight this seems impossible. The very first assumption of the mean-field approximation is that spins are uncorrelated: the probability distribution is factorized. This difficulty may be bypassed by using the linear response theorem,

$$\langle S_i S_k\rangle_{h=0} = -\frac{1}{\beta}\frac{\partial}{\partial h_i}\frac{\partial}{\partial h_k} F[h]\bigg|_{h=0} = \frac{1}{\beta}\frac{\partial m_i}{\partial h_k}\bigg|_{h=0}. \qquad (3.29)$$

We can thus start from the factorized $P[S]$ to compute $F[h]$ for general h, as discussed in the previous section, and obtain the correlation function using Eq. (3.29). The reader may find the argument paradoxical—we use a distribution probability that has zero correlation functions to compute the correlation functions! The solution is evident: Eq. (3.29) holds only for the true F or if $P = P_{eq}$. If $P \neq P_{eq}$ the correlation function and the response function are no longer equal. Now in order to justify the use of Eq. (3.29) we must argue that the error of the response function is smaller than the error of the correlation function when P is near but not equal to P_{eq}. This is true, as can be seen from a quite general argument (which holds for most of the variational problems). If P_{ap} is the approximate distribution probability, we assume that the difference $P_{eq} - P_{ap} \equiv \delta P$ is small in some sense. (If δP is not small, the approximation does not make sense. This assumption is quite natural.) Just to compare orders, let us say that δP is of order ε. It is clear that

$$\langle S_i S_k \rangle_{P_{eq}} = \langle S_i S_k \rangle_{P_{ap}} + 0(\varepsilon)$$
$$\Phi(P_{ap}) = \Phi(P_{eq}) + 0(\varepsilon^2) . \tag{3.30}$$

Indeed, the term proportional to ε vanishes due to the condition $\delta \Phi / \delta P|_{P=P_{eq}} = 0$. Similarly, in the variational approach to quantum mechanics, if there is an error of order ε on the wave function (ψ) of an eigenstate, the error of the corresponding energy ($E = \langle \psi | H | \psi \rangle$) is of order ε^2. This argument justifies the computation of the correlation functions using Eq. (3.29).

For simplicity let us present the computation only in the region $h = 0$, $T > T_c$. Here $m_s = 0$.

If we add an infinitesimal point-dependent magnetic field, the m_i will also be infinitesimally small. Neglecting terms of order m_i^2, we find that Eq. (3.9) reduces to

$$-\sum_k J_{ik} m_k + \frac{m_i}{\beta} = h_i , \tag{3.31}$$

whose solution is in the matrix notation

$$m_i = \sum_k A_{ik} h_k$$
$$A_{ik} = \left(\frac{1}{\beta} - J\right)^{-1}_{i,k} \Leftrightarrow \sum_k \left(\frac{\delta_{ik}}{\beta} - J_{ik}\right) A_{kl} = \delta_{il} . \tag{3.32}$$

In other words, A is the inverse of the matrix $1/\beta - J$.

The solution of Eq. (3.32) can easily be found using the Fourier transform.[9] If the configuration space is a cubic lattice, the Fourier transform is defined in the first Brillouin zone $-\pi \leq p_\nu \leq \pi$, $\nu = 1, \ldots, D$. We obtain

3.4. The Correlation Functions

$$A_{kl} = \frac{\beta}{(2\pi)^D} \int_B d^D p\, G^0(p) \exp[i(\vec{k}-\vec{l})\cdot\vec{p}]$$

$$G^0(p) = \left(1 - 2\beta \sum_\nu^D \cos p_\nu\right)^{-1} \tag{3.33}$$

$$\int_B d^D p \equiv \prod_\nu^D \int_{-\pi}^{\pi} dp_\nu \;.$$

Now it is evident that

$$\langle S_i S_k \rangle = \frac{\frac{\partial m_i}{\partial h_k}}{\beta} = \frac{A_{ik}}{\beta} \;. \tag{3.34}$$

Using this result for the correlation functions, we can compute the corrections to the mean-field result $u = 0$ for $T > T_c$. We find

$$u = -\frac{1}{2} \sum_k \langle S_i S_k \rangle J_{ik} \simeq \frac{-\frac{1}{2}\sum_k A_{ik} J_{ki}}{\beta} \;. \tag{3.35}$$

Using the fact that J connects spins only at distance one, we obtain

$$u = \frac{-1}{(2\pi)^D} \int_B d^D p\, \frac{\sum_\nu^D \cos p_\nu}{1 - 2\beta \sum_\nu^D \cos p_\nu}$$

$$C = \frac{du}{dT} = \frac{2\beta^2}{(2\pi)^D} \int_B d^D p \left(\frac{\sum_\nu^D \cos p_\nu}{1 - 2\beta \sum_\nu^D \cos p_\nu}\right)^2 \;. \tag{3.36}$$

Near the phase transition one finds that $G^0(p)$ develops a pole at $p^2 = 0$:

$$G^0(p) = [\beta p^2 + (1 - 2D\beta) + 0(p^4)]^{-1} \tag{3.37}$$

and

$$A_{k,0} \underset{k\to\infty}{\propto} k^{-D+2} \quad \text{at exactly} \quad 2D\beta = 1.$$

Now the integration region near $p = 0$ may give a large contribution to the specific heat near the transition. Indeed, we find for small $1 - 2D\beta$:

$$C \simeq \text{const.} \qquad D > 4$$

$$C \simeq -\ln(1 - 2D\beta) \qquad D = 4$$

$$C \simeq (1 - 2D\beta)^{D/2 - 2} \qquad D < 4 \qquad (3.38)$$

$$A_{k,0} \underset{k \to \infty}{\propto} \exp\left(-\frac{k}{\xi}\right)$$

$$\xi \propto \frac{1}{(1 - 2D\beta)^{1/2}}.$$

Whereas at high temperatures the specific heat is small (proportional to T^{-2}), in four or fewer dimensions it diverges at the phase-transition point. We can expect that the mean-field approximation is good only if the first corrections are small. However, this is not true near the critical point in four or fewer dimensions, due to the large contribution to the fluctuations of the small p region. A similar computation shows that in the low-T region, $T < T_c$, the corrections are small if we are not too near the critical temperature. The correlation length ξ always diverges at $T = T_c$.

As we said before, the mean-field approximation is exact when $D \to \infty$, at fixed $\tilde{\beta} = 2\beta D$; we are not going to present here the full proof but merely check that the correlation function of two different spins vanishes at fixed $\tilde{\beta}$ when $D \to \infty$ as a power of D^{-1}. (In this limit each spin is coupled to a very large number of spins.) We now compute in the limit $D \to \infty$ the correlation function at fixed distance n (for definiteness, we assume that $i_x - k_x = n$, $i_\nu = k_\nu$ for $\nu \neq 1$). We find it useful to write

$$\tilde{G}^0(n) = \frac{1}{(2\pi)^D} \int_B d^D p \, \exp(i p_x n) G^0(p)$$

$$= \frac{1}{(2\pi)^D} \int d^D p \int_0^\infty d\alpha \, \exp\left[-\alpha\left(1 - \frac{\tilde{\beta}}{D} \sum_1^D \cos p_\nu\right) + i p_x n\right]. \qquad (3.39)$$

The integrals over the different momenta can be calculated independently from one another. It is convenient to use the formula[10]

$$\frac{1}{2\pi} \int_{-\pi}^{\pi} dp \, \exp(\lambda \cos p + inp) = I_n(\lambda). \qquad (3.40)$$

One finds that for $\tilde{\beta} < 1$

3.4. The Correlation Functions

$$\tilde{G}^0(n) = \int_0^\infty d\alpha \, \exp(-\alpha) I_n\left(\frac{\alpha\tilde{\beta}}{D}\right) \left[I_0\left(\frac{\alpha\tilde{\beta}}{D}\right)\right]^{D-1}$$

$$\sim \left(\frac{\tilde{\beta}}{D}\right)^n, \tag{3.41}$$

as can be seen using the following property of the Bessel functions:[11]

$$I_0(\lambda) = 1 - \lambda^2 + 0(\lambda^4)$$

$$I_n(\lambda) = \frac{\lambda^n}{n!} + 0(\lambda^{n+1}) \tag{3.42}$$

$$I_n(\lambda) \sim \frac{\exp \lambda}{\lambda} \quad \text{for } \lambda \to \infty.$$

Indeed, at fixed $\tilde{\beta}$ the integral is dominated by the region of integration in α near $\alpha = 0$, when $D \to \infty$. The possible dangerous region $\alpha\tilde{\beta}/D \gg 1$ does not contribute as long as $\tilde{\beta} \leq 1$. When $\tilde{\beta} > 1$, the α integral is divergent, but for these values of $\tilde{\beta}$ we stay in the low-temperature phase and Eq. (3.29) is not valid.

We have already stated that for large n we have an exponential decay of the correlation functions at distances larger than the correlation length ξ [Eq. (3.38)]. Although this result may be obtained directly from the integral representation (3.41), it is simpler to use the following theorem: If $f(p)$ is an analytic function (with period 2π) and the nearest singularity to the real axis is at $p = ip_s$, we have[11]

$$\int_{-\pi}^{\pi} dp \, f(p) \exp(inp) \underset{n \to \infty}{\approx} \exp(-np_s). \tag{3.43}$$

The proof is very simple, and it is based on the deformation of the integration path in the complex p plane. Now the function

$$f(p) = \int_{-\pi}^{\pi} \prod_{2}^{D} dp_\nu \left[1 - 2\beta\left(\cos p + \sum_{2}^{D} \cos p_\nu\right)\right]^{-1} \tag{3.44}$$

will be singular at the point where

$$1 - 2\beta(\cos(ip_s) + D - 1) = 0. \tag{3.45}$$

For example, in even dimensions we have

$$f(p) \sim (p - ip_s)^{D/2 - 3/2} + \text{regular term}. \tag{3.46}$$

The singular term comes from the integration region where all the p_ν

($\nu = 2, \ldots, D$) are near zero. Equation (3.45) implies

$$p_s \equiv \tilde{\xi}^{-1} = \text{arccosh}\left[\frac{1}{2\beta} - D + 1\right] \simeq \left(\frac{1-2D\beta}{\beta}\right)^{1/2} \quad \text{for } 2D\beta \simeq 1. \tag{3.47}$$

The correlation length goes to infinity at the critical temperature. Indeed, we already know that at the critical temperature the correlation function decays like a power, i.e., much slower than an exponential. In principle we could define the correlation length in more general terms: if $\vec{v}$ is a rational unit vector [i.e., $(\frac{3}{5}, \frac{4}{5}, 0, 0, \ldots)$] the correlation length in the direction $\vec{v}$ is defined by

$$G^0(n\vec{v}) \sim \exp[-n\xi(\vec{v})] \tag{3.48}$$

when n goes to infinity (for those n such that $n\vec{v}$ has integer components). Clearly we recover the previous definition of ξ when $\vec{v}$ is in the direction of one of the axes of the lattice.

The generalization of the previous argument tells us that

$$\sum_{\nu}^{D} \cosh[\xi^{-1}(\vec{v}) v_\nu] = \frac{1}{2\beta}. \tag{3.49}$$

Using this definition, we find that the correlation length depends on the direction v; fortunately near the transition point we have

$$\xi^{-1}(\vec{v}) = \left(\frac{1-2D\beta}{\beta}\right)^{1/2}\left[1 - \frac{1-2D\beta}{6\beta}\vec{v}^4\right]$$

$$\vec{v}^4 \equiv \sum_{\nu}^{D} v_\nu^4. \tag{3.50}$$

The divergent term of the correlation length is independent of v; indeed, near the transition point the correlation function is isotropic at large distances, i.e.,

$$A_{0,k} \simeq A(k^2), \qquad k^2 \equiv \sum_\nu k_\nu^2 \tag{3.51}$$

in the region of large k. The dominant integration region is at small p, where $\tilde{G}(p)$ can be approximated as

$$\beta^{-1}\left[\frac{1-2\beta D}{\beta} + p^2 + 0(p^4)\right]^{-1}. \tag{3.52}$$

3.5. Second-Order Phase Transitions

As we shall see later in the general case as well, the correlation functions will be isotropic at the transition point. We notice *en passant* that Eq. (3.47) is in apparent contradiction with Eq. (3.41): the latter does not reduce to the former when D goes to infinity. The reason for this disagreement is clear: The limits $D \to \infty$ and $n \to \infty$ do not commute, and it is possible that

$$\lim_{D \to \infty} \left\{ \lim_{n \to \infty} \frac{1}{n} \ln[G(n)] \right\} \neq \lim_{n \to \infty} \left\{ \frac{1}{n} \ln[\lim_{D \to \infty} G(n)] \right\}. \quad (3.53)$$

Indeed, Eq. (3.41) has been obtained in the limit $D \to \infty$ at fixed n. A different formula is obtained if we study the large-D limit in the region n much larger than D.

3.5. Second-order phase transitions

We can now use the results we have obtained in the study of the mean-field approximation to get some insight into the general properties of second-order phase transition.[12]

Near the critical point the correlation function behaves as

$$\tilde{G}^0(n) \sim \exp\left(-\frac{n}{\xi}\right) \Big/ \left(n^{(D-1)/2} \cdot \xi^{(D-3)/2}\right) \quad \text{for } \frac{n}{\xi} \to \infty$$

$$\xi \sim |T - T_c|^{-\nu}, \quad \nu = \frac{1}{2} \quad (3.54)$$

$$G^0(n) \sim n^{-(D-2+\eta)}, \quad \eta = 0 \text{ for } T = T_c.$$

In other words, the correlation function is exponentially damped at all temperatures but the critical one, where the correlation length ξ goes to infinity.

Now the correlation function is essentially different from zero at distance ξ only if the system can have a coherent fluctuation of size ξ. The coherence length has thus the meaning of the maximum radius of likely fluctuations, i.e., fluctuations of radius larger than ξ have exponentially small probabilities. It is generally believed that at second-order phase transition point the coherence length goes to infinity. The argument is based on the fact that fluctuations which produce an increase of order $kT = \beta^{-1}$ in free energy are definitely present.

Let us study for simplicity what happens when we reach T_c from below on the line $h = 0$. The critical point ($T = T_c$) is the endpoint of a line of first-order transitions.[13] Now we argue that if the difference between the two phases goes to zero at the critical point (in this case the two phases have a different spontaneous magnetization $\pm m_s$), the correlation length must have gone to infinity. Let us consider the possibility of having a bubble of large radius of phase II (i.e., negative magnetization) while the rest of the system is in phase I (i.e., positive magnetization). See Fig. 3.4 for such a configuration. The only part out of equilibrium would be the surface of the sphere. The contribution to the free energy coming from the interphase is intuitively proportional[14] to $R^{D-1}m_S^2$. Fluctuations up to a critical radius $R_c = m_S^{-2/(D-1)}$ are allowed. Obviously R_c goes to infinity when $m \to 0$.[15]

The argument is compelling from the physical point of view.[16] As we approach the critical point we can have correlated fluctuations on larger and larger scales; just at the critical point the fluctuation radius becomes infinite, and we have correlation functions that decay like powers at large distances.

Collecting together the predictions of the mean-field approximation, Eqs. (3.12)–(3.15), (3.38), and (3.42), we have

$$\xi(T, h)|_{h=0} \simeq |T - T_c|^{-\nu}, \qquad \chi(T, h)|_{h=0} \simeq |T - T_c|^{-\gamma},$$

$$\tilde{G}(n)|_{T=T_c, h=0} \simeq n^{-(D-2+\eta)}, \qquad C(T, h)|_{h=0} \simeq |T - T_c|^{-\alpha}, \qquad (3.55)$$

$$m(T, h)|_{h=0} \simeq |T - T_c|^{\beta}, \qquad m(T, h)|_{T=T_c} \simeq |h|^{1/\delta}$$

where $\alpha, \beta, \gamma, \delta, \eta, \nu$ are called critical exponents and take the values:[17]

$$\alpha = D - 4, \qquad \beta = \tfrac{1}{2}, \qquad \gamma = 1,$$
$$\delta = 3, \qquad \eta = 0, \qquad \nu = \tfrac{1}{2}. \qquad (3.56)$$

The existence of strong fluctuations makes the predictions of mean-field theory unreliable (they have been derived under the hypothesis of small fluctuations). We shall see later on that Eqs. (3.55) are indeed correct,

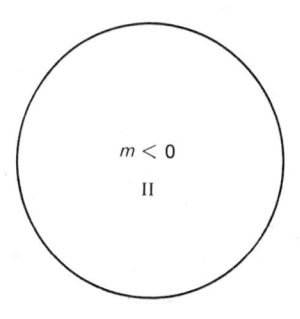

Figure 3.4. A bubble (of radius R) of negative magnetization in a sea of positive magnetization.

3.6. The Infinite-Volume Limit

but for $D < 4$ the true values of the critical exponent are not given by Eq. (3.56).[18]

There is strong experimental and theoretical evidence that the values of the exponents have a high degree of universality (that is, they are the same for different lattice structures). This result should not be considered too surprising because the exponents give us information about fluctuations on a scale much larger than the lattice spacing and at such large distances that the detailed form of the interaction should be an irrelevant detail.

The problems of understanding precisely the reasons for this remarkable universality, the determination of relations among critical exponents [e.g., $\gamma = (2 - \eta)\nu$], and the very computation of the exponents, were the object of very intensive studies in the years 1964–1976. Nowadays, from a physical point of view, the problem is essentially solved, although in most cases only approximate evaluations of the critical exponents are available.

The general strategy for obtaining these results is described in Chapter 7 dedicated to the renormalization group. Before reaching this stage, we must develop tools (e.g., perturbative expansion, diagrammatical techniques) that will allow us to compute the corrections to the mean-field approximation in a systematic way.

3.6. The infinite-volume limit

As we have already stressed, we want to compute the free-energy density in the infinite-volume limit. If the problem is well defined, the free-energy density of a system in a finite box must have a limit when the size of the box goes to infinity, and the limit should not depend on the boundary conditions we have used. We now present a rigorous elementary proof of this fact for the Ising model at zero magnetic fields with nearest-neighbor interaction; the extension to an arbitrary finite-range model is trivial.

We first note that the total energy of a system of N spins satisfies the obvious bounds

$$-ND < E < ND. \tag{3.57}$$

The number of configurations is 2^N, so that the partition function Z_N and the free-energy density f_N satisfy the bounds

$$2^N \exp(-\beta ND) \leq Z_N \leq 2^N \exp(\beta ND)$$

$$-D + \frac{\ln 2}{\beta} \leq f_N \leq D + \frac{\ln 2}{\beta}. \tag{3.58}$$

These bounds are very important. They exclude the most dangerous possibility (i.e., $f_N \to \infty$ when $N \to \infty$). For simplicity, we consider the case of cubic boxes $N = L^D$ (where L can take only the values 2^n) and use free boundary conditions. We denote by Z_N^F the corresponding partition function.

It is easy to see that

$$(Z_{L^D}^F)^{2^D} \exp(-D(2L)^{D-1}\beta)$$
$$\leq Z_{(2L)^D}^F \leq (Z_{L^D}^F)^{2^D} \exp(D(2L)^{D-1}\beta). \qquad (3.59)$$

Indeed, we can glue together 2^D systems of side L to give one system of side $2L$. (See Fig. 3.5 in two dimensions.) The only difference between $(Z_L)^{2^D}$ and Z_{2L} will come from the contribution to the energy coming from spins near the boundary, but their contribution cannot exceed $D(2L)^{D-1}$.

Equation (3.47) implies

$$|f_{2L} - f_L| \leq \frac{C}{L}, \qquad (3.60)$$

where C is a constant independent of L. A standard theorem implies the convergence of the sequence $f_{2^n L}$ for $n \to \infty$. A similar technique can be used to prove the independence of f on the boundary conditions in the infinite-volume limit. Indeed, the effect of the boundary conditions on the total free energy is proportional to L^{D-1} (a surface effect), and it cannot change the value of $f \equiv F_L/L^D$ in a significant way.[19]

A more elegant proof of the existence of the infinite-volume limit for the Ising system may be obtained by proving that the free-energy density

Figure 3.5. Four regions of size L^D glued together to produce a region of size $(2L)^D$ in two dimensions.

with periodic[20] and free[21] boundary conditions is a monotonically increasing or decreasing function, respectively:[22] A monotonic bounded sequence has a limit. The proof, unfortunately, is not so simple and general.

Notes for Chapter 3

1. Although the origin of the force among spins is clearly quantum mechanical, for many purposes it is sufficient to use a classical effective Hamiltonian. In a real material there is always an effect of the lattice structure on the spin-spin forces. If this effect can be neglected, there is no preferred direction for the magnetization, and we obtain the Heisenberg model (full isotropy). In the opposite situation (strong anisotropy), the spins can orient themselves only in one direction, and we obtain the Ising model. Of course in real materials there is also a weak long-range force, the dipole interaction, which plays an important role in the formation of magnetic domains and block walls; for more details see C. Kittel, *Introduction to Solid State Theory*, John Wiley and Sons, New York (1966).
2. For an introduction to frustration see G. Toulouse, *Commun. Math. Phys.* **2**, 115 (1977); the behavior of some fully frustrated (antiferromagnetic) Ising models is studied in G. H. Wannier, *Phys. Rev.* **79**, 3757 (1950) (triangular lattice), M. K. Phani, J. L. Lebowitz, and M. H. Kalos, *Phys. Rev. B* **21**, 4027 (1980) (face-centered cubic lattice).
3. This procedure is very similar to the computation of the ground-state energy of a quantum system using a variational principle, i.e., $E_0 = \min_\psi \langle \psi | H | \psi \rangle$.
4. This identity has been derived by H. B. Callen, *Phys. Lett. B* **4**, 161 (1963). It has been used to put an upper bound on the critical temperature by F. C. Sá Barreto and M. L. O'Carrol, *J. Phys. A* **16**, 1035 (1983).
5. R. L. Dobrushin, *Theory Prob. Appl.* **13**, 197 (1969); O. E. Lanford, III, and D. Ruelle, *Commun. Math. Phys.* **13**, 194 (1969).
6. One of the most popular metastable states is supercooled water (a more dangerous example is a mixture of H_2 and O_2).
7. The method of steepest descent says that $\lim_{N \to \infty} (1/N) \ln[\int dx \exp(Ng(x))] = \max[g(x)]$ if $g(x)$ is a real function; for further information see Chapter 14.

8. A proof of the exactness of the mean-field approximation in infinite dimensions can be found in M. E. Fisher and D. S. Gaunt, *Phys. Rev.* **133**, A224 (1964) and R. Abe, *Prog. Theor. Phys.* **47**, 62 (1972).
9. For the Fourier transform of functions defined on a lattice and the related definition of Brillouin zones, see C. Kittel, op. cit.
10. The I_n are integer order Bessel functions of imaginary arguments; here (and in the rest of the book) we follow for special functions the notation of I. S. Gradshteyn and I. M. Ryzhik, *Table of Integrals, Series and Products*, Academic Press, New York (1965).
11. See, for example, P. M. Morse and H. Feshbach, *Methods of Theoretical Physics*, McGraw-Hill, New York (1953), Chapter 4.
12. A famous review of systems that undergo a second-order phase transition is L. P. Kadanoff *et al.*, *Rev. Mod. Phys.* **39**, 395 (1967).
13. The critical point of the liquid-gas transition is very similar to the Curie point in this respect.
14. The actual power of m_S^2 is not important; we require only that it vanish at the Curie point.
15. This approach may be pushed further. A first attempt in the sixties was not fully successful in obtaining quantitative predictions for the critical exponents. For a review see M. F. Fisher, *Physica* **3**, 255 (1967). Nowadays it is understood that this approach may become quantitative for the Ising case in $1 + \varepsilon$ dimensions (ε being a small number); for more details see the review of D. Wallace in *Proceedings of the 1982 Les Houches Summer School*, ed. by J. B. Zuber and R. Stora, North-Holland, Amsterdam (1984), p. 173.
16. We remark that this result is automatic if m can be found by solving an equation of the form $\varphi'(m, T) = h$, where $\varphi(m, T)$ is a sufficient smooth function. Indeed, the coalescence of the two solutions, that for m different from zero with that for $m = 0$, implies that $\varphi''(0, T_c) = 0$. The magnetic susceptibility $\chi \propto (\partial^2 \varphi / \partial m^2)^{-1}$ is thus infinite. This can happen only if the fluctuations go to zero slowly at large distances [each term in Eq. (2.11) is bounded].
17. We have added the corrections to the mean-field theory coming from fluctuations [Eq. (3.38)], which gives the leading contribution for $D \leq 4$. The specific heat of mean-field theory without corrections would only be discontinuous. (The case $\alpha = 0$ is normally interpreted as a logarithmic singularity.)
18. Quite often the values of the exponents predicted by Eq. (3.44) are called the classical exponents.
19. The argument is quite general; for an application to other systems see K. Huang, *Statistical Mechanics*, John Wiley and Sons, New York (1963).

20. In a box of side L (the D coordinates satisfy the bounds $1 \leq i_\nu \leq L$), the points of would-be coordinates 0 and $L+1$ are identified with those of coordinates L and 1, respectively. The system has the geometry of a D-dimensional torus; the box does not have a surface (i.e., a boundary), and no point is privileged with respect to the other points.
21. If free boundary conditions are used, those terms in the Hamiltonian that connect spins inside and outside the box are omitted, or equivalently, the spins outside the box are set to zero.
22. The increase of the free energy with the size of the box for periodic boundary conditions can be easily proven for systems with a positive transfer matrix (see Chapter 12). The decrease of the free energy for free boundary conditions is a consequence of the so-called Griffith's inequalities for ferromagnetic systems [R. R. Griffiths, *J. Math. Phys.* 8, 478 (1967)].

CHAPTER 4

The Low-Temperature and High-Temperature Expansions

4.1. The low-temperature expansion

The low- and high-temperature expansions make it possible for us to study the free energy systematically, as a function of the temperature and of the magnetic field; for a long time they have been the main tools for investigating the critical properties of a spin system. We consider the low-temperature expansion[1] first.

In the Ising model at low temperature most of the spins are oriented in one direction (let us say conventionally the positive one); only a few spins will be oriented in the negative direction. While at exactly zero temperature all the spins are positive, at a very low temperature we can compute the partition functions in a finite volume by including only those configurations that have up to k reversed spins. For large k the computation is long, but accurate; for small k it is simpler, but less accurate.

Let us see how it works. We consider the partition function of N spins with periodic boundary conditions in a cubic box of side L ($N = L^D$). At the end we shall send L to infinity.

We begin by doing the computation for $k = 2$: the combinatorics are simple enough if $L > k$: there is only one configuration with all the spins up (positive), N configurations with one spin down, $\binom{N}{2} = N(N-1)/2$ configurations with two spins down, and DN configurations with two nearest-neighbor down spins. When all the spins are up, the energy of the configuration is $E_0 = -ND$; if only one spin is down, the energy gap is given by $E - E_0 = 4D$; if two spins are down and if they are nearest neighbors $E - E_0 = 8D - 4$; if they are not nearest neighbors, $E - E_0 = 8D$.

We finally obtain Z,

$$Z = \exp \beta E_0 \left[1 + Nt^D + NDt^{2D-1} + \frac{N(N-1-2D)}{2} t^{2D} + \cdots \right] \tag{4.1}$$

$t = \exp - 4\beta$.

4.1. The Low-Temperature Expansion

In the limit $\beta \to \infty$ ($t \to 0$) all the terms with two spins down are much smaller (for $D > 1$) than those having only one spin down. This observation is crucial. Indeed, Eq. (4.1) tells us that

$$f = -D + \frac{1}{\beta N} \ln\left[1 + \sum_{1}^{k} C_{k'}\right], \qquad (4.2)$$

where the $C_{k'}$, i.e., the contribution from k' flipped spins) are of order $N^{k'}$. Now in order to avoid a trivial zero on the r.h.s. of Eq. (4.2), we first expand the finite volume f in powers of t, and only at the end do we take the $N \to \infty$ limit. In other words, we consider the $C_{k'}$ to be of order $\varepsilon^{k'}$, and we collect in f all terms of order ε^k, using the Taylor expansion for the logarithm.

Using this prescription we would get for $k = 2$ or $k = 3$, respectively,

$$f = -D + \frac{1}{\beta N}\left[C_1 + C_2 - \frac{C_1^2}{2}\right]$$

$$f = -D + \frac{1}{\beta N}\left[C_1 + C_2 - \frac{C_1^2}{2} + C_3 + \frac{C_1^3}{3} - C_1 C_2\right]. \qquad (4.3)$$

Similar formulae hold for higher k's. If we compute the C_k's, we discover that all terms proportional to N cancel in Eq. (4.3), and f is N independent (strictly this happens only if we stick to boxes such that $L > k$). This phenomenon is quite general: If the interaction is short range, the low-temperature expansion for the free-energy density never contains terms proportional to the volume for a ferromagnetic Ising model. Before presenting the general argument, it is useful to consider the following simplified problems. We suppose that the energies of configurations of k down spins are given by $E_k - E_0 = 4kD$ (we forget that the energy changes when two down spins are in contact) and that their number is exactly given by $N^k/k!$ (we neglect excluded volume effects).[2] With these hypotheses we find

$$f = -D + \frac{1}{\beta N} \ln \sum_{0}^{\infty} k \frac{N^k t^{Dk}}{k!} = -D + \frac{t^D}{\beta}. \qquad (4.4)$$

Similarly, if we use the same form for the energies but take care of the excluded volume effects, i.e., instead of $N^k/k!$ we write $\binom{N}{k} = N!/k!(N-k)!$, we find

$$f = -D + \frac{1}{\beta N} \ln\left[\sum_{0}^{\infty}\binom{N}{k} t^{Dk}\right] = -D + \frac{1}{\beta N} \ln(1 + t^D)^N$$

$$= -D + \frac{1}{\beta} \ln(1 + t^D). \qquad (4.5)$$

The general argument runs as follows: if a given cluster of down spins gives a contribution NA to the partition function, the repetition of the same cluster in n other different separated regions gives a contribution $(NA)^n/n!$, where $N^n/n!$ is the usual combinatorial factor); the sum over all these contributions exponentiates to $\exp(NA)$, so that the total free energy is given by $-(1/\beta)NA$. The only nontrivial contributions to the free energy come from excluded volume effects and from the difference in energy between two separated clusters and two clusters in contact. This argument implies that one can bypass Eq. (4.3) and simply pick up the contribution to the free-energy density as the coefficient linear in N of the partition function. We can thus restrict ourselves to considering only connected clusters.

For $k = 2$ the final result is

$$f = -D + \frac{1}{\beta}\left\{t^D - \frac{1}{2}t^{2D} + D(t^{2D-1} - t^{2D})\right\}, \qquad (4.6)$$

where the various terms have the following origin: t^D is the contribution of the single flipped spin, $-\frac{1}{2}t^{2D}$ comes from excluded volume effects (the same spin cannot be down twice), $-Dt^{2D}$ is the contribution of two nearby spins which must be removed because the energy of these configurations is $8D - 4 + E_0$ and not $8D + E_0$, Dt^{2D-1} is just the correct contribution of these last configurations. We leave as an exercise for the willing reader the derivation of the result for the spontaneous magnetization m at this order $k = 2$ and for the free-energy density at the next order:

$$m = 1 - 2t^D - 4t^{2D}\left[-\frac{1}{2} + D(t^{-1} - 1)\right] + \cdots$$

$$f = -D + \frac{1}{\beta}\left\{t^D + t^{2D}\left[D(t^{-1} - 1) - \frac{1}{2}\right]\right. \qquad (4.7)$$

$$\left. + t^{3D}[2D^2(t^{-1} - 1)^2 + D(1 - t^{-2}) + \frac{1}{3}] + \cdots\right\}.$$

It is interesting to note that if we send D to infinity at fixed $\tilde{\beta} = 2D\beta$, the expression for the magnetization [Eq. (4.7)] reduces to

$$m = 1 - 2\exp(-2\tilde{\beta}) - 4\exp(-4\tilde{\beta})\left[-\frac{1}{2} + \frac{\tilde{\beta}}{2}\right] + \cdots \qquad (4.8)$$

Equation (4.8) coincides with the mean-field equations up to the order $\exp(-4\tilde{\beta})$: this coincidence is not surprising. It is a manifestation of the exactness of the mean-field approximation in infinite dimensions.

This analysis may be extended, and one could derive simple rules for

4.2. The Gaussian Model

obtaining the coefficient of the low-temperature expansion starting from the enumeration of connected clusters of negative spins; in this way the computation may be pushed to very high orders; e.g., in three dimensions for the cubic lattice it is known that[3]

$$m = 1 - 2t^3 - 12t^5 + 14t^6 - 90t^7 + 192t^8 - 492t^9 + 2148t^{10} - 7716t^{11}$$
$$+ 23{,}262t^{12} - 79{,}512t^{13} + 252{,}054t^{14} - 846{,}628t^{15} + 2{,}753{,}520t^{16}$$
$$- 9{,}205{,}800t^{17} + 30{,}371{,}124t^{18} - 101{,}585{,}544t^{19} + 338{,}095{,}596t^{20}$$
$$+ O(t^{21}) . \qquad (4.9)$$

The motivations for performing such a long computation stem from the analyticity of the thermodynamic functions: Knowledge of all the terms of the low- (or high-) β expansion is equivalent to knowledge of the functions themselves. The knowledge of a high number of coefficients is enough to compute the thermodynamic functions with very high precision not too near the critical temperature and to extract information on the critical behavior and on the critical exponents. The same motivations are valid also for the high-temperature expansion, which will be described in the next subsections.

4.2. The Gaussian model

Before tackling the construction of the high-temperature expansion of the Ising model, it is convenient to analyze two simpler related cases: the Gaussian model and the random-walk problem.

The Gaussian model is a variation of the Ising model that is exactly soluble and has the same correlation functions as the Ising model in the mean-field approximation in the high-temperature phase.

Let us start by defining a more general model: At any point of the lattice the magnetization S_i can take any value between $-\infty$ and $+\infty$. In the absence of spin-spin interactions (i.e., at $\beta = 0$), the probability of having $S < S_i < S + dS$ is given by $P_0(S)\, dS \equiv dP_0(S)$. The Hamiltonian is the same as before:

$$H[S] = -\frac{1}{2} \sum_{i,k} J_{i,k} S_i S_k - \sum_i h_i S_i . \qquad (4.10)$$

The partition function is given by

$$Z[h] = \int \prod_i dP_0(S_i) \exp[-\beta H[S]] . \qquad (4.11)$$

If $P_0(S) = \delta(1 - S^2)$ we recover the Ising model; if $P_0(S) = (2\pi)^{-1/2} \times \exp(-S^2/2)$ we obtain the Gaussian model; if $P_0(S) \propto \exp[-g(S^2 - 1)^2 - S^2/2]$ we have the Ginsburg-Landau model, which interpolates between the Gaussian model ($g = 0$) and the Ising model $g = \infty$, and whose properties will be discussed in the next chapter.

The Gaussian model can be solved by using the very useful identity

$$\int_{-\infty}^{+\infty} \prod_a^N dy_a \exp\left[-\frac{1}{2}\sum_a^N \sum_b^N A_{ab} y_a y_b + \sum_a^N y_a z_a\right]$$

$$= \frac{(2\pi)^{N/2}}{(\det A)^{1/2}} \exp\left[+\frac{1}{2}\sum_a^N \sum_b^N A_{ab}^{-1} z_a z_b\right] \quad (4.12)$$

$$\sum_b A_{ab} A_{bc}^{-1} = \delta_{ac},$$

where A is a symmetric real positive $N \times N$ matrix; y_a and z_a are N component vectors. The condition of positivity of the eigenvalues of A is crucial for the convergence of the integrals. Quite often the evaluation of det A is simplified by using the two identities

$$\det A = \exp\{\text{Tr}[\ln(A)]\}$$

$$\frac{d}{d\lambda} \ln(A + \lambda) = (A + \lambda)^{-1}. \quad (4.13)$$

If we apply Eqs. (4.12) and (4.13) to the computation of the partition function, we get

$$Z[h] = (\det A)^{-1/2} \exp\left[\frac{\beta^2}{2} \sum_{l,k} h_l A_{lk}^{-1} h_k\right]$$

$$A_{kl} = \delta_{kl} - \beta J_{kl}$$

$$A_{kl}^{-1} \equiv G_{kl} = \frac{1}{(2\pi)^D} \int_B d^D p \exp[i(k - l) \cdot p] G_0(p) \quad (4.14)$$

$$G_0(p) = \frac{1}{1 - \beta \sum_\nu^D \cos p_\nu},$$

where we have used the same procedure as in Eq. (3.33).

The two-spin correlation functions can easily be computed from the linear response theorem:

4.2. The Gaussian Model

$$\langle S_k S_l \rangle = \frac{1}{\beta^2 Z} \frac{\partial}{\partial h_k} \frac{\partial}{\partial h_l} Z[h] . \qquad (4.15)$$

In order to apply Eq. (4.15) we do not need to know det A. We easily obtain

$$\langle S_k S_l \rangle_{h=0} = G_{kl} . \qquad (4.16)$$

The free-energy density at $h = 0$ is simply given by $(1/2N\beta)\, \mathrm{Tr}\ln A$. By diagonalizing A in momentum space we get

$$f(\beta) = \frac{1}{2\beta(2\pi)^D} \int_B d^D p \, \ln\left[1 - \beta \sum_\nu^D \cos p_\nu \right]. \qquad (4.17)$$

The correctness of Eq. (4.17) can be checked by computing the internal energy $u(\beta) = (\partial/\partial\beta)[\beta f(\beta)]$. This can be done by using Eq. (4.17) or Eq. (3.35). The two results coincide. Up to now the Gaussian model reproduces the results of the mean-field approximation. Let us study in detail the behavior of the many-spin correlations. The correlation functions of many spins can be obtained using the identity

$$\left\langle \prod_1^n S_{i_a} \right\rangle\bigg|_{h=0} = \frac{1}{\beta^n} \frac{1}{Z} \prod_1^n \frac{\partial}{\partial h_{i_a}} Z[h]\big|_{h=0} . \qquad (4.18)$$

A tedious but elementary computation tells us that

$$\langle S_i S_k S_l S_j \rangle|_{h=0} = G_{ik} G_{lj} + G_{il} G_{kj} + G_{ij} G_{lk}$$

$$\langle S_i S_k S_l S_j S_m S_n \rangle = G_{ik} G_{lj} G_{mn} + G_{il} G_{kj} G_{mn} + G_{im} G_{lj} G_{kn}$$
$$+ G_{in} G_{lj} G_{km} + G_{ij} G_{kl} G_{mn} + 10 \text{ permutations} \qquad (4.19)$$

$$\langle S_i S_k S_l S_j S_m S_n S_p S_q \rangle = G_{ik} G_{lj} G_{mn} G_{pq} + 104 \text{ permutations} .$$

Generally speaking, the number of terms that appear on the r.h.s. of Eq. (4.19) is equal to the number of ways that n objects may be paired in $n/2$ sets of two objects, i.e., $(2n-1)!!$. This combinatorial result may be checked by considering a very elementary case: a single-site lattice at $\beta = 0$. If $\langle S^2 \rangle = 1$, $\langle S^n \rangle$ is equal to the number of terms on the r.h.s. of Eq. (4.19); on the other hand, we easily obtain

$$M_n \equiv \frac{1}{(2\pi)^{1/2}} \int_{-\infty}^{+\infty} dS \exp\left(-\frac{1}{2} S^2\right) S^n = 0, \quad n \text{ odd}$$

$$M_n = \frac{1}{(2\pi)^{1/2}} \int_0^\infty dz\, z^{n-1/2} \exp\left(-\frac{z}{2}\right) = \frac{2^n}{\pi^{1/2}} \Gamma\left(n + \frac{1}{2}\right) \quad (4.20)$$

$$= (2n-1)!!, \quad n \text{ even},$$

where we have used two properties of the Euler Γ function,

$$\Gamma\left(\frac{1}{2}\right) = \pi^{1/2}, \quad \Gamma(x+1) = x\Gamma(x). \quad (4.21)$$

Equation (4.19) has a nice diagrammatic interpretation, as can be seen in Fig. 4.1, where the two-point correlation is represented by a line. We have inserted small bridges to stress that the eventual crossing of two lines does not have any special meaning.

This diagrammatic representation suggests to us that we can introduce the connected components of the correlation functions by generalizing the definition of the previous chapter. Indeed, the diagrams in Fig. 4.1 are disconnected if $n > 2$, and the only connected blocks are the two-spin correlations. In the region where the correlation functions of an odd number of spins are zero (i.e., $h = 0$ and $T > T_c$), we can define a 4-spin connected correlation function as

$$\langle S_i S_k S_l S_j \rangle = \langle S_i S_k \rangle \langle S_l S_j \rangle + \langle S_i S_l \rangle \langle S_k S_j \rangle$$

$$+ \langle S_i S_j \rangle \langle S_k S_l \rangle + \langle S_i S_k S_l S_j \rangle_c. \quad (4.22)$$

(a)

(b)

(c)

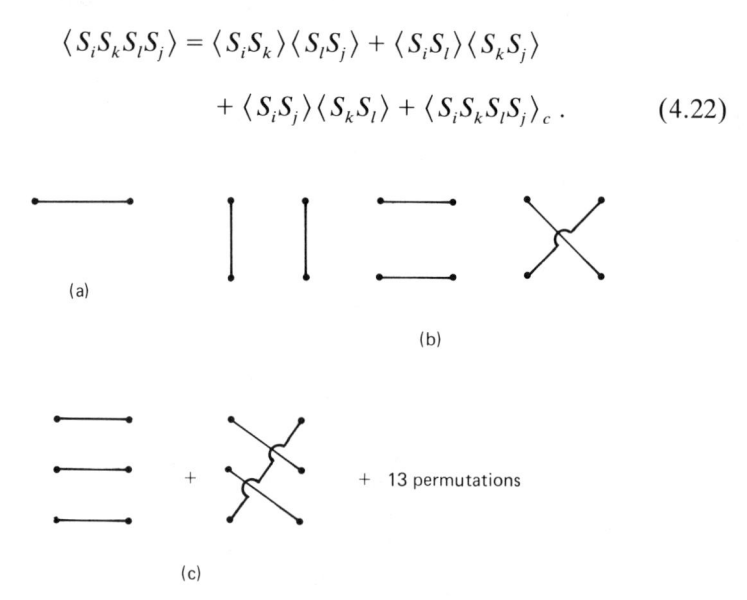

Figure 4.1. Correlation-function diagrams for the Gaussian model: (a) two-field; (b) four-field; (c) six-field.

4.2. The Gaussian Model

The n-point connected correlation function can be defined recursively in the following way:

$$\langle S_i \rangle = \langle S_i \rangle_c$$

$$\langle S_i S_k \rangle = \langle S_i S_k \rangle_c + \langle S_i \rangle_c \langle S_k \rangle_c$$

$$\langle S_i S_k S_l \rangle = \langle S_i S_k S_l \rangle_c + \langle S_i \rangle_c \langle S_k S_l \rangle_c$$
$$+ \langle S_k \rangle_c \langle S_i S_l \rangle_c + \langle S_l \rangle_c \langle S_i S_k \rangle_c + \langle S_i \rangle_c \langle S_k \rangle_c \langle S_l \rangle_c \quad (4.23)$$

$$\left\langle \prod_1^n S_{i_a} \right\rangle = \sum_p \prod_\alpha \left\langle \prod_{i_a \in \mathcal{S}_\alpha} S_{i_a} \right\rangle_c ,$$

where in the last formula Σ_p stands for the sum of all the partitions of the set of n spins in subsets $\mathcal{S}_\alpha$, and α labels the subsets. Equations (4.23) give the general correlation functions in terms of the connected ones, but they can obviously be inverted recursively. We could say that a knowledge of the correlation functions of $n - 1$ spins can be used to guess the correlation function of n spins, and the n-spin connected correlation function is the difference between this guess and the exact answer. Equation (4.19) tells us that at $h = 0$, the two-spin correlation is the only nonzero connected correlation function in the Gaussian model.

Equation (4.18) can be modified to give

$$\left\langle \prod_1^n S_{i_a} \right\rangle_c = \frac{1}{\beta^n} \prod_1^n \frac{\partial}{\partial h_{i_a}} \ln[Z(h)]$$

$$\equiv -\frac{1}{\beta^{n-1}} \prod_1^n \frac{\partial}{\partial h_{i_a}} F(h) . \quad (4.24)$$

The proof is not too complicated.[4] Equation (4.24) clearly implies in a direct way the triviality of higher-order connected correlation functions in the Gaussian model [cf. Eq. (4.14)].

Connected correlation functions are important for three reasons:

a) They are directly related to the free energy by Eq. (4.24).
b) They satisfy a generalized cluster theorem: In a pure thermodynamic state, connected correlation functions vanish when one or more points go to infinity.
c) In all the diagrammatical approaches, connected correlation functions can be written in terms of connected diagrams.

The Gaussian model is very simple; it can be completely solved and has a transition at $\beta = \beta_c \equiv \frac{1}{2}D$. Unfortunately it does not exist for $\beta > \beta_c$

because the matrix A is no longer positive definite in this region. As we shall see later, this model is a good starting point for constructing a perturbative expansion.

4.3. The random walk

Let us suppose that a traveler moves randomly on a lattice, taking steps of one lattice spacing. If the traveler starts at point 0, we want to compute the probability $P_n(k)$ of finding him at point k after n steps.

It is readily apparent that this probability satisfies the recursion relation

$$P_{n+1}(k) = \frac{1}{2D} \sum_l J_{kl} P_n(l) \quad n \geq 0,$$

$$P_0(k) = \delta_{k,0},$$

(4.25)

J_{kl} being the usual matrix [see Eq. (3.10)] with $J = 1$.

It is useful to introduce the generating function of the P_n's:

$$G(k \mid q) = \sum_0^\infty q^n P_n(k).$$

(4.26)

The recursion relation (4.25) implies that

$$\frac{1}{2D} \sum_l J_{kl} G(l \mid q) = \sum_0^\infty q^n P_{n+1}(k) = \frac{G(k \mid q) - \delta_{k,0}}{q},$$

(4.27)

which can be written

$$G(k \mid q) - \frac{q}{2D} \sum_l J_{kl} G(l \mid q) = \delta_{k,0}.$$

(4.28)

Comparing Eqs. (4.14) and (4.28), we see that $G(k \mid q)$ is the correlation function of the Gaussian model at $q = \tilde{\beta} = 2D\beta$. In the next subsection we shall perform the warming up exercise of deriving this relation between the Gaussian model and the random walk, using a diagrammatical approach. Here we want only to show how Eqs. (4.26)–(4.28) can be used to compute the average square distance. It is easy to check that

4.3. The Random Walk

$$\sum_{0}^{\infty}{}_n q^n \sum_k P_n(k) = \sum_k G(k \mid q) = G_0(p)|_{p=0} = \frac{1}{1-q} = \sum_{0}^{\infty}{}_n q^n.$$

$$\sum_{0}^{\infty}{}_n q^n \sum_k \left(\sum_{1}^{D}{}_\nu k_\nu^2\right) P_n(k) = \sum_k \left(\sum_{1}^{D}{}_\nu k_\nu^2\right) G(k \mid q) \qquad (4.29)$$

$$= -\left(\sum_{1}^{D}{}_\nu \frac{\partial^2}{\partial p_\nu^2}\right) G_0(p)|_{p=0} = \frac{q}{(1-q)^2} = \sum_{0}^{\infty}{}_n n q^n.$$

$$G_0(p) \equiv 1 \bigg/ \left(1 - \frac{q}{2D} \sum_{1}^{D}{}_\nu \cos p_\nu\right).$$

By comparing the two sides of Eq. (4.29) we get

$$\sum_k P_n(k) = 1$$
$$\sum_k \left(\sum_{1}^{D}{}_\nu k_\nu^2\right) P_n(k) = n. \qquad (4.30)$$

The first equation tells us that the probability is well normalized, whereas the second tells us that the average square distance is equal to n, as it should be. The distance is the sum of n independent random variables. We leave as an exercise for the reader the proof that for large n (in agreement with the central-limit theorem) the probability distribution is well approximated by a Gaussian distribution,

$$P_n(k) \sim \left(\frac{2\pi n}{D}\right)^{-D/2} \exp\left[\frac{-D\left(\sum_{1}^{D}{}_\nu k_\nu^2\right)}{2n}\right]. \qquad (4.31)$$

We note that

$$P_n(k-l) = \frac{1}{(2D)^n} R_n(l, k) \equiv \frac{1}{(2D)^n} R_n(k-l), \qquad (4.32)$$

where $R_n(k, l)$ is the number of paths on the lattice that go from k to l in n steps: the random-walk problem is also a random-path problem.

It would be interesting to compute, as well, the number of closed paths of length n ($R_n^c(N)$) on a lattice of size N.

Now it is evident that

$$\sum_k R_n(k, k) = 2n R_n^c(N). \qquad (4.33)$$

Indeed, each closed path of length n appears $2n$ times on the l.h.s. of Eq. (4.33).[5] We finally get

$$r_n^c \equiv \lim_{N\to\infty} \frac{1}{N} R_n^c(N) = \frac{1}{2n} R_n(k,k) = \frac{1}{2n} R_n(0). \qquad (4.34)$$

The generating function of the r_n^c is given by

$$r^c(q) \equiv \sum_1^\infty q^n r_n^c = \sum_1^\infty \frac{1}{2n} \frac{1}{(2\pi)^D} \int_B d^Dp \Big(\sum_1^D \cos p_\nu\Big)^n \Big(\frac{q}{2D}\Big)^n$$

$$= \frac{-1}{2(2\pi)^D} \int_B d^Dp \ln\Big[1 - \frac{q}{2D}\Big(\sum_1^D \cos p_\nu\Big)\Big], \qquad (4.35)$$

which is proportional to the free energy of the Gaussian model for $\beta = q/2D$.

If in the finite-volume case we consider a box of size $L^D = N$ and we use periodic boundary conditions (the box has the geometry of a torus), we should substitute $\int_B d^Dp$ with $L^{-D} \prod_\nu^D \sum_{l_\nu}^L$, where $p_\nu = (2\pi/L)l_\nu$ (momenta are quantized on a box). Trigonometric identities may be used to prove the evident fact that the r_n are L independent for $n < L$; for $n \geq L$ the r_n also get contributions from paths that wind around the box and that disappear when $n < L$.

4.4. The high-temperature expansion of the Gaussian model

In this subsection we want to recompute the free energies and the correlation functions of the Gaussian model by expanding them in powers of β, using a diagrammatical approach [i.e., forgetting the existence of the powerful identity (4.12)]. We want to do such a computation because a modification of the same technique will allow us to find the high-temperature expansion of the Ising model.

Let us start from

$$\langle S_a S_b \rangle = \frac{\int \prod_i dS_i \exp\Big[-\sum_i \frac{S_i^2}{2} + \frac{\beta}{2}\sum_i\sum_k J_{ik}S_iS_k\Big] S_a S_b}{\int \prod_i dS_i \exp\Big[-\sum_i \frac{S_i^2}{2} + \frac{\beta}{2}\sum_i\sum_k J_{ik}S_iS_k\Big]} \qquad (4.36)$$

and expand both the denominator and the numerators in powers of β. We now put the expansion in a graphical form by using

4.4. The High-Temperature Expansion of the Gaussian Model

$$\exp \beta \sum_{(i,k)} S_i S_k = \prod_{(i,k)} \exp \beta S_i S_k = \prod_{(i,k)} \left[\sum_{0}^{\infty} \frac{1}{n_{i,k}!} \beta^{n_{i,k}} (S_i S_k)^{n_{i,k}} \right]$$

$$= \sum_{\{n_{i,k}\}} \beta^{\left[\sum_{(i,k)} n_{i,k}\right]} \prod_{(i,k)} \frac{1}{(n_{i,k})!} (S_i S_k)^{n_{i,k}}. \tag{4.37}$$

In other words, with each link of the lattice (i, k) we associate an integer number $n_{i,k}$. $\Sigma_{\{n_{i,k}\}}$ indicates the sum of all possible values of the $n_{i,k}$'s. At a finite order in β, only a finite number of $n_{i,k}$ will be different from zero. Pictorially, in order to compute the order β^r we must lay down r sticks on the links of the lattice (we are allowed to put more than one stick on the same link) in all possible ways. If we draw the lattice by representing these sticks as lines, we could say that we can compute the order β^r by summing over all possible graphs with r lines both for the denominator and the numerator. The weight of each graph can easily be found:

$$\langle S_a S_b \rangle = \frac{\sum_r \beta^r \left\{ \sum_{\{n_{i,k}\}}^{(r)} \left\{ \left[\prod_i' M_{n_i} \right] M_{n_a+1} M_{n_b+1} \cdot \prod_{(i,k)} \frac{1}{n_{i,k}!} \right\} \right\}}{\sum_r \beta^r \left\{ \sum_{\{n_{i,k}\}}^{(r)} \left[\prod_i M_{n_i} \prod_{(i,k)} \frac{1}{n_{i,k}!} \right] \right\}} \tag{4.38}$$

$$n_i = \sum_k n_{i,k}$$

where the M_n are defined in Eq. (4.20), $\Sigma_{\{n_{i,k}\}}^{(r)}$ is the sum over all possible configurations of the $n_{i,k}$ such that $\Sigma_{(i,k)} n_{i,k} = r$, $\prod_i'$ indicates that the product runs over all i different from a and b (let us assume $a \neq b$).

In other words, we must sum all possible graphs of order r with a weight M_{n_i} for each node of the graph, and $1/n_{ik}!$ for each multiple occupancy of the bonds. Typical graphs with nonzero weight are shown in Fig. 4.2 for both the denominator and the numerator: some of these graphs are connected, some are disconnected. In order to reduce the number of diagrams, it would be much better if we could study only connected diagrams.

To achieve this aim let us decompose each graph as an ensemble of paths. For the denominator all paths will be closed; for the numerator only one path will be open. If $n_i = 2$ at each node of the graph, the decomposition of the graph in the path is unique. If we have nodes with $n_i > 2$, the situation changes as shown in Fig. 4.3: A single graph having an $n_i = 4$ can be decomposed in a path in three different ways. If we forget problems connected with multiple occupancy of the same link (e.g., if we suppose that $n_{(i,k)} = 0$ or 1), the weights appearing in Eq.

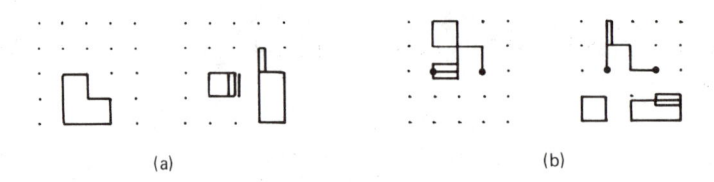

The Low-Temperature and High-Temperature Expansions 4.4

Figure 4.2. Some nonzero diagrams contributing to the denominator (a) and the numerator (b) of Eq. (4.38).

(4.38) just count the ways in which the graph may be decomposed into paths. At each node we must associate the lines arriving at that point in pairs of 2, and this can be done in $(2n-1)!!$ different ways. Similar results hold if we take care of multiple occupancy (we prefer not to enter into the details—the combinatorics start to be complicated).

The final result is

$$Z = \sum_0^\infty \frac{1}{n!} \left[\sum_{\{C\}} \beta^{l_C}\right]^n = \exp\left[\sum_{\{C\}} \beta^{l_C}\right] \quad (4.39)$$

$$Z\langle S_a S_b \rangle = \sum_{\{O_{ab}\}} \sum_0^\infty \frac{1}{n!} \left[\sum_{\{C\}} \beta^{l_C}\right]^n \beta^{l_{O_{ab}}} = Z \sum_{\{O_{ab}\}} \beta^{l_{O_{ab}}}$$

where $\Sigma_{\{C\}}$ and $\Sigma_{\{O_{ab}\}}$ denote the sum over all the closed paths, and the open paths (having a and b as endpoints) of length l_C and $l_{O_{ab}}$, respectively.

Indeed Eq. (4.39) for Z can be written

$$Z = \sum_0^\infty \sum_{\{C_1,\ldots,C_n\}} w(C_1,\ldots,C_n) \beta^{(\Sigma_i^n l_{C_i})}, \quad (4.40)$$

where the weight $w(C_1, \ldots, C_n)$ is one if all the paths C_i are different. If there are repetitions of the same path, $w(C_1, \ldots, C_n)$ is smaller. (If there are k_1 repetitions of one path, k_2 of another path, up to k_m, ($\sum_1^m k_j = n$), $w(C_1, \ldots, C_n) = \prod_1^m 1/k_j!$, as can be seen from the multinomial formula.

Figure 4.3. The three different decompositions of a graph in a path at a node of order 4.

If we neglect both multiple occupancy and the repetition of the same path (the two problems go together), the denominator of Eq. (4.38) is 4.40.

In this way we have reduced the Gaussian model to the random-walk problem which was solved in the previous section. The reader can see that the graphical method is much more complex than the algebraic one. The key point is that, after we go from graphs to paths, disconnected diagrams exactly cancel between numerator and denominator, leaving no excluded volume factor. The aim of this section is to let the reader become familiar with the combinatoric techniques of the high-temperature expansion in an "easy" case, where the results can be algebraically derived.

4.5. The high-temperature expansion of the Ising model

In the Ising model we can proceed as in the previous section. However, the M_n are given by

$$M_n = 1, \quad n \text{ even},$$
$$M_n = 0, \quad n \text{ odd}. \tag{4.41}$$

If we translate Eq. (4.38) in terms of paths, we clearly see that we must add some extra weight when the paths cross. To do this at the diagrammatical level is rather tricky; we shall see later on how to do it algebraically. In order to compute the first orders of the high-temperature expansion it is convenient to use a slightly modified approach. The key step is the identity

$$\exp(\beta S_i S_j) = \cosh \beta [1 + k S_i S_j] \quad k = \tanh \beta, \tag{4.42}$$

which holds if S_i and S_j can take only the values ± 1.

The partition function can now be written as

$$Z = [2 \cosh \beta]^N \sum_{\{n_{i,j}\}} k^{\sum_{(i,j)} n_{i,j}}, \quad n_i = \sum_j n_{i,j} \tag{4.43}$$

with the constraint that all the $n_{i,j}$ are 0 or 1 and the n_i are even. In this way the problem of multiple occupancy is avoided. Disconnected diagrams cancel, but we have to take care of excluded volume effects.

At the order k^4 the only diagram different from zero is the square; we get

$$f(\beta) = -\frac{1}{\beta} \ln[2\cosh(\beta)] + \tilde{f}(\beta)$$

$$\tilde{f}(\beta) = -\frac{1}{\beta} \frac{D(D-1)}{2} k^4 .$$
(4.44)

With some effort one can get the expansion up to quite high orders: for example, it is known that on the cubic lattice[6]

$$-\beta \tilde{f}(\beta) = 3k^4 + 22k^6 + 187\tfrac{1}{2}k^8 + 1980k^{10} + 24{,}044k^{12}$$

$$+ 319{,}170k^{14} + 4{,}514{,}757\tfrac{3}{4}k^{16} + 67{,}003{,}469\tfrac{1}{3}k^{18} + \cdots .$$
(4.45)

As we said, this kind of expansion can be used to find the critical temperature and the critical exponents. Indeed, if we assume that the critical temperature is the nearest singularity to the origin in the complex k^2 plane, and that the singularity at $k^2 = k_c^2$ has the form $(k_c^2 - k^2)^\omega$, Appel's comparison theorem tells us that

$$f(k^2) = \sum_{n}^{\infty} f_n (k^2)^n$$

$$k_c^2 = \lim_{n \to \infty} \left(\frac{f_n}{f_{n+1}} \right)$$
(4.46)

$$\omega = \lim_{n \to \infty} \left[-n^2 \left(\frac{f_n}{f_{n+1}} - \frac{f_{n-1}}{f_n} \right) \right] k_c^{-2} - 1 .$$

It is always tricky to estimate the asymptotic value of a sequence from the knowledge of a finite number of terms. The error that must be attached to such a value is sometimes rather subjective. For a reasonable hypothesis,[7] one finds that Eq. (4.45) implies that

$$k_c \simeq 0.2181$$

$$\omega \simeq 1.9$$
(4.47)

$$\alpha = 2 - \omega \simeq 0.1 .$$

Similar results may be obtained for different lattices and different dimensions; the quality of the results clearly depends on the number of computed coefficients.

We can extract from the high-temperature expansion the exact solution of the one-dimensional Ising problem at $h = 0$:

4.6. The Free Energy of the Two-Dimensional Ising Problem

$$f(\beta) = -\frac{1}{\beta} \ln[2 \cosh \beta]$$

$$\langle S_a S_b \rangle = k^{|a-b|}.$$
(4.48)

No closed graph can be constructed on the infinite line. However, if the size of the system is L and periodic boundary conditions are used, it is easy to check that

$$f_L(\beta) = f_\infty(\beta) - \frac{1}{\beta L} \ln(1 + k^L).$$
(4.49)

In the next section we shall obtain the free energy of the two-dimensional Ising system using the same method.

4.6. The free energy of the two-dimensional Ising problem

In two dimensions the expansion for the free energy may be written in a compact way. Let us work out what happens up to the order k^8. The only nontrivial diagrams are shown in Fig. 4.4, their contribution to Z being given by

$$Z(\beta) = [2 \cosh \beta]^N \tilde{Z}(\beta)$$
(4.50)

$$\tilde{Z}(\beta) = 1 + N[c_a k^4 + c_b k^6 + (c_c + c_d + c_e)k^8] + \frac{N^2 - 5N}{2} c_a^2 k^8 + 0(k^{10})$$

where $Nc_a, Nc_b, \ldots, Nc_e$ are the numbers of ways in which the graph $a, b, \ldots, e$ can be embedded in the lattice ($c_a = 1$, $c_b = 2$, $c_c = 2$, $c_d = 1$, $c_e = 4$). We obtain

$$\frac{1}{N} \ln \tilde{Z}(\beta) = c_a k^4 + c_b k^6 + (c_c + c_d + c_e)k^8 - \frac{5}{2} c_a^2 k^8 + 0(k^{10})$$

$$= k^4 + 2k^6 + \frac{g}{2} k^8 + 0(k^{10}).$$
(4.51)

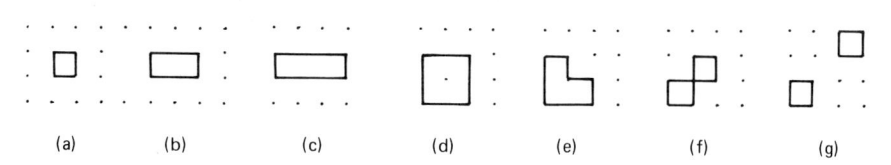

(a) (b) (c) (d) (e) (f) (g)

Figure 4.4. Diagrams contributing to the high-temperature expansion of the Ising model up to the order k^8.

A similar result can be obtained in any dimension (the c's being polynomial in D).

It is clear that we could also write

$$\tilde{Z}(\beta) = \sum_0^\infty \frac{1}{n!}\left[\sum_{\{C\}} k^{l_C} p_C\right]^n \tag{4.52}$$

with an appropriate extra weight p_C; a remarkable theorem states that in two dimensions p_C is given by

$$p_C = (-1)^{w_C - 1}, \tag{4.53}$$

for nonbacktracking paths, $p_C = 0$ if the walk is backtracking, where w_C is the winding number (see Fig. 4.5). The winding number is defined as the number of rotations of a unit vector tangent to the path when transported along the path. This number is an integer (there is an ambiguity of sign for nonoriented paths). We could also write $w_C = (N_L - N_R)/4$, N_L and N_R being the number of left and right turns. The contributions of diagrams 4.4(a)–(e) and 4.4(g) are not changed with respect to Eq. (4.50) and have positive weight. One gets as extra contributions the diagrams shown in Fig. 4.6.

We readily see that Eq. (4.53) gives the correct result at the order k^8 (the sum of all the new diagrams reduces to f); a topological theorem, which we do not prove,[8] tells us that this is true at all orders in k. An example of cancellations of unwanted diagrams at a high order is shown in Fig. 4.7.

We finally get

$$\tilde{Z}(\beta) = \exp\left[\sum_{\{C\}} k^{l_C}(-1)^{w_C - 1}\right]$$
$$N\tilde{f}(\beta) = -\frac{1}{\beta}\sum_{\{C\}} k^{l_C}(-1)^{w_C - 1}. \tag{4.54}$$

To evaluate $\tilde{Z}(\beta)$ it is convenient to define the amplitude $A_l^{\alpha,\beta}(j)$:

$$A_l^{\alpha,\beta}(j) = \sum_{\{O_{0,j}\}} \exp\left[\frac{i\pi}{2}(N_L - N_R)\right] k^{l_{O_{0,j}}}, \tag{4.55}$$

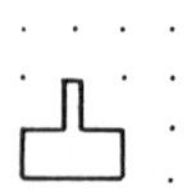

Figure 4.5. An example of a backtracking walk.

4.6. The Free Energy of the Two-Dimensional Ising Problem

Figure 4.6. New diagrams with their weights contributing to Eq. (4.52).

where the sum is over all open paths $(O_{0,j})$, nonbacktracking, of length l starting from 0 and arriving at j from the direction β, the weight having been computed by supposing that the paths arrived at 0 coming in the direction α (α and β label the four directions of the lattice, which can be called N, W, S, E, or 1, 2, 3, 4).

The same arguments as in Section 4.3 tell us that[9]

$$-\beta f(\beta) = \frac{1}{2} \sum_{\alpha}^{4} \sum_{l}^{\infty} \frac{1}{l} A_l^{\alpha,\alpha}(0) . \tag{4.56}$$

It is evident that $A_l^{\alpha,\beta}(j)$ satisfies the following recursion equation:

$$A_{l+1}^{\alpha,N}(j_1, j_2) = k\left[A_l^{\alpha,N}(j_1, j_2 - 1) + \exp\left(-\frac{i\pi}{4}\right) A_l^{\alpha,W}(j_1, j_2 - 1) \right.$$
$$\left. + \exp\left(+\frac{i\pi}{4}\right) A_l^{\alpha,E}(j_1, j_2 - 1) \right]$$

$$A_{l+1}^{\alpha,W}(j_1, j_2) = k\left[\exp\left(\frac{i\pi}{4}\right) A_l^{\alpha,N}(j_1 + 1, j_2) + A_{l+1}^{\alpha,W}(j_1 + 1, j_2) \right. \tag{4.57}$$
$$\left. + \exp\left(-\frac{i\pi}{4}\right) A_l^{\alpha,S}(j_1 + 1, j_2) \right]$$

$$\vdots$$

$$A_0^{\alpha,\beta}(j_1, j_2) = \delta_{\alpha,\beta} \delta_{j_1,0} \delta_{j_2,0} ,$$

where we denote the two coordinates of j by j_1, j_2.

In momentum space we obtain

$$A_{l+1}^{\alpha,\beta}(p) = \sum_{\gamma} k A_l^{\alpha,\gamma}(p) Q^{\gamma\beta}(p)$$
$$A_0^{\alpha,\beta}(p) = \delta^{\alpha,\beta} , \tag{4.58}$$

Figure 4.7. An example of cancelling diagrams at the order k^{12}.

where the matrix Q has the form

$$\begin{bmatrix} \exp(-ip_y) & \exp\left(ip_x + \dfrac{i\pi}{4}\right) & 0 & \exp\left(-ip_x - \dfrac{i\pi}{4}\right) \\ \exp\left(-ip_y - \dfrac{i\pi}{4}\right) & \exp(ip_x) & \exp\left(ip_y + \dfrac{i\pi}{4}\right) & 0 \\ 0 & \exp\left(ip_x - \dfrac{i\pi}{4}\right) & \exp(ip_y) & \exp\left(-ip_x + \dfrac{i\pi}{4}\right) \\ \exp\left(-ip_y + \dfrac{i\pi}{4}\right) & 0 & \exp\left(ip_y - \dfrac{i\pi}{4}\right) & \exp(-ip_x) \end{bmatrix}$$
(4.59)

The final result is

$$\tilde{f}(\beta) = -\frac{1}{2\beta(2\pi)^2} \int_B d^2p \, \mathrm{Tr} \ln[1 - kQ(p)]$$

$$= -\frac{1}{2\beta(2\pi)^2} \int_{-\pi}^{\pi} dp_x \int_{-\pi}^{\pi} dp_y \, \ln[(1 + k^2)^2 - 2k(1 - k^2)$$

$$\times (\cos p_x + \cos p_y)], \qquad (4.60)$$

as can be seen from Eq. (4.13) or by computing the eigenvalues of Q.

The free-energy density is singular at $k = k_c = \sqrt{2} - 1$; the specific heat has a logarithmic singularity ($\alpha = 0$) indicating a second-order phase transition.

A much more complete computation tells us that the other critical exponents are given by[10]

$$\beta = \tfrac{1}{8}, \qquad \gamma = \tfrac{7}{4}, \qquad \delta = 15, \qquad \eta = \tfrac{1}{4}, \qquad \nu = 1. \qquad (4.61)$$

In particular, the spontaneous magnetization for $T < T_c$ and the coherence length for $T > T_c$ are given by

$$m_s = (1 + z^2)^{1/4}(1 - 6z^2 + z^4)^{1/8}(1 - z^2)^{-1/2}, \qquad z = \exp(-2\beta),$$

$$k = \frac{1 - z}{1 + z}, \qquad \xi^{-1} = -\ln\left(\frac{k}{k_c}\right). \qquad (4.62)$$

Only quite recently have all the correlation functions been computed in a closed form[11] at zero magnetic field.

For years the Ising model in two dimensions has been the only soluble nontrivial model with a second-order phase transition. It has played and it still plays a crucial role as a testing ground of new ideas and approximation schemes.

4.7. Self-duality

Let us look again at the low-temperature expansion in two dimensions: the relevant configurations contain a few down spins within a sea of up spins. We can define the boundaries between up and down spins as done in Fig. 4.8.

We remark that

a) There is a one-to-one correspondence between boundaries and spin configurations; we can substitute the sum over the spin configurations for the sum of all possible boundaries.
b) Not all possible sets of closed curves may be a boundary: two boundaries cannot overlap.
c) The weight for each spin configuration is $\exp(-2\beta l)$, l being the length of the boundary.

Points (a), (b), and (c) together imply that the low-temperature expansion in powers of $\exp(-2\beta)$ coincides with the high-temperature expansion in powers of k.[12]

More precisely, if we write

$$f(\beta) = -D + \frac{1}{\beta} \tilde{g}_L(\exp -2\beta)$$
$$f(\beta) = -\frac{1}{\beta} \ln[2\cosh\beta] + \frac{1}{\beta} \tilde{g}_H(\tanh\beta),$$
(4.63)

we find that the functions $\tilde{g}_L$ and $\tilde{g}_H$ coincide. In other words, if we know the free energy at k, Eq. (4.61) allows us to compute it at $k' = (1-k)/(1+k)$. This is the celebrated self-duality relation,[13] which is satisfied by Eq. (4.60). This relation is enough to fix the transition temperature (under the hypothesis of only one transition point) and the internal energy (if the transition is of second order) at the transition point:

$$\exp(-2\beta_c) = \tanh\beta_c \Rightarrow \beta_c = \operatorname{arctanh}(\sqrt{2}-1)$$
$$u(\beta_i) = \frac{1}{\sqrt{2}}.$$
(4.64)

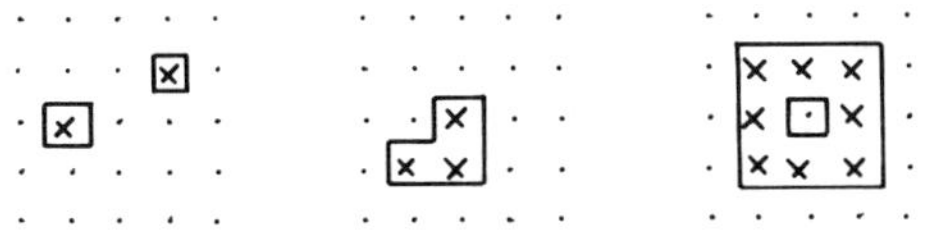

Figure 4.8. Configurations with down spins (crosses) in a sea of up spins (points). The solid lines are the corresponding boundaries between up spins and down spins.

The reader can check that Eqs. (4.7) and (4.51) agree with the duality relation up to the order $k^6(t^3)$.

In three dimensions, the boundaries of the low-temperature expansion are closed surfaces. The three-dimensional Ising model is dual to a model whose high-temperature expansion is constructed from closed surfaces. If we pursued this approach, we would construct theories invariant under the Z_2 gauge group, but this task goes beyond the aims of this book; we can only say that a closed expression for the free energy of the three-dimensional Ising model has not yet been found.[14]

Notes for Chapter 4

1. For a review of the techniques used to derive the low-temperature expansion in the Ising model, see C. Domb in *Phase Transitions and Critical Phenomena*, C. Domb and M. S. Green, eds., Vol. III, Academic, London (1974).
2. This rule for the energies corresponds to decoupled spins ($J = 0$) in a magnetic field equal to $2D$.
3. C. Domb and A. J. Guttmann, *J. Phys. C* **3**, 1652 (1970).
4. Equation (4.24) can be written

$$\sum_{n=0}^{\infty} \frac{1}{n!} \sum_{i_1} \cdots \sum_{i_n} \langle S_{i_1} \cdots S_{i_n} \rangle h_{i_1} \cdots h_{i_n} \equiv \sum_{n=0}^{\infty} \frac{1}{n!} \left\langle \left(\sum_i S_i h_i \right)^n \right\rangle$$

$$\equiv Z(h) \equiv \exp[-\beta F(h)]$$

$$= \exp\left\{ \sum_{n=0}^{\infty} \frac{1}{n!} \sum_{i_1} \cdots \sum_{i_n} \langle S_{i_1} \cdots S_{i_n} \rangle_c h_{i_1} \cdots h_{i_n} \right..$$

The correctness of this equation may be verified if we expand the exponential in powers of its argument and make use of the multinomial formula. See P. M. Morse and H. Feshbach, *Methods of Theoretical Physics*, McGraw-Hill, New York (1953), p. 412.
5. The paths in $R_n(k, k)$ have a starting point and an ending point; therefore they are oriented. This is not the case for the paths in $R_n^c(N)$. This difference explains the origin of the factor 2.
6. For a review of the high-temperature expansion for the Ising model, see M. Wortis, in *Phase Transitions and Critical Phenomena*, C. Domb and M. S. Green, eds., Academic, London (1974).

7. For a lucid presentation of various computations of the critical exponents using series extrapolation, see D. S. Gaunt and A. J. Gurrmann, in *Phase Transitions and Critical Phenomena*, C. Domb and M. S. Green, eds., Academic, London (1974).
8. A nice presentation of the solution of the Ising model at $h = 0$ in two dimensions can be found in R. P. Feynman, *Statistical Mechanics*, Benjamin (1972); the proof of the topological theorem is due to S. Sherman, *J. Math. Phys.* *1*, 202 (1960), *4*, 213 (1963); see also N. V. Wdovichenko, *Sov. Phys. JEPT 20*, 477 (1965), *21*, 350 (1965).
9. The factor $\frac{1}{2}$ has the same origin as in Eqs. (4.34) and (4.35).
10. A review of the critical properties of the two-dimensional Ising model can be found in Kadanoff *et al.*, *Rev. Mod. Phys. 39*, 395 (1967) or B. M. McCoy and T. T. Wu, *The Two-Dimensional Ising Model*, Harvard University Press, Cambridge, MA (1978).
11. B. M. McCoy and T. T. Wu, *Phys. Rev. Lett. 45*, 675 (1980).
12. The boundaries can be considered as links of a new lattice whose points are in the center of the squares of the old lattice.
13. H. A. Kramers, G. H. Wannier, *Phys. Rev. 60*, 252 (1941).
14. F. Wegner, *J. Math. Phys. 12*, 2259 (1971) and R. Balian, J. M. Drouffe, and C. Itzykson, *Phys. Rev. D 11*, 2098 (1975) study the dual theory of the three-dimensional Ising model. Recent attempts at finding the analytic solution of the three-dimensional Ising model are described in C. Itzykson, *Nucl. Phys. B 210* [FS6], 477 (1982) and references therein.

CHAPTER 5

The Landau-Ginsburg Model

5.1. The model

In the preceding chapters we have seen some properties of the Ising model and of the Gaussian model; here we shall study the Landau-Ginsburg model, which was devised with the aim of providing a simple general form of the effective Hamiltonian for magnetic systems.

In Chapter 2 we saw how we could write an effective Hamiltonian for the local magnetization, which is the relevant variable for the study of magnetic phenomena. At the phase transition point the singular terms in the thermodynamic functions are produced by fluctuations on large scales (i.e., at small momenta in the Fourier space); the fine details of the probability distribution of the magnetization on a small scale (e.g., whether S^2 is equal to a fixed value, the shape of the lattice, the existence of a nonzero second-neighbor interaction) seem not to be very important for deriving general properties, such as the value of the critical exponents (we shall come back to this point later). We can thus carry Eq. (3.2) a step further: we introduce as a relevant variable the field $\varphi(x)$, which is the average of the magnetization around the point x (e.g., $\varphi(x) \propto \Sigma_i \exp[-(x-i)^2/2\delta^2] S_i$), and we consider a φ-dependent effective model Hamiltonian. The radius δ of the region over which the microscopic magnetization S_i is averaged must be large enough, that $\varphi(x)$ is a sufficiently smooth function of x, but it should not be larger than the correlation length, otherwise trivial results are obtained; in particular, if we want to keep a simple form for the effective action, δ should not diverge at the transition point; δ equal to a few lattice spacings is likely to be a good choice. The probability distribution of the Landau-Ginsburg model is by definition

$$dP[\varphi] \propto \exp[-\beta H_{\text{ef}}[\varphi]] \, d[\varphi]$$

$$\beta H_{\text{ef}}[\varphi] = \int d^D x \left\{ \frac{1}{2} \mu \varphi^2(x) + \frac{1}{2} \sum_{\nu}^{D} (\partial_\nu \varphi(x))^2 + \frac{g}{4!} \varphi^4(x) + h(x)\varphi(x) \right.$$
$$\left. + \frac{1}{2\Lambda^2} (\Delta \varphi(x))^2 \right\}. \tag{5.1}$$

5.1. The Model

Let us explain the roles of the terms in the effective Hamiltonian. The first four are essential, whereas the last one is not so crucial and can be omitted if some precautions are taken. If $\mu > 0$ the minimum of the effective Hamiltonian is just at $\varphi = 0$; if $\mu < 0$ there are two minima at $\varphi = \pm\sqrt{-6\mu/g}$; if we approximate the free energy with the minimum of $H_{\text{ef}}[\varphi]$, i.e.,

$$\beta F = \min_{\{\varphi\}} [\beta H_{\text{ef}}[\varphi]], \qquad (5.2)$$

we find that spontaneous magnetization is present as soon as $\mu < 0$. In this approximation the critical temperature (T_c) is at $\mu = 0$; we therefore have

$$\mu = \alpha(T - T_c) + 0(T - T_c)^2, \qquad (5.3)$$

where α is a constant. Here, g cannot be negative, otherwise the functional integral would diverge. It must be nonzero or we would recover the Gaussian model, which is not definite for $T < T_c$. The term $\frac{1}{2}\sum_{\nu}^{D}(\partial_{\nu}\varphi)^2$ describes a ferromagnetic interaction, so its coefficient must be positive, since the effective energy must increase (and consequently the probability decrease) if φ changes with x (i.e., if spins are less aligned). Conventionally its coefficient can be set to $\frac{1}{2}$ by a multiplicative redefinition of the φ field. The term $1/2\Lambda^2(\Delta\varphi)^2$ is very large when the function φ is not smooth. It is needed in two and higher dimensions (at least at an intermediate stage) to suppress those configurations for which φ changes too fast. Of course, the term $\frac{1}{2}(\partial\varphi)^2$ also suppresses these configurations, but it is not so efficient as the term proportional to $1/2\Lambda^2$. As we shall see, at $\Lambda = \infty$ $\langle\varphi^2\rangle$ is divergent as soon as $D \geq 2$.

It should be clear that if the field φ is the average value of the magnetization, the true effective Hamiltonian would be much more complex. The main advantage of Eq. (5.1) is its extreme simplicity: no term may be removed from it without changing the physics of the problem in a deep way. We shall see in Section 11.3 that the insertion of other terms in the effective Hamiltonian does not change the results for the critical exponents: From our point of view these other terms are irrelevant.

It is normally assumed that the parameters g, Λ, and μ are smooth functions of the temperature near the critical point; since the change in sign of μ drives the transition in the approximation (5.2), the variation of μ are the most important. In computing the singular terms of the thermodynamic functions, it is therefore sufficient to consider g, Λ, and β as constants and to concentrate all the temperature dependence in μ. In approximation (5.2), the transition point corresponds to $\mu = 0$; if this approximation is removed, the critical value of μ (μ_c) will change. Then

Eq. (5.3) is replaced by

$$\mu - \mu_c(g, \Lambda) = \alpha(T - T_c), \tag{5.4}$$

where we have neglected the higher orders in $T - T_c$. At this point, in order to lighten the notation, we set conventionally $\alpha = 1$ and $\beta = 1$.

The reader may ask what is the precise mathematical meaning of $d[\varphi]$. Moreover, in a finite volume the space of all functions $\varphi(x)$ is huge: if we consider L^2 functions,[1] the space is still infinite dimensional, and the generic L^2 function is not differentiable. For most of the functions $\varphi(x)$ belonging to L^2, $H_{ef}[\varphi]$ is infinite. However, these questions are not directly related to the physical problem, i.e., the study of fluctuations on a large scale. They can be bypassed by defining the field φ only on the points of a lattice (which may or may not coincide with the original lattice). The effective model Hamiltonian then becomes

$$H_{ef}[\varphi] = a^D \left[\sum_{i,k} \frac{1}{2a^2} J_{i,k}(\varphi_i - \varphi_k)^2 + \sum_i \left(\frac{\mu}{2} \varphi_i^2 + \frac{g}{4!} \varphi_i^4 + h_i \varphi_i \right) \right], \tag{5.5}$$

where a is the lattice spacing and $J_{i,k}$ are defined on p. 25.

If the φ_i's are the values of a smooth function $\varphi(x)$ on the points of the lattice (i.e., $\varphi_i = \varphi(i)$), Eq. (5.5) reduces to

$$H_{ef}[\varphi] = \int d^D x \left[\frac{1}{2} \sum_\nu (\partial_\nu \varphi)^2 + \frac{\mu}{2} \varphi^2 + \frac{g}{4!} \varphi^4 + h\varphi + 0(a^2(\partial^2\varphi)^2) \right]. \tag{5.6}$$

If the lattice spacing becomes very small, or if the scale of variation of the function $\varphi(x)$ becomes very large (as happens at the critical point), the terms of order $a^2(\partial^2\varphi)^2$ can be neglected. The introduction of a lattice clarifies the mathematical structure by reducing the problem to a finite number of degrees of freedom. The continuum limit can be considered as the limit $a \to 0$ of the lattice theory. In the following we shall present a unified treatment of the lattice and the continuum theory as far as possible.

5.2. The Gaussian model again

If we set $g = 0$ we recover the Gaussian model; we can formally perform the integral over the φ field using Eq. (4.12). We thus find[2] that in a box of size L ($V = L^D$) for large L

5.2. The Gaussian Model Again

$$F = -\frac{V}{2(2\pi)^D} \int d^D p \ln \tilde{G}_\Lambda(p) - \frac{1}{2} \int dx\, dy\, h(x)h(y)G_\Lambda(x-y)$$
$$+ 0(\exp(-L\mu))$$

$$G_\Lambda(x) = \frac{1}{(2\pi)^D} \int d^D p\, \tilde{G}_\Lambda^2(p) \exp(ipx) \tag{5.7}$$

$$\tilde{G}_\Lambda(p) = 1 \bigg/ \left(p^2 + \mu + \frac{(p^2)^2}{\Lambda^2} \right) \qquad p^2 \equiv \sum_\nu p_\nu^2 .$$

The h-dependent part can be obtained by noting that, for any reasonable definition, the measure $d\mu[\varphi]$ must be invariant under translations in functional space: $d[\varphi] = d[\psi]$, if $\psi(x) = \varphi(x) + \omega(x)$. By an appropriate choice of ω, i.e., $\omega(x) = \int dy\, G_\Lambda(x-y)h(y)$, the h-dependent part may be factorized.

A similar formula [Eq. (4.14)] holds on the lattice; using the same normalizations as in Eq. (5.6) we get

$$F_L \simeq -\frac{V}{2(2\pi)^D} \int_B dp \ln \tilde{G}_L(p) - \frac{1}{2} a^2 \sum_l \sum_k h_l h_k G_L(l-k)$$

$$G_L(k) = \frac{1}{(2\pi)^D} \int_B d^D p\, \tilde{G}_L(p) \exp(iap \cdot k) \tag{5.8}$$

$$\tilde{G}_L(p) = \frac{1}{\mu + \sum_\nu (2 - 2\cos(p_\nu a))/a^2} \qquad \int_B dp = \int_{-\pi/a}^{\pi/a} \prod_1^D dp_\nu .$$

Let us compare Eqs. (5.7) and (5.8); we begin with the h-dependent part. We consider a smooth (analytic) function $h(x)$. Its Fourier transform $\tilde{h}(p)$ decreases very fast (exponentially) at large moments.[3] If the lattice spacing is smaller than the range of variation of the function $h(x)$, the Fourier transform at $p = \pi/a$ is very small; we now want to see the difference between Eqs. (5.7) and (5.8) if we set $h_i = h(i)$.

The Fourier transform of h_i ($\tilde{h}_L(p)$) is given by

$$\tilde{h}_L(p) = \prod_1^D \sum_{n_\nu = -\infty}^{+\infty} \tilde{h}\left(p + \frac{2\pi}{a} n_\nu \right) \simeq \tilde{h}(p) \qquad \text{for } |p| \ll 0\left(\frac{1}{a}\right). \tag{5.9}$$

Now Eqs. (5.7) and (5.8) can be written, respectively, as

$$\int_{-\infty}^{+\infty} \tilde{h}(p)\tilde{h}(-p)\tilde{G}_\Lambda(p)\, d^D p$$
$$\int_{-\pi/a}^{\pi/a} \tilde{h}_L(p)\tilde{h}_L(-p)\tilde{G}_L(p)\, d^D p . \tag{5.10}$$

The integral in Eq. (5.10) is dominated by the contribution of small momenta $p \sim 0$, where (if $\Lambda \sim a^{-1}$) $\tilde{G}_\Lambda(p)$ and $\tilde{G}_L(p)$ coincide with good accuracy, $[\tilde{G}_\Lambda(p) - \tilde{G}_L(p) = 0(a^2 p^4)]$. The difference between the two expressions in Eq. (5.10) is of order a^2.

For the h-independent part of the free energy things are more complex. The integral in Eq. (5.7) is divergent at large p. This divergence originates in the fact that the effective Hamiltonian $H_{\text{eff}}[\varphi]$ is infinite for a general function of L^2; the probability measure $d\mu[\varphi]$, defined by

$$d\mu[\varphi] \equiv dP[\varphi] \tag{5.11}$$

is therefore concentrated in a "small" subspace of the total space. It is not strange that the entropy (and consequently the partition function) of such a distribution is divergent. However, for a continuous distribution the absolute normalization of the entropy (and of the free energy) is arbitrary, and this divergence is harmless.

When we consider the internal energy and the specific heat, we find

$$u(\mu) \equiv -\frac{d}{d\mu} f(\mu) = -\frac{1}{2(2\pi)^D} \int d^D p \, \tilde{G}_\Lambda(p) = -\frac{\langle \varphi^2(0) \rangle}{2}$$

$$c(\mu) \equiv \frac{du}{d\mu} = -\frac{d^2 f}{d\mu^2} = \frac{1}{2(2\pi)^D} \int d^D p (\tilde{G}_\Lambda(p))^2$$

$$= \frac{\int dx \langle \varphi^2(x) \varphi^2(0) \rangle_c}{4}. \tag{5.12}$$

For $D < 4$, but $u(\mu)$ and $c(\mu)$ are finite; however, when $\Lambda \to \infty$, the internal energy divergence in $D \geq 2$ as

$$\begin{aligned} \ln \Lambda, & \quad D = 2 \\ \Lambda, & \quad D = 3. \end{aligned} \tag{5.13}$$

The absolute value of the internal energy is somewhat conventional; the crucial physical requirement is that the specific heat be finite. If we do not care about the normalization of the internal energy, we can send $\Lambda \to \infty$ and $g = 0$ (for $D < 4$) without further problems; however, the divergence of $\langle \varphi^2(0) \rangle$ and of $\langle \varphi^4(0) \rangle$ ($\langle \varphi^4(0) \rangle \geq \langle \varphi^2(0) \rangle^2$) suggests that new divergences will appear as soon as g is different from zero. In the following sections we shall see that these new divergences can be removed by considering the Hamiltonian

$$H_{\text{ef}}^R = H_{\text{ef}} + \frac{1}{2}\int d^D x\, c(g,\Lambda)\varphi^2(x)$$

$$c(g,\Lambda) \simeq c_1 g \ln \Lambda^2 \qquad D = 2,\ c_1 = -\frac{1}{8\pi} \tag{5.14}$$

$$c(g,\Lambda) \sim c_1 g\Lambda + c_2 g^2 \ln \Lambda^2 \qquad D = 3,\ c_1 = -\frac{1}{8\pi},\ c_2 = \frac{1}{192\pi},$$

where H_{ef} is given by Eq. (5.1).

The correlation functions and the specific heat of the Hamiltonian (5.14) have a good limit when $\Lambda \to \infty$. In other words, for the Hamiltonian H_{eff} the critical value of $\mu(\mu_c)$ goes to infinity with Λ. The extra term in Eq. (5.14) is an additive renormalization of μ: For the Hamiltonian H_{eff}^R, μ_c remains finite when Λ goes to infinity.

5.3. On the zero lattice spacing limit

The origin of the divergences mentioned in the previous paragraph can be intuitively understood from the following considerations. We want to obtain a theory in which the coherence length ξ is finite and the lattice spacing a is zero; we can change our point of view by considering the lattice spacing as fixed and sending the coherence length ξ to infinity in lattice spacing units. Quantities that are not dimensionless will behave very differently if measured in units of ξ or of a, respectively.[4]

In order to implement this new point of view it is convenient to remove the a^2 factor in front of the kinetic term $J_{ik}\varphi_i\varphi_k$ by introducing a new field ψ_i equal to $a^{(D-2)/2}\varphi_i$.

The effective Hamiltonian is thus

$$H = \sum_{i,k} \frac{1}{2} J_{i,k}(\psi_i - \psi_k)^2 + \sum_i \left(\frac{M}{2}\psi_i^2 + \frac{\lambda}{4!}\psi_i^4\right)$$

$$M = \mu a^2 \qquad \lambda = g a^{4-D}. \tag{5.15}$$

If we send a to zero in Eq. (5.15), we obtain in fewer than four dimensions the critical Gaussian model, whose correlation functions are well defined in more than two dimensions. In other words, the probability distribution of the field becomes Gaussian in the limit a going to zero, if observed on a scale that goes to zero with a.

If we use the Hamiltonian (5.15), the value of ξ/a is a function only of M and λ; it becomes infinite on a critical line $M = M_c(\lambda)$. Obviously $M_c(0) = 0$. If g measured in units of ξ remains finite when a goes to zero,

λ also goes to zero in fewer than four dimensions; in this case we can estimate the critical value of M by using perturbation theory. We get (as expected) $M_c(\lambda) = A\lambda$ (A being a constant) plus high-order corrections; moreover, we find no indications of ψ_i^2 vanishing at the phase transition point. These results are rather natural; however, they imply that in three dimensions $\mu_c \sim (1/a)g$, $\langle \varphi^2 \rangle \sim 1/a$, i.e., both μ_c and $\langle \varphi^2 \rangle$ diverge in the limit $a \to 0$. In other words, when ξ goes to infinity, all the quantities that receive contributions mainly from the large-scale fluctuations are finite in units of ξ, while other quantities that receive contributions mainly from the short-scale fluctuations are finite in their natural unit a, but divergent if measured in units of ξ. It is evident that a here plays the same role as Λ^{-1} in the previous section.

The above considerations rely implicitly on dimensional analysis. Indeed, we can measure everything in units of inverse length. The requirement that the Hamiltonian, being the argument of an exponential, must be dimensionless, implies

$$[a] = [\Lambda^{-1}] = [x] = -1$$

$$[\psi] = 0 \quad [\varphi] = \frac{D-2}{2} \tag{5.16}$$

$$[M] = 0 \quad [\mu] = 2 \quad [g] = 4 - D$$

where the brackets, [ ], stand for "dimension of."

We want to study the (carefully defined) zero lattice spacing (or infinite Λ) theory, because in this way we automatically concentrate our attention on quantities that are dominated by the critical fluctuations, and we neglect all short-distance phenomena.

In this limit the only dimensionless parameter in the Hamiltonian is $g/(\mu - \mu_c)^{(4-D)/2}$ (we shall see later that this ratio may remain finite in the limit $a \to 0$). Dimensional analysis implies that the effective exponent

$$\gamma(g, \mu - \mu_c) \equiv \frac{d \ln \chi}{d \ln(\mu - \mu_c)} \quad (\gamma_c = \gamma(g, 0)) \tag{5.17}$$

satisfies the following relations:

$$\gamma(g, \mu - \mu_c) = \gamma\left(\frac{g}{(\mu - \mu_c)^{(4-D)/2}}\right) = \gamma(u),$$

$$\tag{5.18}$$

$$u = \frac{g}{(\mu - \mu_c)^{(4-D)/2}}, \quad \gamma_c = \gamma(u)|_{u=u_c} \quad \begin{array}{l} u_c = 0, \; D > 4 \\ u_c = \infty, \; D < 4 \end{array}.$$

5.4. The Perturbative Expansion

In the next section we shall see how to construct a perturbative expansion in g (or in u). If $D>4$ we expect trivial critical exponents [the classical ones, Eq. (3.44)], with u going to zero at the transition point. If $D<4$, u_c goes to infinity at the transition, and we face the problem of studying the very strong coupling limit. In the next chapters we shall see how this can be done in a natural way by introducing the so-called renormalized coupling constant.

5.4. The perturbative expansion

We want to show how to compute the correlation functions for the Hamiltonian (5.1) or (5.5) as an expansion in series[5] of powers of g. For the time being it is convenient to suppose that $D<4$ and Λ is finite, so no divergences are present in the correlation functions. In the next subsection we shall study the limit $a \to 0$ or $\Lambda \to \infty$. We shall study together the cases of the lattice and of the continuum. In order to avoid duplications, we use only the continuum notation; the reader can switch to the lattice using the correspondence formulae:

$$x \leftrightarrow i$$

$$\int d^D x \leftrightarrow a^D \sum_i$$

$$\delta(x) \leftrightarrow a^{-D} \delta_{i,0}$$

$$\Delta \varphi = \sum_\nu^D \partial_\nu^2 \varphi \leftrightarrow \frac{\sum_k J_{ik}(\varphi_k - \varphi_i)}{2a^2} \qquad (5.19)$$

$$\tilde{G}_0(p) = \frac{1}{p^2 + m^2 + p^4/\Lambda^2} \leftrightarrow \frac{1}{\left[\sum_\nu^D \frac{4\sin^2(p_\nu a/2)}{a^2} + m^2\right]}$$

$$\int d^D p \leftrightarrow \int_B d^D p$$

$$\delta(p) \leftrightarrow \sum_k \delta(p - 2\pi k).$$

Let us study the correlation function of two φ fields; we get

$$G(x) = \langle \varphi(x)\varphi(0) \rangle \equiv \frac{\int d[\varphi]\exp[-H[\varphi]]\varphi(x)\varphi(0)}{\int d[\varphi]\exp[-H[\varphi]]}$$

$$= \lim_{V\to\infty} \frac{\sum_n \left(-\frac{g}{4!}\right)^n \frac{1}{n!} \left\langle \left[\int_V dy\, \varphi^4(y)\right]^n \varphi(x)\varphi(0)\right\rangle_0}{\sum_n \left(-\frac{g}{4!}\right)^n \frac{1}{n!} \left\langle \left[\int_V dy\, \varphi^4(y)\right]^n \right\rangle_0} \qquad (5.20)$$

where $\langle\ \rangle_0$ denote the $g=0$ expectation values, i.e., those of the Gaussian model. The expectation values appearing in Eq. (5.19) can be computed using Eq. (4.19). We have thus reduced the problem of computing the coefficient of the Taylor expansion in g of Eq. (5.19) to the evaluation of some integrals.

Let us see how this works in practice by computing the first nontrivial contributions. At the first order in g we have

$$G(x) = \frac{\langle \varphi(x)\varphi(0)\rangle - (g/4!)\int_V dy\,\langle \varphi^4(y)\varphi(x)\varphi(0)\rangle}{1 - \frac{g}{4!}\int_V dy\,\langle \varphi^4(y)\rangle}$$

$$= \frac{G_0(x) - (g/4!)\int_V dy[3(G_0(0))^2 G_0(x) + 12G_0(x-y)G_0(0)G_0(y)]}{1 - (g/4!)\int_V dy[3G_0(0)^2]}$$

$$= G_0(x) - \frac{g}{2}\int_V dy\, G_0(x-y)G_0(y)G_0(0). \qquad (5.21)$$

It is easy to check that $\langle \varphi^4(x)\rangle_0 \equiv \langle \varphi(x)\varphi(x)\varphi(x)\varphi(x)\rangle_0$ is given by $3G_0(0)^2$. Unfortunately, the direct application of Eq. (4.19) to the evaluation of $\langle \varphi(x)\varphi(0)\varphi^4(y)\rangle$ will give rise to 5!! (15) different terms; this would become very tedious at high orders. It is convenient to use the equivalent relation

$$\langle \varphi(x_1)\varphi(x_2)\cdots\varphi(x_N)\rangle = G_0(x_1-x_2)\langle \varphi(x_3)\cdots\varphi(x_N)\rangle_0$$
$$+ G_0(x_1-x_3)\langle \varphi_1(x)\cdots\varphi(x_N)\rangle_0 + \cdots$$
$$+ G(x_1-x_N)\langle \varphi(x_2)\cdots\varphi(x_{N-1})\rangle_0$$
$$= \sum_{2}^{N} G(x_1-x_a)\left\langle \prod_{2\,b\neq a}^{N} \varphi(x_b)\right\rangle_0 \qquad (5.22)$$

5.4. The Perturbative Expansion

and to collect equal terms at each step,

$$\langle \varphi(x)\varphi(0)\varphi^4(y)\rangle_0 = G_0(x)\langle \varphi^4(y)\rangle_0 + 4G_0(x-y)\langle \varphi^3(y)\varphi(0)\rangle_0$$

$$= 3G_0(x)G_0(0)\langle \varphi^2(y)\rangle_0 + 12G_0(x-y)G_0(y-0)$$

$$\times \langle \varphi^2(y)\rangle_0$$

$$= 3G_0(x)G_0(0)^2 + 12G_0(x-y)G_0(y)G_0(0). \quad (5.23)$$

Using the diagrammatical representation introduced in the previous chapter, Eq. (5.20) can be visualized in Fig. 5.1. It is important to note that in the final result only the connected diagram survives, and the disconnected ones cancel (as usual) between numerator and denominator.

This cancellation of disconnected diagrams is present at all orders; if it were not, there would be a serious disaster: disconnected diagrams are proportional to the volume V, and noncanceling disconnected diagrams would appear as terms proportional to V^n in the connected correlation functions, thus destroying the possibility of a perturbative expansion in powers of g. If μ is different from zero, connected diagrams cannot diverge when the volume goes to infinity (they are free of infrared singularities): $G_0(x) \sim \exp(-\mu^{1/2}|x|)$, and the contribution of any point y_i to the integral in Eq. (5.20) is exponentially damped when y_i goes to infinity. If Λ is finite, $G_0(x)$ is bounded and no short-distance (ultraviolet) divergences are present. In this situation all connected diagrams are well defined (divergence free).

At the next order we obtain

$$\langle \varphi(x)\varphi(0)\rangle$$

$$= \frac{\langle \varphi(x)\varphi(0)\rangle_0 - \frac{g}{4!}\int dy_1 \langle \varphi^4(y_1)\varphi(x)\varphi(0)\rangle_0 + \frac{1}{2}\frac{g^2}{(4!)^2}\int dy_1\,dy_2 \langle \varphi^4(y_1)\varphi^4(y_2)\varphi(x)\varphi(0)\rangle_0}{1 - \left(\frac{g}{4!}\right)\int dy_1 \langle \varphi^4(y_1)\rangle_0 + \frac{1}{2}\left(\frac{g}{4!}\right)^2 \int dy_1\,dy_2 \langle \varphi^4(y_1)\varphi^4(y_2)\rangle_0},$$

(5.24)

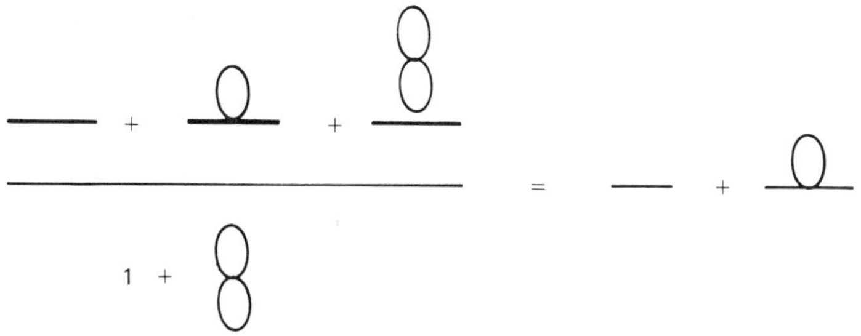

Figure 5.1. The two-point correlation function [Eq. (5.19)] at order g: After the cancellations between the numerator and the denominator only connected diagrams survive.

where we have expanded the result up to the order g^2. After some simple algebra we find that the coefficient of g^2 is given by

$$\frac{g^2}{(4!)^2} \int dy_1\, dy_2 [144 G_0(x-y_1) G_0(0) G_0(y_1-y_2) G_0(0) G_0(y_2)$$

$$+ 144 G_0(x-y_1) G_0(y_1-y_2)^2 G_0(0) G_0(y_1)$$

$$+ 96 G_0(x-y_1) G_0(y_1-y_2)^3 G_0(y_2)]. \tag{5.25}$$

Equation (5.24) is represented diagrammatically in Fig. 5.2.

There is a faster way to obtain the result (5.25). We suppose that disconnected diagrams can be neglected and retain only the connected ones,[6]

$$\langle \varphi(x)\varphi(0) \rangle$$
$$= \sum_n \left(\frac{-g}{4!}\right)^n \frac{1}{n!} \int dy_1 \cdots dy_n \langle \varphi(x)\varphi(0)\varphi^4(y_1) \cdots \varphi^4(y_n) \rangle_0^c. \tag{5.26}$$

A further simplification can be achieved by noticing that the r.h.s. of (5.26) contains $n!$ equal terms that differ from the permutations of the y's; we can consider only one of such contributions, $\langle\ \rangle^{cp}$, and cancel the $n!^{-1}$. We finally get

$$\langle \varphi(x)\varphi(0) \rangle = \sum_n \left(\frac{-g}{4!}\right)^n \int dy_1 \cdots dy_n \langle \varphi(x)\varphi(0)\varphi^4(y_1) \cdots \varphi^4(y_n) \rangle^{cp}. \tag{5.27}$$

Let us recompute Eq. (5.25): We start from

$$\langle \varphi(x)\varphi(0)\varphi^4(y_1)\varphi^4(y_2) \rangle_0 = 4 G_0(x-y_1)\langle \varphi(0)\varphi^3(y_1)\varphi^4(y_2) \rangle_0$$

$$+ 4 G_0(x-y_2)\langle \varphi(0)\varphi^3(y_2)\varphi^4(y_1) \rangle_0$$

$$+ G_0(x)\langle \varphi^4(y_1)\varphi^4(y_2) \rangle_0. \tag{5.28}$$

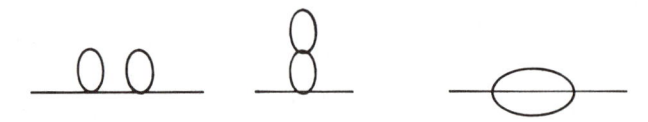

Figure 5.2. Connected diagrams for the two-point correlation function at the order g^2 [Eq. (5.24)].

5.4. The Perturbative Expansion

The last term is disconnected and can be neglected; we can retain only one of the first two terms. We finally get

$$4G_0(x - y_1)\langle \varphi(0)\varphi^3(y_1)\varphi^4(y_2)\rangle_0 = 4G_0(x - y_1)$$
$$\times [3G_0(y_1)\langle \varphi^2(y_1)\varphi^4(y_2)\rangle_0$$
$$+ 4G_0(y_2)\langle \varphi^3(y_1)\varphi^3(y_2)\rangle_0]. \quad (5.29)$$

The first term arises by connecting 0 with y_1 (3 ways), the second by connecting 0 with y_2 (4 ways).

We can go on: If in the first term we connect y_1 with y_1 [extracting $G_0(y_1 - y_1) \equiv G_0(0)$] we are left with $\langle \varphi^4(y_2)\rangle_0$, which yields a disconnected piece. Therefore we must connect y_1 with y_2 (4 ways). In the second term, either we connect y_1 with y_1 (3 ways) or we connect y_1 with y_3 (3 ways). We finally get

$$4G_0(x - y_1)\{3G_0(y_1)G_0(y_1 - y_2)\langle \varphi(y_1)\varphi^3(y_2)\rangle_0$$
$$+ 4G_0(y_2)[3G_0(0)\langle \varphi(y_1)\varphi^3(y_2)\rangle_0 + 3G(y_1 - y_2)\langle \varphi^2(y_1)\varphi^2(y_2)\rangle_0]\}$$
$$= 4G_0(x - y_1)\{3G_0(y_1)G_0(y_1 - y_2)3G(y_1 - y_2)\langle \varphi^2(y_2)\rangle_0$$
$$+ 4G_0(y_2)[3G_0(0)G(y_1 - y_2)\langle \varphi^2(y_2)\rangle_0$$
$$+ 3G(y_1 - y_2)(G_0(0)\langle \varphi^2(y_2)\rangle_0$$
$$+ 2G_0(y_1 - y_2)\langle \varphi(y_1)\varphi(y_2)\rangle_0)]\}. \quad (5.30)$$

The general diagrammatical rules for computing the order g^n are as follows: we draw all connected diagrams with n vertices with 4 lines and two external points; diagrams equivalent by the interchange of the vertices must be identified; with each diagram we associate an integrand which is the product of the G_0's corresponding to the lines of the diagrams; we finally integrate over the coordinates of the vertices. The result must be multiplied by $(-g/24)^n$ and by the multiplicity factor of the diagram [how many times the diagram can be obtained by using Eq. (5.22)]; the multiplicity factors for the diagrams in Fig. 5.2 are, respectively, 144, 144, 96. The computation of the correct multiplicity factor is normally a tedious job. It can be simplified by using the following rule: The multiplicity of a diagram of order n for the φ^4 interaction, divided by $(24)^n$, is equal to one divided by the number of symmetries of the diagram (i.e., the number of permutations of the lines of the diagrams that do not change the diagram). The diagrams of Fig. 5.2 have 4, 4, and 6 symmetries, respectively. It is usually a good idea not to trust the results and to double-check the multiplicities quite carefully. It is very easy to make mistakes, as shown by the large number of preprints (and papers) with wrong multiplicities.

We leave as an exercise for the reader the derivation of the results for the order g^3 shown in Fig. 5.3.

The evaluation of the integrals is simpler in momentum space; we can use the relations

$$\int d^D x \exp(ipx) \int d^D y \, f(y)g(x-y) = \tilde{g}(p)\tilde{f}(p)$$

$$\tilde{g}(p) = \int d^D x \exp(ipx) g(x)$$

$$g(x) = \int \frac{d^D p}{(2\pi)^D} \exp(-ipx) \tilde{g}(p) \quad (5.31)$$

$$\delta^D(x) = \int \frac{d^D p}{(2\pi)^D} \exp(ipx) \,.$$

After having transcribed a few diagrams from position to momentum space, we arrive at the following general rules: Each external leg of the diagram is associated with a momentum p. Each internal line (α) is associated with a momentum q_α, which runs along the line in a given direction. We must impose the condition that the sum of all incoming minus all outgoing momenta be equal to zero at each vertex. At the order g^n there remain n unfixed momenta (let us call them k's) on which the q_α depend linearly. The contribution of the diagram is given by the integral over the k's of the product of all $\tilde{G}(q_\alpha)$'s. These rules are clarified by the

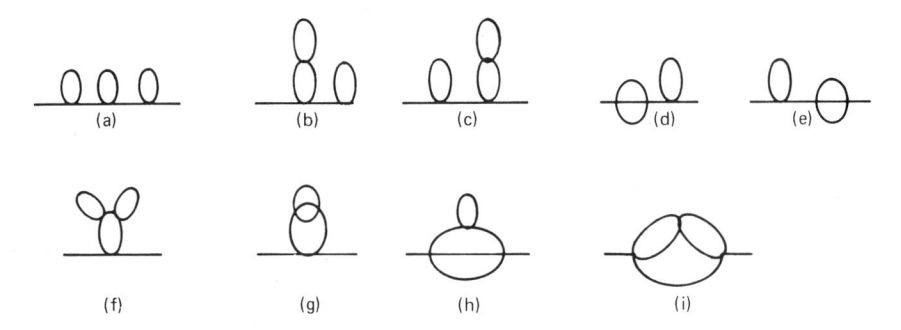

Figure 5.3. Connected diagrams at the order g^3 for the two-point correlation function: Only diagrams (f)–(i) are one-line irreducible. See evaluations in text, Eq. (5.32).

5.4. The Perturbative Expansion

explicit evaluations of the diagrams of Fig. 5.3:

$$a = -\left(\frac{g}{(2\pi)^D}\right)^3 \frac{1}{8} [\tilde{G}_0(p)]^4 \left[\int d^D k_1 \, \tilde{G}_0(k_1)\right]^3$$

$$b = c = -\left(\frac{g}{(2\pi)^D}\right)^3 \frac{1}{8} [\tilde{G}_0(p)]^3 \left[\int d^D k_1 \, \tilde{G}_0(k_1)\right]^2 \int d^D k_2 (\tilde{G}_0(k_2))^2$$

$$d = e = -\frac{1}{12} \left(\frac{g}{(2\pi)^D}\right)^3 [\tilde{G}_0(p)]^3 \int d^D k_1 \, d^D k_2 \, \tilde{G}_0(k_1)\tilde{G}_0(k_2)$$

$$\times \tilde{G}_0(p - k_1 - k_2) \int d^D k_3 \, \tilde{G}_0(k_3)$$

$$f = -\frac{1}{8} \left(\frac{g}{(2\pi)^D}\right)^3 [\tilde{G}_0(p)]^2 \int d^D k_1 [\tilde{G}_0(k_1)]^3 \left[\int d^D k_2 \, \tilde{G}_0(k_2)\right]^2 \quad (5.32)$$

$$g = -\frac{1}{12} \left(\frac{g}{(2\pi)^D}\right)^3 [\tilde{G}_0(p)]^2 \int d^D k_1 \, d^D k_2 \, d^D k_3 (\tilde{G}_0(k_1))^2$$

$$\times \tilde{G}_0(k_2)\tilde{G}_0(k_3)\tilde{G}_0(k_1 - k_2 - k_3)$$

$$h = -\frac{1}{4} \left(\frac{g}{(2\pi)^D}\right)^3 [\tilde{G}_0(p)]^2 \int d^D k_1 \, d^D k_2 \, d^D k_3 [\tilde{G}_0(p - k_1 - k_2)]^2$$

$$\times \tilde{G}_0(k_1)\tilde{G}_0(k_2)\tilde{G}_0(k_3)$$

$$i = -\frac{1}{4} \left(\frac{g}{(2\pi)^D}\right)^3 [\tilde{G}_0(p)]^2 \int d^D k_1 \, d^D k_2 \, d^D k_3 \, \tilde{G}_0(k_1)\tilde{G}_0(k_2)$$

$$\times \tilde{G}_0(k_3)\tilde{G}_0(p - k_1 - k_3)\tilde{G}_0(p - k_2 - k_3).$$

In order to decrease the proliferation of diagrams with order, it is convenient to introduce diagrams that are one-line irreducible, i.e., they cannot be divided into two disconnected diagrams by cutting only one internal line [the diagrams in Fig. 5.3(f)–(i) are one-line irreducible]. These diagrams can be written as $(\tilde{G}_0(p))^2 \Sigma(p)$, where $\Sigma(p)$ is called the self-energy and is the contribution of the diagrams neglecting the two external lines (amputated diagrams). We readily obtain

$$\tilde{G}(p) = \tilde{G}_0(p) + \tilde{G}_0(p)\Sigma(p)\tilde{G}_0(p) + \tilde{G}_0^3(p)\Sigma^2(p) + \cdots$$

$$= \frac{1}{[\tilde{G}_0^{-1}(p) + \Sigma(p)]}. \quad (5.33)$$

Equation (5.31) is shown diagrammatically in Fig. 5.4.

Figure 5.4. The graphical representation of Eq. (5.31) of the text. The square and the blob denote, respectively, the self-energy and the exact two-point correlation.

The only delicate point in proving Eq. (5.33) is to check that the weights of the diagrams are correct.

We can now develop in powers of g the four-field correlation functions $\langle \varphi(x)\varphi(y)\varphi(z)\varphi(t) \rangle$.

We have three classes of diagrams shown in Fig. 5.5. In the first class we find the fully disconnected diagrams, i.e., there are vertices that are not connected to the external points. These diagrams cancel with the corresponding ones coming from the expansion of the denominator. In the second class all vertices are connected to the external points, but the diagrams can be divided into two disconnected parts. These diagrams are

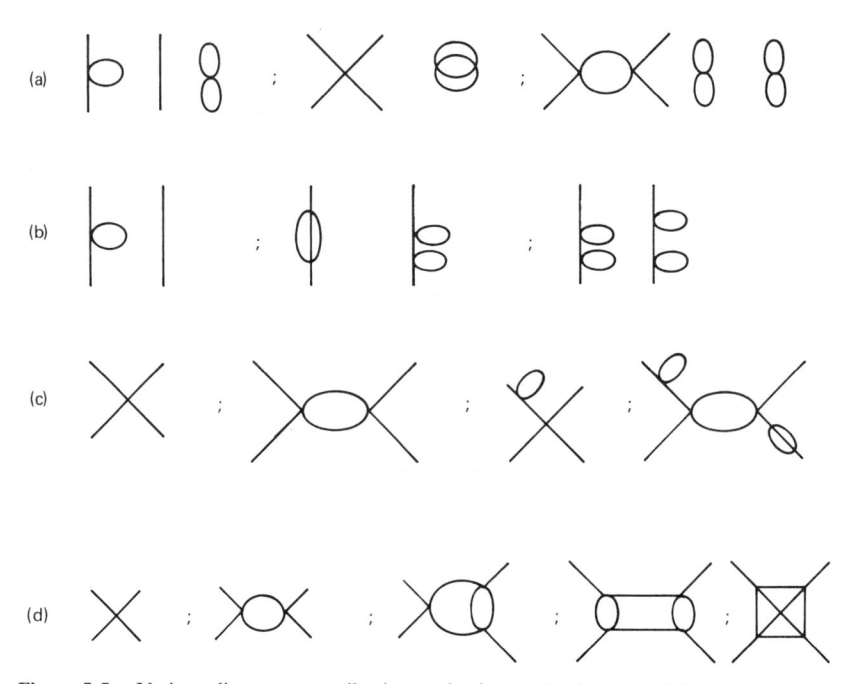

Figure 5.5. Various diagrams contributing to the four-point function: (a) fully disconnected diagrams; (b) partially connected diagrams; (c) connected diagrams; (d) some connected one-line irreducible diagrams contributing to Γ.

the same as those of the two-field correlation functions, and their contributions can be summed to $G(x-y)G(z-t) + G(x-z)G(y-t) + G(x-t)G(y-z)$. In the third class we find fully connected diagrams. They are the only ones that contribute to the connected correlation functions.

In order to simplify the analysis it is convenient to introduce one-line irreducible amputated diagrams. Their sum, neglecting the external leg, is denoted $\Gamma(p_1, p_2, p_3, p_4)$; at the order g^2 we are left with the diagrams shown in Fig. 5.5(d).

In momentum space we find

$$-\Gamma(p_1, p_2, p_3, p_4) = \delta(p_1 + p_2 + p_3 + p_4)\left[g - \frac{g^2}{2}(I(p_1 + p_2)\right.$$

$$\left. + I(p_1 + p_3) + I(p_1 + p_4)) + O(g^3)\right]$$

$$I(p) = \frac{1}{(2\pi)^D}\int d^D k\, \tilde{G}_0(k)\tilde{G}_0(p-k) \quad (5.34)$$

$$G(p_1, p_2, p_3, p_4) = \tilde{G}(p_1)\tilde{G}(p_3)\delta(p_1 + p_2)\delta(p_3 + p_4)$$

$$+ \tilde{G}(p_1)\tilde{G}(p_2)\delta(p_1 + p_3)\delta(p_2 + p_4)$$

$$+ \tilde{G}(p_1)\tilde{G}(p_3)\delta(p_1 + p_4)\delta(p_4 + p_3)$$

$$+ \tilde{G}(p_1)\tilde{G}(p_2)\tilde{G}(p_3)\tilde{G}(p_4)\Gamma(p_1, p_2, p_3, p_4).$$

The δ function in Eq. (5.34) is the consequence of the translational invariance of the correlation functions. It is clear that if we already know the two-field correlation functions, in order to get the four-field correlation functions we need only compute Γ, which receives contributions from only "a few" diagrams. A similar analysis can be performed for the many-field correlation functions.

5.5. The removal of ultraviolet divergences

We want to study what happens when $\Lambda \to \infty$ or $a \to 0$. Some diagrams will be divergent, some will remain finite; we have to take care of the first and to compensate for their divergence by the introduction of appropriate counterterms [Eq. (5.14)].

The structure of ultraviolet divergences can be derived from general theorems that will be discussed in Chapter 9. Here we merely remark that a diagram whose expression is given by

$$\int \prod_a^L d^D k_a \prod_\alpha^N \tilde{G}_0[q_\alpha(k)], \tag{5.35}$$

the q_α being linear functions of the k's, is certainly divergent (for $\Lambda \to \infty$) if the number of powers of k coming from the integrations (DL) is greater than or equal to the number of powers coming from the denominator $(2N)$, as can be seen from rescaling all the k's together. If the opposite happens, i.e.,

$$DL < 2N \Leftrightarrow \mathcal{D} < 0, \tag{5.36}$$

the diagram is said to be superficially convergent; $\mathcal{D} = DL - 2N$ is the degree of divergence of a diagram. It is possible that a superficially convergent diagram is divergent because a subdiagram (a subintegration) is divergent. The necessary condition for convergence is that the diagram and all the subdiagrams be superficially convergent.[7]

We must therefore classify all the superficially divergent diagrams. In fewer than four dimensions this task is quite easy; there are the following relations between the number of external lines E, the number of vertices V, the number of internal lines N, and the number of loops (k integrations):

$$E + 2N = 4V \qquad V = L + \frac{E-2}{2}$$
$$\mathcal{D} = (D-4)L + 4 - E. \tag{5.37}$$

We see that if $D < 2$ no divergences are present, as expected; for $2 \le D < 4$ only the self-energies ($E = 2$), may be divergent[8]—at the first order for $D = 2$ and at the second order in g for $D = 3$.

In $D = 2$ only the divergent self-energy diagram is shown in Fig. 5.6(a). Its value is given by

$$\Sigma(p, \mu) = -\frac{g}{(2\pi)^D} \frac{1}{2} \int d^D k \frac{1}{(k^2 + \mu + k^4/\Lambda^2)} \tag{5.38}$$

where conventionally $k^4 \equiv (k^2)^2 = (\sum_\nu^D k_\nu^2)^2$. Let us decide that when we evaluate a diagram we substitute the following expression for $\Sigma(p, \mu)$:

$$\Sigma_R(p, \mu) = \Sigma(p, \mu) - \Sigma(0, \mu_N), \tag{5.39}$$

(a)　　　(b)

Figure 5.6. Divergent diagrams for $D = 3$.

5.6. A Constructive Approach

where μ_N is a given normalization point. The substitution (5.39) corresponds to adding a counterterm[9] in the Hamiltonian equal to $\frac{1}{2} \int d^2x\, \varphi^2(x)$. Here Σ_R is finite for any μ [for $\mu = \mu_N$ the contribution of Fig. 5.6(a) is zero because this diagram produces a momentum-independent $\Sigma(p, \mu)$]. In the same way in $D = 3$, the only new divergent diagram is shown in Fig. 5.6(b); its contribution to Σ is given by

$$\Sigma(p, \mu) = \frac{1}{6} \left(\frac{g}{(2\pi)^2}\right)^2 \int d^D k_1\, d^D k_2\, \frac{1}{k_1^2 + \mu + k_1^4/\Lambda^2} \cdot \frac{1}{k_2^2 + \mu + k_2^4/\Lambda^2}$$

$$\cdot \frac{1}{(p - k_1 - k_2)^2 + \mu + (p - k_1 - k_2)^4/\Lambda^2}. \qquad (5.40)$$

As will be shown in the appendix, $\Sigma(p, \mu)$ has a divergence proportional to $\ln \Lambda^2$, which is p independent.

If the renormalized self-energy is defined in the same way as in Eq. (5.39), all the divergences disappear; the corresponding counterterm in the Hamiltonian behaves like $\ln \Lambda/\mu_N$. We have thus recovered the results announced in Eq. (5.14).[10]

In the foregoing arguments it is crucial that the divergences in Σ be removable by the addition of a momentum-independent term. In the general case the divergences may be polynomials in the external momenta. If that happens (e.g., for $D = 4$), we must add to the Hamiltonian counterterms containing derivatives of φ as $\int d^D x (\partial_\nu \varphi)^2$.

5.6. A constructive approach

The problem of removing the cutoff (e.g., sending the lattice spacing to zero) has been the subject of many physical mathematical studies; it has been rigorously proven that the limit $a \to \infty$ or $\Lambda \to \infty$ of the correlation functions exists (including in the infinite-volume limit) after the introduction of the counterterms if $D < 4$.[11]

Here we want only to discuss the mathematical meaning of the functional integral in a box with $\Lambda < \infty$. Rather than define $d[\varphi]$ or $\exp[-H]$, we start from the free case and define a probability measure $d\mu[\varphi]$, which is formally given by

$$d\mu[\varphi] \propto d[\varphi] \exp\left[\int -\left(\frac{1}{2}(\partial_\nu \varphi)^2 + \frac{1}{2}(\Delta \varphi)^2 + \frac{\mu}{2} \varphi^2\right) d^D x\right] \qquad (5.41)$$

in terms of its functional Fourier transform

$$\int d\mu[\varphi]\exp\left[i\int d^Dx\, h(x)\varphi(x)\right] = \int \exp\left[\frac{-1}{2}\int dx\, dy\, h(x)h(y)G(x-y)\right]. \tag{5.42}$$

The result (5.42) could obviously be derived by Eq. (5.41) in a nonrigorous way. From the mathematical point of view it is convenient to consider Eq. (5.42) as the basic definition. Now it can be proved that if[12]

$$A[\varphi] \geq 0, \qquad \int d\mu[\varphi]A[\varphi] = \langle A\rangle_0 < \infty \tag{5.43}$$

the functional integral $Z(g) \equiv \int d\mu[\varphi]\exp[-gA[\varphi]]$ exists and has the following properties:

$$1 \geq \int d\mu[\varphi]\exp[-gA[\varphi]] \geq \exp(-g\langle A\rangle_0)$$

$$Z(g) = \int_0^\infty d\nu(a)\exp[-ga]. \tag{5.44}$$

$d\nu(a)$ is the probability measure[13] for having $A[\varphi] = a$; $d\mu(a) = \int d\mu[\varphi]\delta(A[\varphi] - a)$.

The condition (5.43) is obviously satisfied by the interaction term $\frac{1}{24}\int d^Dx\, \varphi^4 \equiv A[\varphi]$ but fails when $\Lambda \to \infty$ because of the additional counterterms, which ruin the positivity of the perturbation. The representation (5.44) is very powerful if combined with the perturbative expansion of $Z(g)$ in powers of g, which we have derived in the preceding sections,[14]

$$Z(g) \underset{A}{=} \sum_0^\infty Z_n g^n = \exp[-F(g)]. \tag{5.45}$$

We use the symbol $\underset{A}{=}$ to indicate that the equality is true in the sense of the expansion in powers of g:

$$Z(g) - \sum_0^N Z_n g^n = 0(g^{N+1}). \tag{5.46}$$

The expansion of Z in powers of g is not convergent, and it is only an asymptotic series, as we shall see soon.

It is well known that Eqs. (5.45) and (5.46) do not fix the function $Z(g)$ in terms of the Z_n [$\exp(-1/g^2)$ is a well-known example].[15] However, Eqs. (5.45) and (5.46) with the integral representation (5.44) tell us that the generalized convexity inequalities hold;

5.7. The Borel Transform

$$L_N(g) \leq L_{N+1}(g) \leq Z(g) \leq U_{N+1}(g) \leq U_N(g), \quad (5.47)$$

where

$$\lim_{N \to \infty} L_N(g) = \lim_{N \to \infty} U_N(g) = Z(g). \quad (5.48)$$

The upper and lower bounds U_N and L_N can be constructed using the conditions

$$U_N(g) = \sum_{k=1}^{N} a_k^{(N)} \exp(-b_k^{(N)} g) + c^{(N)}$$

$$L_N(g) = \sum_{k=1}^{N} \alpha_k^{(N)} \exp(-\beta_k^{(N)} g) \quad (5.49)$$

$$U_N(g) - I(g) = 0(g^{2N+1}), \qquad L_N(g) - I(g) = 0(g^{2N}).$$

Here a, b, c, α, and β are given as nonlinear functions of the Z_n.

In this case not only does the perturbative expansion fix the function $Z(g)$ (ambiguities are removed), but we can construct a convergent sequence of upper and lower bounds: we have the errors under control.[16] Unfortunately, such a wonderful result is lost if we consider the correlation functions, or if we send the ultraviolet cutoff or the volume of the box to infinity. In these cases weaker results can be obtained. They will be presented in the next section.

5.7. The Borel transform

In general it is not possible to transform the Taylor expansion (5.46) in convergent sequences of upper and lower bounds. We may thus ask under which conditions the asymptotic expansion fixes the function $Z(g)$ in a unique way.[17] The answer is given by Watson's theorem:[18] We need to study $Z(g)$ in the complex plane.

If the functions $Z(g)$ and $H(g)$ satisfy in the sector $|g| < R$

$$|\arg(g)| < \frac{\pi}{2} + \delta, \qquad \delta > 0,$$

the inequalities

$$|Z(g) - H(g)| < A g^N r^N N! \quad (5.50)$$

for any N, the functions $H(g)$ and $Z(g)$ coincide. This means that there can be only one function $Z(g)$ that is the best approximated by the asymptotic expansion, in the sense that

$$\left| Z(g) - \sum_0^N Z_n g^n \right| \leq A g^{N+1} r^{N+1} (N+1)! \ . \tag{5.51}$$

If Eq. (5.51) is satisfied, we can apply the Borel theorem: The function $B(b)$ [the Borel transform of $Z(g)$] defined by

$$B(b) = \sum_0^\infty \frac{1}{n!} Z_n b^n \tag{5.52}$$

is an analytic function of b for $|b| < r^{-1}$, and it may be extended analytically in the sector $0 \leq |b| < \infty$, $|\arg(b)| < \delta$. The function $Z(g)$ is thus given by the convergent integral

$$Z(g) = \int_0^\infty db \, \exp(-b) B(gb) = \frac{1}{g} \int_0^\infty db \, \exp(-b/g) B(b) \ . \tag{5.53}$$

The compatibility of Eq. (5.45) with (5.53) can be checked by expanding the left-hand side and the right-hand side in powers of g, using the relation $\int_0^\infty db \, \exp(-b) b^n = n!$. Vice versa, if the perturbative expansion is such that the function $B(b)$ has good analyticity properties and does not increase too fast at infinity—Eq. (5.53) can be used to sum the perturbative expansion, $B(b)$ being said to be the Borel sum of the expansion.

It is known that in fewer than four dimensions the perturbative expansion for the correlation functions described in the previous subsections can be Borel summed and that the Borel sum is the correct result: the proof, not so simple, goes through the following steps.[19]

a) One proves that the correlation functions exist and are analytic in the region Re $g \geq 0$. This is the definition of the function $Z(g)$.
b) One notices that in order to prove Eq. (5.50) it is necessary only to show that

$$\left| \left(\frac{d}{dg} \right)^k Z(g) \right| < A(k!)^2 r^k R^{-k} \tag{5.54}$$

is in the sector.
c) One shows by direct evaluation, using intelligent upper bounds, that Eq. (5.54) holds at $g = 0$.
d) One writes the left-hand side of (5.54) in the case of the two-point function as

$$\left| \left\langle \varphi(x) \varphi(0) \left[\int d^D x \, \varphi^4(x) \right]^k \right\rangle_c \right| \ . \tag{5.55}$$

5.7. The Borel Transform

It is possible to show that Eq. (5.55) decreases when g becomes different from zero if Re $g \geq 0$ and it is bounded by the value at $g = 0$.

e) In this way one obtains Eq. (5.54) for Re $g \geq 0$. We need, however, to get it in a larger sector: here $\delta = 0$, and we need $\delta > 0$. This last difficulty may be resolved by giving a small imaginary part to μ and noticing that the parameter expansion is g/μ^{4-D}. This last trick would not work in four dimensions.

f) All these steps must be done controlling the infinite-volume limit and the divergences coming from the zero lattice spacing.[20]

This theorem is very important because it tells us that the perturbative expansion contains all the information on the true correlation functions. We can now try to construct convergent sequences of approximation to $Z(g)$ starting from the perturbative coefficients.

The main difficulty in using Eq. (5.53) is connected with the fact that the integral over b extends to infinity: We need to construct "uniformly" convergent approximations to the function $B(b)$ on the whole positive real line. If, for example, the Borel transform has a singularity on the real negative axis at $b = -b_s$ ($b_s > 0$), the Z_n behaves as

$$Z_n \simeq n!(-b_s)^{-n} \qquad n \to \infty \qquad (5.56)$$

and the Taylor expansion for $B(b)$ is divergent for $|b| > b_s$. This difficulty may be bypassed by a conformal mapping:[21] Let us suppose for simplicity that the only singularities of the Borel transform are a cut running from $-\infty$ to $-b_s$. We define

$$y = \frac{b}{b_s + b} \qquad b = b_s \frac{y}{1-y}$$

$$\tilde{B}(y) = B\left(b_s \frac{y}{1-y}\right). \qquad (5.57)$$

We have, therefore,

$$Z(g) = \frac{b_s}{g} \int_0^1 \frac{dy}{(1-y)^2} \exp\left[-\frac{b_s y}{g(1-y)}\right] \tilde{B}(y). \qquad (5.58)$$

The function $\tilde{B}(y)$ is now analytic inside the unit circle with a cut from one to infinity. [The coefficients of the Taylor expansion of $\tilde{B}(y)$ can be obtained from those of $B(b)$ (or $Z(g)$) in a simple way].[22] Its Taylor expansion converges as soon as $y < 1$. At the point $y = 1$ there may be some difficulty, but the whole region near $y = 1$ is exponentially damped. If in Eq. (5.58) we perform the approximation

$$\tilde{B}_N(y) = \sum_0^N \tilde{B}_k y^k, \qquad (5.59)$$

the corresponding $Z_N(g)$ form a convergent sequence of approximants to $Z(g)$.

It is clear that this strategy can be implemented only if we know something of the singularity of the Borel transform. We can either use the rigorous information coming from the theorems or develop an heuristic approach (which is not fully rigorous) to compute the position and the shape of the singularities of the Borel transform. These results are interesting because they give information on the large-k behavior of the perturbative coefficients Z_k. Before examining the singularities of the Borel transform in the next section, we observe that the property of being Borel summable is remarkably stable: if $f(g)$ and $h(g)$ are Borel summable and $h(0) = 0$, then $f(h(g))$ is also Borel summable; if $f(g) = g + O(g^2)$ is Borel summable, then $g(f)$ (i.e., the inverse function) is also Borel summable. Most manipulations (e.g., writing the connected correlations as functions of the correlations) do not destroy Borel summability. This kind of stability is clearly not valid for all kinds of resummation techniques based on positivity conditions, like those of the previous section.

5.8. The singularities of the Borel transform

Let us consider an elementary case: the partition function of the Ginsburg-Landau model on a lattice with only one point. We have for $N = 2$ (see pp. 23 and 108)

$$I(g) = \int_{-\infty}^{+\infty} d\varphi_1 \, d\varphi_2 \, \exp\left[-\frac{\mu}{2}(\varphi_1^2 + \varphi_2^2) - \frac{g}{24}(\varphi_1^2 + \varphi_2^2)^2\right]$$

$$= \pi \int_0^\infty dr \, \exp\left[-\frac{\mu}{2} r - g r^2\right]$$

$$= \frac{\pi}{g} \int_0^\infty ds \, \exp\left[-\frac{(\mu/2)s + \frac{1}{24} s^2}{g}\right] = \frac{1}{g} \int_0^\infty db \, B(b) \exp\left(-\frac{b}{g}\right)$$

$$B(b) = \pi \int ds \, \delta\left(\frac{\mu}{2} s + \frac{1}{24} s^2 - b\right)$$

$$= \pi \left(\frac{db}{ds}\bigg|_{s=s(b)}\right)^{-1} = \frac{\pi}{(\mu + s(b)/6)} \tag{5.60}$$

where $(\mu/2)s(b) + \frac{1}{24} s^2(b) = b$, i.e., $s(b) = -6\mu + \sqrt{36\mu^2 + 24b}$. Equation (5.60) implies that $B(b)$ is singular when

5.8. The Singularities of the Borel Transform

$$\frac{dA(\psi)}{d\psi} = 0 \qquad A(\psi) = \frac{\mu}{2}\psi^2 + \frac{\psi^4}{24} \Rightarrow \psi^2 = -6\mu \qquad b = -b_s = -\frac{3}{2}\mu^2, \tag{5.61}$$

the singularity being a cut running from $-b_s$ to $-\infty$.

It is easy to see that the argument can be generalized to any finite-dimensional integral of the form

$$Z(g) = \int d\psi_i \exp\left[-\frac{A[\psi]}{g}\right]$$

$$\langle f \rangle Z = \int d\psi_i \exp\left[-\frac{A[\psi]}{g}\right] f(\psi). \tag{5.62}$$

The singularities of the Borel transform are given by the solutions of the equations[23]

$$\frac{\partial A}{\partial \psi_i} = 0 \qquad b_s = A[\psi], \tag{5.63}$$

the positions of the singularities being given by the value of the function $A[\psi]$ evaluated on the solutions of Eq. (5.63). The Ginsburg-Landau partition function can be written in the form (5.62) by rescaling the field φ ($\psi = g^{1/2}\varphi$).

It is normally assumed that Eq. (5.63) survives to the limits $a \to 0$ and $V \to \infty$. Then the nearest singularity of the Borel transform is given for the Hamiltonian (5.1) ($\Lambda = \infty$) by the minimum value of $A[\psi]$ computed on the nontrivial ($\psi \neq 0$) solutions of the equations[24]

$$\frac{\delta A}{\delta \psi(x)} = (-\Delta + \mu)\psi(x) + \psi^3(x)/6 = 0$$

$$A[\psi] = \int d^D x \left[\frac{1}{2}(\partial_\nu \psi)^2 + \frac{\mu}{2}\psi^2 + \frac{\psi^4}{4!}\right]. \tag{5.64}$$

The problem (5.64) can be solved numerically to high accuracy with a pocket computer; one finds that $i\psi(x)$ is a positive spherically symmetric function without zeros and that b_s is given by

$$\begin{aligned} b_s &= \tfrac{3}{2} & D &= 0 \\ b_s &= 8 & D &= 1 \\ b_s &= 25.1405 & D &= 2 \\ b_s &= 40.6025 & D &= 3 \end{aligned} \tag{5.65}$$

While the position of the singularity of the Borel transform is universal, i.e., it does not depend on the particular correlation function we consider, the form and the residue of the singularity depend on the particular Green's function. Their computation is somewhat more complex, but it is feasible.

If the function $Z(g)$ is analytic in the cut plane and it can be written as

$$Z(g) = \int_0^\infty \frac{dt}{g+t} D(t), \qquad (5.66)$$

with $D(t)$ having arbitrary sign [the representation (5.66) is supposed to hold in many cases], the following three statements are equivalent:

a) $D(t) \equiv \dfrac{1}{2\pi i} \text{Disc } Z(g)_{g=-t} = \dfrac{1}{\pi} \text{Im } Z(g)|_{g=-t} \sim t^{-\gamma} \exp\left(-\dfrac{b_s}{tb_s}\right)$

$$(5.67)$$

b) $Z_n \sim \Gamma(n+\gamma) b_s^{-n}$ \hfill (5.68)

c) The nearest singularity of the Borel transform has the form

$$B(b) \sim \frac{1}{(b_s + b)^\gamma}. \qquad (5.69)$$

The singularities of the Borel transform for negative b are thus connected to the nonconvergence of the integral representation (5.62) for negative g. There are theories for which the Borel transform has a singularity on the real positive axis:[25] in this case perturbation theory alone is not sufficient to reconstruct the correct result, and we must add new information coming from a nonperturbative analysis.

Appendix to Chapter 5

In Sec. 5.4 we saw how to compute a diagram, i.e., how to associate with a diagram an integral that must be evaluated either in position space or, more practically, in momentum space [Eq. (5.30)]. Unfortunately, the final expression is rather complex: a diagram with n loops requires n integrations over D-dimensional variables. Although in most cases (ex-

Appendix to Chapter 5

cept for $D = 1$) it is impossible to obtain an explicit evaluation of the integrals in terms of known functions, it is convenient to manipulate integrals of the kind in Eq. (5.30) in order to get a more handy representation.

A standard approach is based on the following formula:

$$\frac{1}{A^n} = \frac{1}{\Gamma(n)} \int_0^\infty d\alpha \, \alpha^{n-1} \exp(-\alpha A) \quad (n > 0) \qquad (A5.1)$$

where $\Gamma(n)$ is the Euler function that for integer n satisfies $\Gamma(n) = (n-1)!$. If we apply this formula to the propagator $G_0(k)$, we get

$$\frac{1}{k^2 + m^2} = \int_0^\infty \exp[-\alpha(k^2 + m^2)] \, d\alpha . \qquad (A5.2)$$

We can now use equation (A5.2) for all propagators that must be integrated. At this point the integration over the k's is Gaussian and can be easily performed by using the well-known results

$$\int d^D k \, \exp[-\alpha k^2 + k \cdot p] = \frac{1}{(\alpha \pi)^{D/2}} \exp\left[-\frac{p^2}{4\alpha}\right]. \qquad (A5.3)$$

Let us see how all this works for the integrals in Eq. (5.32a–f).

a) $\displaystyle \int d^D k (k^2 + m^2)^{-1} = \int_0^\infty d\alpha \int d^D k \, \exp[-\alpha(k^2 + m^2)]$

$\displaystyle \qquad = \pi^{-D/2} \int_0^\infty d\alpha \, \alpha^{-D/2} \exp[-\alpha m^2]$

$\displaystyle \qquad = \pi^{-D/2} \Gamma\left(1 - \frac{D}{2}\right) m^{(D-2)}$

b) $\displaystyle \int d^D k (k^2 + m^2)^{-2} = \pi^{-D/2} \int_0^\infty d\alpha \, \alpha^{-(D/2-1)} \exp[-\alpha m^2]$

$\displaystyle \qquad = \pi^{-D/2} \Gamma\left(2 - \frac{D}{2}\right) m^{(D-4)}$

c) $\displaystyle \int d^D k (k^2 + m^2)^{-3} = \frac{1}{2} \pi^{-D/2} \Gamma\left(3 - \frac{D}{2}\right) m^{(D-6)}$

d) $\int d^D k_1 \, d^D k_2 (k_1^2 + m^2)^{-1}(k_2^2 + m^2)^{-1}((p - k_1 - k_2)^2 + m^2)^{-1}$

$= \int d\alpha \, d\beta \, d\gamma \int d^D k_1 \, d^D k_2 \exp[-\alpha(k_1^2 + m^2) - \beta(k_2^2 + m^2)$
$\quad - \gamma((p - k_1 - k_2)^2 + m^2)]$

$= \pi^{-D/2} \int \frac{d\alpha \, d\beta \, d\gamma}{(\alpha + \gamma)^{D/2}} \int d^D k_2 \exp\left\{-\left[(\beta + \gamma)k_2^2 - \frac{(p + k_2)^2 \gamma^2}{\alpha + \gamma}\right.\right.$
$\quad \left.\left. + 2\gamma k_2 \cdot p + (\alpha + \beta + \gamma)m^2\right]\right\}$

$= \frac{1}{\pi^D} \int \frac{d\alpha \, d\beta \, d\gamma}{(\alpha\beta + \beta\gamma + \gamma\alpha)^{D/2}} \exp\left[-\frac{\alpha\beta\gamma}{\alpha\beta + \beta\gamma + \gamma\alpha} p^2\right.$
$\quad \left. - (\alpha + \beta + \gamma)m^2\right]$

$= \frac{1}{\pi^D} \int_0^1 dx \int_0^1 dy \int_0^1 dz \, \delta(x + y + z - 1) \int_0^\infty d\Sigma$
$\quad \times \frac{1}{(xy + yz + zx)^{D/2}} \Sigma^{2-D} \exp\left[-\Sigma \cdot \left(\frac{xyz}{xy + yz + zx} p^2 + m^2\right)\right]$

$= \frac{\Gamma(3 - D)}{\pi^D} \int_0^1 dx \int_0^1 dy \int_0^1 dz (xy + zx + yz)^{-D/2}$
$\quad \times \left[\frac{xyz}{xy + yz + zx} p^2 + m^2\right]^{D-3}$ \hfill (A5.4)

The integrals (a), (b), (c), and (d) are divergent, respectively, for $D \geq 2$, $D \geq 4$, $D \geq 6$, $D \geq 3$, as can be seen from the relation

$$\Gamma(x) \sim \frac{1}{x} \quad \text{for } x \sim 0. \tag{A5.5}$$

More generally the Γ function has simple poles at negative integer values, as can be seen from the identity

$$\Gamma(-x) = \frac{\pi}{\sin(\pi x)\Gamma(x + 1)}. \tag{A5.6}$$

Higher-order diagrams can be computed in the same way: After some algebra we find for the diagrams of Eq. (5.32g), (5.32h), and (5.32i), respectively,

Appendix to Chapter 5

a) $\pi^{-3D/2} \int_0^\infty \dfrac{d\alpha\, d\beta\, d\gamma\, d\delta\, \alpha}{(\alpha\beta\gamma + \alpha\gamma\delta + \alpha\delta\beta + \beta\gamma\delta)^{D/2}} \exp[-m^2(\alpha+\beta+\gamma+\delta)]$

$= \pi^{-3D/2} \Gamma\left(5 - \dfrac{3}{2}D\right) m^{(-10+3D)} \int_0^1 dx_1\, dx_2\, dx_3\, dx_4$

$\times \dfrac{\delta(x_1 + x_2 + x_3 + x_4 - 1)}{(x_1 x_2 x_3 + x_1 x_3 x_4 + x_1 x_3 x_2 + x_2 x_3 x_4)^{5-3D/2}}$

b) $\pi^{-3D/2} \int_0^\infty \dfrac{d\alpha\, d\beta\, d\gamma\, d\delta\, \beta}{[\alpha(\beta\gamma + \gamma\delta + \delta\beta)]^{D/2}} \exp\left[-\dfrac{p^2 \beta\gamma\delta}{\beta\gamma + \delta\gamma + \delta\beta}\right.$

$\left. - m^2(\alpha + \beta + \gamma + \delta)\right]$

$= \pi^{-3D/2} \Gamma\left(5 - \dfrac{3}{2}D\right) \int_0^1 \dfrac{dx_1\, dx_2\, dx_3\, dx_4}{[x_1(x_2 x_3 + x_3 x_4 + x_4 x_2)]^{D/2}}$

$\times \dfrac{1}{[p^2(x_2 x_3 x_4 / (x_2 x_3 + x_3 x_4 + x_4 x_2)) + m^2]^{5-3D/2}} \qquad (A5.7)$

c) $\pi^{-3D/2} \int \dfrac{d\alpha\, d\beta\, d\gamma\, d\delta\, d\varepsilon}{[(\alpha+\beta)(\gamma+\delta)\varepsilon + (\alpha+\beta)\gamma\delta + (\gamma+\delta)\alpha\beta]^{D/2}}$

$\times \exp\left\{-\dfrac{\varepsilon[(\alpha+\beta)\delta\gamma + (\delta+\gamma)\alpha\beta]p^2}{(\alpha+\beta)(\gamma+\delta)\varepsilon + (\alpha+\beta)\delta\gamma + (\gamma+\delta)\alpha\beta}\right.$

$\left. - (\alpha + \beta + \gamma + \delta + \varepsilon)m^2\right\}$

$= \pi^{-3D/2} \Gamma\left(5 - \dfrac{3}{2}D\right) \int_0^1 dx_1\, dx_2\, dx_3\, dx_4\, dx_5$

$\times \dfrac{\delta(x_1 + x_2 + x_3 + x_4 + x_5 - 1)}{[(x_1+x_2)(x_3+x_4)x_5 + (x_1+x_2)x_3 x_4 + (x_3+x_4)x_1 x_2]^{D/2}}$

$\times \left[\dfrac{x_5[(x_1+x_2)x_3 x_4 + (x_3+x_4)x_1 x_2]p^2}{(x_1+x_2)(x_3+x_4)x_5 + (x_1+x_2)x_3 x_4 + (x_3+x_4)x_1 x_2} + m^2\right]^{(3D/2-5)}$

In both Eqs. (A5.4) and (A5.7) the last integration has been carried out by this or similar transformations

$$\alpha = x_1 \Sigma, \qquad \beta = x_2 \Sigma, \qquad \gamma = x_3 \Sigma, \qquad \alpha + \beta + \gamma = \Sigma$$
$$d\alpha\, d\beta\, d\gamma = dx_1\, dx_2\, dx_3\, \Sigma^2\, d\Sigma\, \delta(x_1 + x_2 + x_3 - 1) \qquad (A5.8)$$

and computing the integral over Σ using Eq. (A5.1).

The algebra for computing high-order integrals may become rather long if many loops are present; there are fortunately some simple graphical rules for obtaining the result directly under the form in Eq. (A5.7) just by looking at the diagrams.[26] As usual it is easy to make mistakes, so it is better, especially at the beginning, to derive the results using the two different approaches.

Let us now consider the problem of convergence of the diagram. It is easy to check that the argument of the Γ function in the final result is just $\mathcal{D}/2$, so that the diagram is divergent as soon as its degree of divergence $\mathcal{D}$ is nonnegative. This divergence is connected to the integration region in the space where all the variables go to zero simultaneously (i.e., superficial divergence); in this region the integrand weakly depends on p^2. Consequently it may be approximated by a polynomial in p^2. We finally find that for superficially divergent diagrams the divergent part is a polynomial in p^2. For example, in the case of Eq. (A5.4d) the result can be written as

$$\frac{1}{\pi^D} \sum_n \frac{(p^2)^n}{n!} \Gamma(3 - D + n) \int_0^1 dx \int_0^1 dy \int_0^1 dz$$

$$\times \frac{(xyz)^n}{(xy + yz + zy)^{D/2+n}} \delta(x + y + z - 1). \tag{A5.9}$$

The term proportional to $(p^2)^n$ is divergent as soon as the degree of divergence of the diagram ($\mathcal{D} = 2D - 6$) becomes larger than or equal to $2n$, whereas high-order terms in the Taylor expansion are finite. The fact that the only possible divergences of a superficially divergent diagram are polynomial in the external momenta will be of great utility in Chapters 8 and 9. Here we only need to remark that the only superficially divergent diagrams in two and three dimensions are those of Eqs. (A5.4a) and (A5.4d) (the last only in $D = 3$), and their divergence is momentum independent [cf. Eq. (A5.13)]. Thus we can reabsorb their divergences in a renormalization of the coefficient of the term proportional to φ^2.

As can be checked, the diagrams in Eq. (A5.7a) and (A5.7b) contain subdivergent diagrams: although these are superficially divergent only for $D \geq \frac{10}{3}$, they are divergent in $D = 3$; in the second case (b) the divergence is connected to the integration over α, while in the first case (a) the divergence comes from the integration region where β, γ, and δ become simultaneously small. This last divergence may be exposed by setting

$$\beta = \lambda x_1 \qquad \gamma = \lambda x_2 \qquad \delta = \lambda x_3. \tag{A5.10}$$

One finds

Appendix to Chapter 5

$$\pi^{-3D/2} \int_0^\infty d\lambda \int_0^\infty d\alpha \int_0^1 dx_1 \int_0^1 dx_2 \int_0^1 dx_3$$

$$\times \frac{\alpha \lambda^2 \delta(x_1 + x_2 + x_3 - 1)}{\lambda^D [\alpha(x_1 x_2 + x_2 x_3 + x_3 x_1) + \lambda x_1 x_2 x_3]^{D/2}} . \quad (A5.11)$$

In three dimensions the integral over λ is logarithmically divergent at $\lambda = 0$.

As we have said in the text, these divergences may be removed if we subtract from all the self-energy subdiagrams their value computed at zero external momentum. In these two cases we find

a) $\pi^{-3D/2} \int d\alpha\, d\beta\, d\gamma\, d\delta\; \alpha \cdot \exp[-m^2(\alpha + \beta + \gamma + \delta)]$

$\cdot \{(\alpha\beta\gamma + \alpha\gamma\delta + \alpha\delta\beta + \beta\gamma\delta)^{-D/2} - [\alpha(\beta\gamma + \gamma\delta + \delta\beta)]^{-D/2}\}$

b) $\pi^{-3D/2} \int d\alpha\, d\beta\, d\gamma\, d\delta\; \beta \exp\left[-p^2 \frac{\beta\gamma\delta}{\beta\delta + \beta\gamma + \gamma\delta} - m^2(\beta + \gamma + \delta)\right]$

$\cdot \{[\alpha(\beta\gamma + \gamma\delta + \delta\gamma)]^{-D/2} - [\alpha(\beta\gamma + \gamma\delta + \delta\gamma)]^{-D/2}\} = 0 .$
$\hspace{10cm} (A5.12)$

In case (b) the result is obviously finite, whereas in case (a) an explicit computation shows that in the limit $\lambda \to 0$ no divergences are present. The two terms of opposite sign cancel in this limit; indeed, using Eq. (A5.10) we get

$$\pi^{-3D/2} \int_0^\infty d\lambda \int_0^\infty d\alpha \int_0^1 dx_1 \int_0^1 dx_2 \int_0^1 dx_3\, \delta(x_1 + x_2 + x_3 - 1)\alpha$$

$$\lambda^{(2-D)} \{[\alpha(x_1 x_2 + x_2 x_3 + x_3 x_1) + \lambda x_1 x_2 x_3]^{-D/2} \quad (A5.13)$$

$$- [\alpha(x_1 x_2 + x_2 x_3 + x_3 x_1)]^{-D/2}\} .$$

The last task of this appendix is the computation of the dependence of the counterterms upon the cutoff. The propagator with cutoff can be written as

$$\frac{1}{p^2 + m^2 + p^4/\Lambda^2} = Z[(p^2 + \mu_1)^{-1} - (p^2 + \mu_2)^{-1}] \quad (A5.14)$$

where Z, μ_1, and μ_2 are simple algebraic functions of m^2 and Λ^2. In the large Λ^2 limit we have approximately

$$\frac{1}{p^2 + m^2 + p^4/\Lambda^2} \simeq (p^2 + m^2)^{-1} - (p^2 + \Lambda^2)^{-1}. \tag{A5.15}$$

For the diagram in Fig. 5.6(a) we find

$$-\frac{1}{2} \frac{g}{(4\pi)^{D/2}} \int_0^\infty \frac{d\alpha}{\alpha^{D/2}} \{\exp[-m^2\alpha] - \exp[-\Lambda^2\alpha]\}$$

$$-\frac{1}{2} \frac{g}{(4\pi)^{D/2}} \Gamma\!\left(1 - \frac{D}{2}\right)[m^{2-D} - \Lambda^{2-D}].$$

The result is finite in the presence of the cutoff in fewer than four dimensions (in four dimensions a slightly stronger cutoff is needed); in the large cutoff limit we recover the values for the counterterm c_1 of Eq. (5.14) [$\Gamma(\tfrac{1}{2}) = \sqrt{\pi}$ and $\Gamma(x+1) = x\Gamma(x)$].

We notice that in the presence of the cutoff

$$\langle \varphi^2 \rangle = (4\pi)^{-D/2} \Gamma\!\left(1 - \frac{D}{2}\right)[m^{2-D} - \Lambda^{2-D}]. \tag{A5.16}$$

For finite Λ, $\langle \varphi^2 \rangle$ is finite: it diverges when Λ^2 goes to infinity for $D \geq 2$.

We have already noticed that without a cutoff Fig. 5.6(b) is divergent only at zero external momentum; we can thus restrict our considerations to zero external momentum. In this situation the value associated with the diagram is

$$\frac{1}{6}\left(\frac{g}{(4\pi)^{D/2}}\right)^2 A(m^2, \Lambda^2)$$

$$A(m^2, \Lambda^2) \equiv \int_0^\infty \frac{d\alpha\, d\beta\, d\gamma}{(\alpha\beta + \beta\gamma + \gamma\alpha)^{D/2}} [\exp(-\alpha m^2) - \exp(-\alpha \Lambda^2)] \tag{A5.17}$$

$$\times [\exp(-\beta m^2) - \exp(-\beta \Lambda^2)][\exp(-\gamma m^2) - \exp(-\gamma \Lambda^2)].$$

The new terms in the numerator of the integrand become very small as soon as $\Lambda^2 \to \infty$; in three dimensions the integral has a logarithmic divergence around zero: it is obviously finite in the presence of a cutoff and for infinite cutoff diverges as $C \ln \Lambda^2/m^2$.

The direct evaluation of C is rather complex; it is simplified by noticing that (always for $D = 3$) $A[m^2, \Lambda^2] = \tilde{A}[\Lambda^2/m^2]$. We thus obtain

Appendix to Chapter 5

$$C = \lim_{m^2 \to 0} \left[-m^2 \frac{d}{dm^2} A(m^2, \Lambda^2) \right]$$

$$\simeq m^2 \int_0^\infty \frac{d\alpha \, d\beta \, d\gamma (\alpha + \beta + \gamma)}{(\alpha\beta + \beta\gamma + \gamma\alpha)^{3/2}} \exp[-m^2(\alpha + \beta + \gamma)]$$

$$= \int_0^1 dx_1 \int_0^1 dx_2 \int_0^1 dx_3 \frac{\delta(x_1 + x_2 + x_3 - 1)}{(x_1 x_2 + x_2 x_3 + x_3 x_1)^{3/2}}. \quad \text{(A.5.18)}$$

The last integral is not very complex and can be performed analytically. The result may be obtained more quickly if we observe that

$$\int_0^\infty \frac{d\alpha \, d\beta \, d\gamma}{(\alpha\beta\gamma)^{3/2}} \exp\left[-\frac{1}{\alpha} - \frac{1}{\beta} - \frac{1}{\gamma} \right]$$

$$= \int_0^\infty d\lambda \, \lambda^{-5/2} \int_0^1 dx_1 \int_0^1 dx_2 \int_0^1 dx_3 \frac{\delta(x_1 + x_2 + x_3 - 1)}{(x_1 x_2 x_3)^{3/2}}$$

$$\times \exp\left[-\frac{1}{\lambda} \left(\frac{1}{x_1} + \frac{1}{x_2} + \frac{1}{x_3} \right) \right]$$

$$= \Gamma\!\left(\frac{3}{2}\right) \int_0^1 dx_1 \int_0^1 dx_2 \int_0^1 dx_3 \frac{\delta(x_1 + x_2 + x_3 - 1)}{(x_1 x_2 x_3)^{3/2} \left[\frac{1}{x_1} + \frac{1}{x_2} + \frac{1}{x_3}\right]^{3/2}}$$

$$= \frac{\sqrt{\pi}}{2} \int_0^1 dx_1 \int_0^1 dx_2 \int_0^1 dx_3 \frac{\delta(x_1 + x_2 + x_3 - 1)}{(x_1 x_2 + x_2 x_3 + x_3 x_1)^{3/2}} = \frac{\sqrt{\pi}}{2} C.$$

(A5.19)

On the other hand, the first integral in Eq. (A5.19) can be easily evaluated,

$$\left[\int_0^\infty \frac{d\alpha}{\alpha^{3/2}} \exp\!\left(-\frac{1}{\alpha}\right) \right]^3 = [\Gamma(\tfrac{1}{2})]^3 = \pi^{3/2}. \quad \text{(A5.20)}$$

We finally obtain

$$C = 2\pi, \quad c_2 = \frac{1}{192\pi^2}. \quad \text{(A5.21)}$$

An essentially equivalent way to compute C consists in evaluating the diagram 5.6(b) in position space first: we have

$$C = \lim_{m^2 \to 0} (4\pi)^3 m^2 \frac{d}{dm^2} \int d^3x [G_0(x)]^3, \quad \text{(A5.22)}$$

where $G_0(x)$ is the correlation function in position space.

Proceeding in the usual way we get in D dimensions

$$G_0(x) = \frac{1}{(4\pi)^{D/2}} \int_0^\infty \frac{d\alpha}{\alpha^{D/2}} \exp\left[-\alpha m^2 - \frac{x^2}{4\alpha}\right]$$

$$= \frac{1}{(2\pi)^{D/2}} K_{D/2-1}(m|x|) \Big/ \left(|x|^{(D-2)/2} m^{(2-D)/2}\right), \quad \text{(A5.23)}$$

where K_ν is a Bessel function. Bessel functions of half-integer order reduce to elementary functions, so that the final expression is simplified in odd dimensions. For example, in one and three dimensions we obtain

$$G_0(x) = \frac{1}{2m} \exp(-m|x|), \qquad D = 1,$$

$$G_0(x) = \frac{1}{4\pi|x|} \exp(-m|x|), \qquad D = 3. \quad \text{(A5.24)}$$

Substituting Eq. (A5.24) in Eq. (A5.22), we get

$$C = m^2 \frac{d}{dm^2} \int d^3x |x|^{-3} \exp[(3m|x|)]$$

$$= \frac{1}{2} m \frac{d}{dm} \int_0^\infty dr \cdot \frac{4\pi}{r} \exp[-(3mr)] = 2\pi \quad \text{(A5.25)}$$

in perfect agreement with Eq. (A5.21).

Equation (A5.23) may be quite useful for computing the behavior of $G_0(x)$ at small or large x; we easily find

$$G_0(x) \sim \left(\frac{\pi}{2}\right)^{1/2} (2\pi)^{-D/2} |x|^{(-D/2-1/2)} m^{-(3/2-D/2)} \exp(-m|x|) \qquad |x| \to \infty$$

$$G_0(x) \sim \frac{1}{4} (\pi)^{-D/2} \Gamma\left(\frac{D}{2} - 1\right) |x|^{-(D-2)} \qquad x \to 0, D > 2$$

$$G_0(x) \sim (4\pi)^{-D/2} \Gamma\left(1 - \frac{D}{2}\right) m^{(D-2)} \qquad x \to 0, D < 2.$$

$$\text{(A5.26)}$$

The coherence length is given by m^{-1} (as it should be).

Notes for Chapter 5

1. A function $f(x)$ belongs to L^2 if $\int dx\, f^2(x) < \infty$. For more details see any book on functional analysis, e.g., A. F. Taylor and D. C. Coy, *Introduction to Functional Analysis*, John Wiley and Sons, New York (1980).
2. The exponentially small corrections due to the finite size of the box can easily be computed using the relations

$$F(L)|_{h=0} = \sum_{\vec{n}} \ln[\tilde{G}_\Lambda(\vec{p}_n)] \qquad \vec{p}_n = \frac{2\pi}{L}\vec{n}$$

$$\sum_{\vec{n}} \tilde{f}(\alpha\vec{n}) = \frac{1}{(2\pi\alpha)^D} \sum_{\vec{n}} f\left(\frac{2\pi}{\alpha}\vec{n}\right)$$

$$\tilde{f}(p) = \int d^D x\, \exp(ipx) f(x).$$

If $\tilde{f}(p)$ is an analytic function near the real p axis and the nearest singularity is at $|\operatorname{Im} p| = \lambda$, $f(x)$ must decrease as $\exp(-\lambda|x|)$ when x goes to infinity. This result may be proven in one dimension (the proof can be easily extended to higher dimensions) starting from the relation $f(x) = (1/2\pi) \int dp\, \exp(-ipx)\tilde{f}(p)$ and deforming the integration path in the complex p plane and picking the contribution of the singularity. In the extreme case where $\tilde{f}(p)$ is an entire function, $f(x)$ decreases at infinity faster than $\exp(-\lambda|x|)$ for any λ, e.g., $\tilde{f}(p) \propto \exp(-p^2/2)$, $f(x) \propto \exp(-x^2/2)$.
3. Useful theorems on the Fourier transform can be found in P. M. Morse and H. Feshbach, *Methods of Theoretical Physics*, McGraw-Hill, New York (1953), Chapter 4. Nowadays it is clear that the Fourier transform finds its most natural setting in the theory of distributions. A nice introduction to distribution theory and related theorems on Fourier transforms can be found in R. Streater and A. S. Wightman, *PCT, Statistics and all That* (Benjamin, 1964).
4. Review articles that share this point of view are K. Wilson and J. Kogut, *Phys. Rep.* 12C, 77 (1974); J. Kogut, *Rev. Mod. Phys.* 51, 659 (1979), S. Shenker in *Proceedings of the 1982 Les Houches Summer School*, ed. by J. B. Zuber and R. Stora, North-Holland, Amsterdam (1984).
5. The derivation of the diagrammatical rules can be found in any book on quantum field theory. See, for example, S. S. Schweber, *An Introduction to Relativistic Quantum Field Theory*, Harper and Row, New York (1964) Chapter 14 or C. Itzykson and J. B. Zuber, *An Introduction to Field Theory*, McGraw-Hill, New York (1980).

6. We use the definition of connected correlation functions from Chapter 4. Note: $\langle \varphi(x)\varphi(0)\varphi(y)\varphi(y)\varphi(y)\varphi(y)\rangle_c \neq \langle \varphi(x)\varphi(0)\varphi^4(y)\rangle_c$. The first expression is zero in the Gaussian model. The second expression is given by $\langle \varphi(x)\varphi(0)\varphi^4(y)\rangle - \langle \varphi(x)\varphi(0)\rangle\langle \varphi^4(y)\rangle$ (if $\langle \varphi \rangle = 0$), which is nonzero in the Gaussian model, that is, it is equal to $12 G(x-y)G(y)G(0)$.
7. This is essentially the content of the Weinberg theorem: S. Weinberg, *Phys. Rev.* **118**, 838 (1960). See also J. Bjorken and S. D. Drell, *Relativistic Quantum Field Theory*, McGraw-Hill, New York (1965) or C. Itzykson and J. B. Zuber, op. cit. Chapter 8.
8. The BHP theorem presented in Chapter 9 implies that in the region $0 \leq \mathscr{D} < 2$ the divergent part of the diagram does not depend on μ or on the external momentum p. In other words, if we subtract from the diagram its value at a given value of the momenta, a finite nondivergent result is obtained.
9. To add a constant C to Σ is equivalent to changing the two field correlation functions from $G(p)$ to $1/(G^{-1}(p) - C)$, i.e., to modifying the value of μ; this operation corresponds to adding to the Hamiltonian a new term equal to $(C/2) \int d^D x \, \varphi^2(x)$.
10. If we use Eq. (5.39) only for the first two diagrams, we recover Eq. (5.14); we are free, however, to change our definitions and to add other finite terms proportional to $\frac{1}{2} \int d^D x \, \varphi^2(x)$ in H. For example, if we stick to Eq. (5.39) for higher-order diagrams, the coefficient of the term $\frac{1}{2} \int d^D x \, \varphi^2(x)$ will be $\mu - C(\Lambda, \mu_R, g)$, where the function C is constructed in such a way that $G(p, \mu_R)|_{p=0} = 1/\mu_R$. Of course the difference $C(\Lambda, \mu_R, g) - C(\Lambda, g)$ will remain finite when Λ goes to infinity. A very popular procedure consists in requiring that $\Sigma(p, \mu)|_{p=0} = 0$ [i.e., $G(p, \mu)|_{p=0} = 1/\mu$] and developing in powers of g at fixed μ; this corresponds to using the Hamiltonian $\frac{1}{2} \int d^D x \, \varphi^2(x)[\mu - C(\Lambda, \mu, g)]$.
11. An incredible amount of work has been necessary to derive this seemly simple result. For a review of the results obtained in rigorous constructive quantum field theory see J. Glimm and A. Jaffe in *Cargèse Lectures 1976*, M. Levy et al., eds. (Plenum Press, 1974); J. Glimm and A. Jaffe, *Quantum Physics*, Springer-Verlag, Berlin (1981), A. Simon, *The $P(\varphi)_2$ Euclidean (Quantum) Field Theory*, Princeton University Press, Princeton (1974).
12. A proof of the convexity inequality (5.44) and of the more general upper and lower bounds (5.49) can be found in D. Bessis, P. Moussa, and M. Villani, *J. Math. Phys.* **16**, 462 and 2318 (2075).
13. A probability measure is a positive measure, normalized to 1.
14. $Z(g)$ can be computed by looking at the diagrams contributing to the denominator of Eq. (5.20). In this way connected and disconnected diagrams contribute. On the other hand $F(g)$ receives the contribu-

tion of connected diagrams with zero external legs. In order to check the combinatorics it is useful to use the relation $\partial F/\partial g = (1/4!)\langle \int d^D x\, \varphi^4(x)\rangle$.

15. There are many techniques for dealing with divergent (i.e., nonconvergent) series. See, for example, R. B. Dingle, *Asymptotic expansions: their derivation and interpretation*, Academic Press, London (1973).

16. A related case in which similar results are obtained is the following: if the function $A(y)$ can be written as $\int_0^\infty d\mu(t)/(g+t)$ and $d\mu(t)$ is a positive measure which goes to zero sufficiently fast when $t \to 0$, the Padé approximate [i.e., rational function approximants constructed with a recipe similar to Eq. (5.8)] provide a monotonic convergent sequence of lower and upper bounds. A proof of this theorem and of other interesting properties of the Padé approximants can be found in G. Baker, *Essentials of Padé Approximants*, Academic Press, New York (1975).

17. From here on $Z(g)$ denotes a generic function of g (e.g., the free energy or a correlation function).

18. The proof of Watson's theorem and the essential properties of the Borel transform can be found in E. T. Whittaker and G. N. Watson, *A Course in Modern Analysis*, Cambridge University Press, Cambridge (1962).

19. J. P. Eckman, J. Magnen and R. Sénéor, *Commun. Math. Phys.* 39, 251 (1976).

20. For one-dimensional systems (i.e., for quantum mechanics, as we shall see in Chapter 13) the proof is much simpler: S. Graffi, V. Grecchi, and B. Simon, *Phys. Lett. B* 32, 631 (1970).

21. See, for example, J. C. Le Guillou and J. Zinn-Justin, *Phys. Rev. Lett.* 39, 95 (1977), where this technique is used for computing the critical exponents as explained in Chapter 8. An application to a quantum-mechanical problem can be found in G. Parisi, *Phys. Lett. B* 69, 329 (1977).

22. If $B(b) = \sum_{0}^{\infty} B_n b^n$ and $\tilde{B}(y) = \sum_{0}^{\infty} \tilde{B}_n y^n$, we have

$$\tilde{B}_n = \sum_{1k}^{n} B_k b_s^k \frac{(n-1)!}{(k-1)!(n-k)!}.$$

23. The rest of this section is very sketchy: For less compressed expositions see G. Parisi and J. Zinn-Justin in *Cargèse Lectures 1976*, M. Levy et al., eds., Plenum Press, New York (1977) or J. Zinn-Justin in *Proceedings of the 1982 Les Houches Summer School*, ed. by J. B. Zuber and R. Stora, North-Holland, Amsterdam (1984).

24. For the partition function of the Landau-Ginsburg model, a less general argument can be made: indeed, the Borel transform of

$\int d\mu[\varphi]\exp[(-g/24)\int d^D x\, \varphi^4(x)]$ is given by $\int d\mu[\varphi]J_0\{[(b/24)\int d^D x\, \varphi^4(x)]^{1/2}\}$, where the Bessel function $J_0(z)$ is an entire function which for large values of z behaves as $\exp(z)$ and is bounded on the real positive axis. The boundedness of $J_0(z)$ explains the absence of singularities on the real positive b axis, whereas b_s can be characterized as the value of b for which $\int(\tfrac{1}{2}(\partial_\mu\varphi)^2 + (\mu/2)\varphi^2) - [(b_s/24)\int d^D x\, \varphi^4(x)]^{1/2}$ is no longer positive definite, as has been shown by G. Parisi, *Phys. Lett.* **66B**, 167 (1977), C. Itzykson, G. Parisi, and J. B. Zuber, *Phys. Rev. Lett.* **38**, 306 (1977).

25. A typical example of a theory that has a perturbative expansion that is not Borel summable is given by the following one-dimensional Hamiltonian $\int dx\{\tfrac{1}{2}(d\varphi/dx)^2 + (\mu/2)\varphi^2(1-g\varphi)^2\}$, see E. Brézin, J. C. Le Guillou, and J. Zinn-Justin, *Phys. Rev. D* **15**, 1558 (1977). E. Brézin, G. Parisi, and J. Zinn-Justin, *Phys. Rev. D* **15**, 408 (1977).

26. K. Symanzik, *Prog. Theor. Phys.* **20**, 690 (1958).

CHAPTER 6

Near the Transition

6.1. Infrared divergences

Before putting forward the general ideas of the renormalization group, let us see what happens if we proceed in a naive way. As we already said, it is convenient to expand the correlation functions in powers of g at $\tilde{\mu} = \mu - \mu_c(g)$ fixed. Dimensional analysis tells us that the perturbative expansion in powers of g for dimensionless quantities is in reality an expansion in powers of $u = g/(\mu - \mu_c)^{(4-D)/2}$, so that it cannot be used directly at $\mu = \mu_c$. On the other hand, we could try to compute the correlation functions as functions of the momenta p at $\mu = \mu_c$; for the two-point function we get

$$G(p, g) = \frac{1}{p^2} f(g/p^{4-D}). \tag{6.1}$$

Unfortunately this program also fails. Using the identity

$$\frac{\int d^D x \exp(ipx)}{(x^2)^a} = \frac{(4\pi)^{D/2}\Gamma\left(\frac{D}{2} - a\right)}{2^a \Gamma(a)} (p^2)^{(a-D/2)}, \tag{6.2}$$

w see that the diagrams of Fig. 6.1 for Γ take the values[1]

$$g, \quad \frac{-\frac{1}{2}g^2 C}{|p|^{4-D}}, \quad \frac{\frac{1}{4}g^2 C^2}{|p|^{2(4-D)}}, \quad C = \frac{\Gamma^2\left(\frac{D-2}{2}\right)\Gamma\left(2 - \frac{D}{2}\right)}{\pi^{D/2}\Gamma(D-2)}. \tag{6.3}$$

In general, if we have n bubbles we get a contribution proportional to $g^{n+1}/|p|^{n(4-D)}$.

105

Figure 6.1. Diagrams for Γ considered in subsection 6.1.

Figure 6.2. An infrared-divergent diagram in the massless theory in dimensions $D \leq 3$.

The diagram of Fig. 6.2 gives, therefore, in $D = 3$

$$\Sigma(p) \propto \int \frac{d^3k}{|k|^3} \frac{1}{(p+k)^2}, \qquad (6.4)$$

and the integral is divergent at small k (infrared divergence). The function in (6.1) cannot be expanded[2] in powers of g. The perturbative expansion seems to be useless; some work is needed in order to transform it into a powerful tool for the computation of the critical exponents.

6.2. The Hartree-Fock approximation

A possible way to use the perturbative expansion in the nonperturbative region consists in making a selection of an infinite number of diagrams, which are simultaneously evaluated. In the so-called Hartree-Fock approximation[3] one includes for Σ only those diagrams having the topology shown in Fig. 6.3. These diagrams are generated by the recursive solution of the following integral equations:

$$G(k) = \frac{1}{m^2 + k^2} \qquad m^2 = \mu - \Sigma_R$$

$$\Sigma_R = -\frac{g/2}{(2\pi)^D} \int d^D k \left[\frac{1}{k^2 + m^2} - \frac{1}{k^2 + \mu_N} \right]$$

$$= -\frac{g/2}{(4\pi)^{D/2}} \Gamma\left(1 - \frac{D}{2}\right) [m^{D-2} - \mu_N^{D/2-1}] \qquad (D \neq 2) \qquad (6.5)$$

$$\Sigma_R = -\frac{g/2}{4\pi} \ln\left(\frac{\mu_N}{m^2}\right) \qquad (D = 2)$$

which are diagrammatically written in Figs. 6.3 and 6.4.

6.2. The Hartree-Fock Approximation

Figure 6.3. Some of the diagrams considered in the Hartree-Fock approximation.

The solution of the equations is rather simple. The self-energy is momentum independent, and Eq. (6.5) reduces to the "gap" equation,

$$m^2 = \mu + \frac{g/2}{(4\pi)^{D/2}} \Gamma\left(1 - \frac{D}{2}\right)[m^{D-2} - \mu_N^{D/2-1}]. \tag{6.6}$$

By explicitly evaluating the integrals we find in $D = 3$

$$m^2 + \frac{g}{8\pi} m = \mu + \frac{g}{8\pi} \mu_N^{1/2}$$

$$\mu_c = -\frac{g}{8\pi} \mu_N^{1/2} \quad m^2 \sim (\mu - \mu_c)^2$$

$$-\Gamma(p_1 p_2 p_3 p_4) = \delta(p_1 + p_2 + p_3 + p_4)[V(p_1 + p_2)$$
$$+ V(p_1 + p_3) + V(p_1 + p_4)] \tag{6.7}$$

$$V(p) = \frac{g}{1 + g\Pi(p, m)}$$

$$\Pi(p, m) = \frac{1}{2(2\pi)^3} \int d^3k \frac{1}{(p-k)^2 + m^2} \frac{1}{k^2 + m^2}$$

$$\Pi(0, m) \sim \frac{1}{m} \quad (m \sim 0) \qquad \Pi(0, p) \sim \frac{1}{p} \quad (p \sim 0)$$

If $D > 2$ there is a value of μ for which m is zero: $\mu_c \propto -g\mu_N^{(D-2)/2}$, and we have $m \propto (\mu - \mu_c)^{1/(D-2)}$. However, if $D = 2$, μ_c does not exist, and no transition is present ($m \to 0$ for $\mu \to -\infty$).

The magnetic susceptibility is m^{-2}, the correlation length is m^{-1}, and $G(p)$ at the critical temperature is given by p^{-2}. The corresponding critical exponents are

$$\gamma = \frac{2}{D-2}, \qquad \nu = \frac{1}{D-2}, \qquad \eta = 0. \tag{6.8}$$

Figure 6.4. Equation (6.5) in graphical terms.

It is interesting to note that if $g \to \infty$ at fixed $\xi = m^{-1}$, the function Γ goes to a finite limit $[V \to 1/\Pi(p, m)]$. However, if ξ goes to infinity ($m \to 0$), Γ becomes zero at small momenta; we have

$$V(p)|_{g=\infty} = \frac{1}{\Pi(p,m)} \to \frac{1}{p^{4-D}} \quad \text{for } p \to \infty$$
$$V(p)|_{m=0} \sim \frac{1}{p^{4-D}} \quad \text{for } p \to 0. \tag{6.9}$$

The selection of diagrams looks arbitrary. However, we shall see in the next section how to systematize this approach.[4]

6.3. The large-N expansion

It is convenient to generalize the model by considering N-component fields $\varphi_a(x)$, $a = 1, \ldots, N$. The Hamiltonian is supposed to be invariant under the $0(N)$ symmetry group (i.e., under rotations in N dimensions),

$$H[\varphi] = \int d^D x \left[\frac{1}{2} \sum_\nu^D \sum_a^N (\partial_\nu \varphi_a)^2 + \frac{\mu}{2} \sum_a^N \varphi_a^2 + \frac{g}{24N} \left(\sum_a^N \varphi_a^2 \right)^2 \right]. \tag{6.10}$$

From symmetry arguments we know that the two-field correlation function must have the form

$$\langle \varphi_a(x) \varphi_b(0) \rangle = \delta_{ab} G(x), \tag{6.11}$$

at least in the high-temperature phase ($\langle \varphi_a \rangle = 0$), where the symmetry φ is not spontaneously broken.

The diagrams are the same as in the previous chapter; multiplicities are the only differences. In order to keep track of the multiplicities in an efficient way, it is convenient to write the partition function as

$$Z = \int d[\alpha] d[\varphi] \exp -H[\alpha, \varphi]$$
$$H[\alpha, \varphi] = \int d^D x \left[\frac{1}{2} \sum_\nu^D \sum_a^N (\partial_\nu \varphi_a(x))^2 + \left(\frac{\mu}{2} + \frac{i\alpha(x)}{\sqrt{12N}} \right) \right.$$
$$\left. \times \sum_a^N \varphi_a^2(x) + \frac{1}{2g} \alpha^2(x) \right]. \tag{6.12}$$

The α correlation function is $\langle \alpha(x)\alpha(0)\rangle = g\delta(x)$ at $g \simeq 0$, i.e., a constant in momentum space; the Γ function can be written as

6.3. The Large-N Expansion

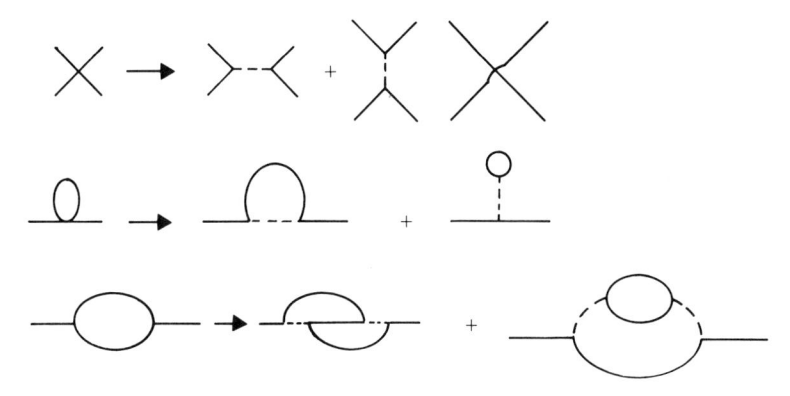

Figure 6.5. Diagrams of the old representation (left) transcribed into the new representation (right).

$$\Gamma_{abcd} = (\delta_{ab}\delta_{cd} + \delta_{ac}\delta_{bd} + \delta_{ad}\delta_{bc})\frac{g}{3N} + 0(g^2). \quad (6.13)$$

We could also introduce a $\varphi_a\varphi_b\alpha$ vertex, which is given by $(ig^{1/2}/\sqrt{12N})\delta_{ab}$ in momentum space.

The new diagrams can be obtained from the old one by substituting as described in Fig. 6.5. This new representation is useful for controlling the limit. Indeed in such a limit we lose a power of N for each coupling constant g, but we gain a power of N for each closed loop of φ lines (the combinatorial factor is proportional to N). It is easy to check that when $N\to\infty$ at fixed g, the only surviving diagrams for the φ and α propagators[5] are those shown in Fig. 6.6.[6] We have thus recovered the results of Section 6.2 for $N\to\infty$, the α propagator being given by $V(p)$ of Eq. (6.7).[7] It is a good exercise for the reader to transcribe some simple diagrams from one representation to the other. After some practice one is able to tell almost immediately which class of diagrams will survive at a given order in $1/N$ when N goes to infinity.

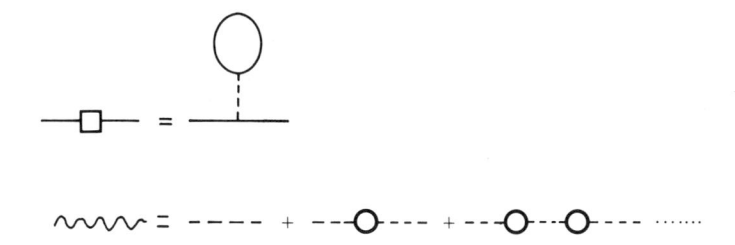

Figure 6.6. The only diagrams surviving in the limit $N\to\infty$ for the two-ϕ correlation function and the two-α correlation function. The wavy line indicates the α correlation function.

In the new representation only a finite number of diagrams survives at a fixed order in $1/N$. We can now expand the correlation functions in inverse powers of $1/N$, generating the diagrams in Fig. (6.7). An explicit computation shows that in $D = 3$ at the critical point in the small momentum region we have[8]

$$G(p) = \frac{1}{p^2}\left[1 + \frac{1}{N}(a_{11} \ln p^2 + b_1)\right.$$

$$\left. + \frac{1}{N^2}(a_{22} \ln^2 p^2 + a_{21} \ln p^2 + b_2) + O\left(\frac{1}{N^3}\right)\right]$$

$$a_{11} = \frac{4}{3\pi^2}$$

$$a_{22} = \frac{1}{2} a_{11}^2 \qquad (6.14)$$

$$a_{21} = b_1 a_{11} - \frac{256}{27\pi^4}$$

If we assume that for small p, $G(p)$ has the form

$$G(p) \sim \frac{B(N)}{p^{2-\eta(N)}}, \qquad (6.15)$$

we find

$$\frac{1}{2}\eta(N) = \frac{\eta_1}{N} + \frac{\eta_2}{N^2} + O\left(\frac{1}{N^3}\right),$$

$$\eta_1 = a_{11}, \qquad \eta_2 = a_{21} - a_{11} b_1 \qquad (6.16)$$

$$B(N) = 1 + \frac{b_1}{N} + b_2/N^2 + O\left(\frac{1}{N^3}\right).$$

Clearly it is consistent to assume that Eq. (6.15) holds only because a_{22} is equal to $a_{11}^2/2$; otherwise Eq. (6.15) would be dead wrong. If Eq. (6.15) is correct, the term proportional to $1/N^3$ must satisfy many more consistency conditions.

In this way one can compute the critical exponents[9] as powers of N^{-1}. Of course some care must be taken to prove that everything goes

Figure 6.7. The only diagram for the φ self-energy surviving in the limit $N \to \infty$.

smoothly, Eq. (6.15) holds, and no infrared divergences are present. In particular, it is not evident that the limits $p\to 0$ and $N\to\infty$ may be exchanged. These problems may be solved by using the technique of the next chapters.[10] The $1/N$ expansion is quite useful; it gives a concrete example for testing explicitly various conjectures on the critical behavior, though it is not so useful for computing critical exponents in the range $N = 1–3$.

Notes for Chapter 6

1. Equation (6.2) implies that $G(x)$ is given by $\Gamma[(D-2)/2]/[2\pi^{D/2}(x^2)^{(D-2)/2}]$. The contribution to $\Gamma_4(p)$ is the Fourier transform of $[G(x)]^2$. We thus obtain Eq. (6.3).
2. Using the approach of the next section, one finds terms proportional to $g^4 \ln g$ in $\Sigma(p)$ at $D = 3$.
3. The same approximation carries different names in different contexts: random-phase approximation (RPA), coherent potential approximation (CPA), and so on.
4. We shall find that the diagrams considered here are the first nontrivial order of the large-N expansion described in subsection 6.3.
5. The two-field correlation function is sometimes called the propagator.
6. There are many papers on the $1/N$ expansion; an arbitrary selection is E. Brézin and D. J. Wallace, *Phys. Rev. B* **7**, 1967 (1973); K. S. Ma, *Rev. Mod. Phys.* **45**, 589 (1973); G. Parisi, *Nucl. Phys. B* **100**, 368 (1975).
7. The generalized Heisenberg model is soluble in the limit $N\to\infty$. A. Stanley, *Phys. Rev.* **176**, 718 (1968); in this limit it becomes identical to the "spherical model," T. H. Berlin and M. Kaz, *Phys. Rev.* **86**, 821 (1952).
8. R. Abe, *Prog. Theor. Phys.* **49**, 1877 (1973).
9. This can be done only if we assume that Eq. (6.15) is correct; the validity of this equation is highly nontrivial. It implies the relation $a_{22} = \frac{1}{2}(a_{11})^2$.
10. The proof of the correctness of Eq. (6.15) is not straightforward; especially if $D \leq 3$ there are some subtle points. The interested reader should look into the original literature, e.g., K. Symanzik DESY Preprint 77/05, unpublished (1977).

CHAPTER 7

The Renormalization Group

7.1. Block variables

We have seen in the preceding chapters that the critical point is characterized by the onset of correlated fluctuations at large distances: their study is particularly difficult, insofar as all the methods we have seen here do not work: the correlation length is zero at $T = 0$ and at $T = \infty$, and at any order in the high-temperature (or low-temperature) expansion we can keep track of the correlated fluctuations of only a finite number of spins.[1] Indeed, far from the critical point thermodynamic quantities may well be estimated by approximating the infinite system with a relatively small finite system. If periodic boundary conditions are used, the finite-volume corrections are proportional to $\exp(-L/\xi)$, where L and ξ are, respectively, the size of the box and the coherence length. In contrast, at the critical point the finite-volume corrections are proportional to a negative power of L and our aim is to study the collective fluctuations that produce these power-law corrections.[2] In other words, at the critical point the number of relevant degrees of freedom is actually infinite; the usual reductionistic procedure of considering only a finite number of degrees of freedom fails here, and a new approach is needed.

The renormalization group approach[3] is based on the recursive introduction of block variables.[4] These can be defined by generalizing the procedure we have used in Chapter 5 to derive the Landau-Ginsburg model (where the magnetization averaged around a point was the relevant variable on which the effective Hamiltonian depended.

Let us consider an infinite (or nearly infinite) system and let us divide the system into boxes of side $L_n = 2^n$ for $n = 0, 1, 2, \ldots$ (as was done in the proof of the infinite-volume limit in Chapter 3). We label each box by k. In this way[5] we construct a sequence of nested lattices having spacing L_n (for $n = 0$ we recover the original lattice). We define the block variables $\varphi_n(k)$ as

7.1. Block Variables

$$\varphi_n(k) = \sum_{i \in k_n} \sigma_i, \tag{7.1}$$

where the sum runs over all the spins σ_i (S_i in the notation of the previous chapters) in the box k_n. Equation (7.1) can be written recursively as[6]

$$\varphi_0(k) = \sigma_k$$
$$\varphi_{n+1}(k) = \sum_{i_n \in k_{n+1}} \varphi_n(i_n). \tag{7.2}$$

There are $2^{Dn} = L_n^D$ and 2^D terms in the sums in Eq. (7.1) and (7.2), respectively.

The probability distribution $\exp[-\beta H(\sigma)]$ of the σ variables induces naturally a probability distribution on the φ variables. As soon as L_n is definitively larger than the coherence length ξ (i.e., $n \gg n_\xi \simeq \ln_2 \xi$),[7] the correlations of block variables belonging to different blocks will be essentially zero.[8] In this case, in the large-n limit, the probability distribution of the $\varphi_n(k)$ is factorized; moreover, each φ_n is the sum of $2^{D(n-n_\xi)}$ practically uncorrelated variables. The central limit theorem tells us that these have a Gaussian distribution and their variance can be easily computed:

$$\langle \varphi_n(k)\varphi_n(k) \rangle = \left\langle \left(\sum_{i \in k_n} \sigma_i\right)\left(\sum_{j \in k_n} \sigma_j\right) \right\rangle$$
$$\simeq \sum_{i \in k_n} \left\langle \sigma_i\left(\sum_j \sigma_j\right) \right\rangle = L_n^D \frac{\chi}{\beta} \equiv L_n^D \tilde{\chi}. \tag{7.3}$$

In deriving Eq. (7.3) we have used the fact that if correlations are exponentially damped, the correlation function of a generic point i of the box (not too close to the boundary) with any point j outside the box is completely negligible: The points outside the box may be added without changing the final result. We thus find

$$dP[\varphi_n] = \prod_{k_n} \{d\varphi_n(k)(2\pi L_n^D \tilde{\chi})^{-1/2} \exp[-\varphi_n^2(k)/(2L_n^D \tilde{\chi})]\}. \tag{7.4}$$

If we introduce rescaled variables $\psi_n = L_n^{-D/2} \varphi_n$, the new variables have finite variance when $n \to \infty$, if we stay at $T \neq T_c$:

$$dP[\psi_n] = \prod_{k_n} \left\{d\psi_n(k)(2\pi\tilde{\chi})^{-1/2} \exp\left[-\frac{\psi_n^2(k)}{(2\tilde{\chi})}\right]\right\}. \tag{7.5}$$

In the presence of a nonzero magnetic field we find

$$dP[\varphi_n] = \prod_{k_n} \left\{ (2\pi L_n^D \tilde{\chi})^{-1/2} \, d\varphi_n(k) \exp\left[-\frac{(\varphi_n(k) - L_n^D m)^2}{2L_n^D \tilde{\chi}} \right] \right\}$$

$$\langle \sigma_i \rangle \equiv m .$$
(7.6)

In this case we have two options: we can define

$$\psi_n = \frac{\varphi_n}{L_n^D}$$

$$\tilde{\psi}_n = \frac{\varphi_n - mL_n^D}{L_n^{D/2}} .$$
(7.7)

The corresponding probability distributions are, for large n:

$$dP[\psi_n] = \prod_{k_n} \{ d\psi_n(k) \delta(\psi_n(k) - m) \}$$

$$dP[\tilde{\psi}_n] = \prod_{k_n} \left\{ d\tilde{\psi}_n(k) (2\pi\tilde{\chi})^{-1/2} \exp\left[-\frac{\tilde{\psi}_n(k)}{2\tilde{\chi}} \right] \right\} .$$
(7.8)

Below T_c at $h = 0$, if we consider the symmetric nonclustering state $\langle \, \rangle_s = \frac{1}{2}[\langle \, \rangle_+ + \langle \, \rangle_-]$, Eq. (7.8) becomes

$$dP[\psi_n] = \frac{1}{2} \sum_{\varepsilon = \pm 1} \prod_{k_n} \{ d\psi_n(k) \delta(\psi_n(k) - \varepsilon m_s) \} ,$$
(7.9)

where m is the spontaneous magnetization.

Let us consider from now on the case $h = 0$. From the preceding equations we see that, if we write

$$\psi_n(k) = \frac{\varphi_n(k)}{L_n^{D\omega}} ,$$
(7.10)

the value of ω, such that the ψ has finite nonzero variance, jumps from $\frac{1}{2}$ to 1 when we cross the critical temperature.[9] At the critical temperature ω cannot be $\frac{1}{2}$ if $\chi = \infty$, nor can it be 1 if $m = 0$ at $T = T_c$. Let us assume that there is one intermediate value of ω (ω_c), such that the ψ_n are well normalized when $n \to \infty$. The crucial hypothesis, which is at the basis of the renormalization group approach, is that for $\omega = \omega_c$ at the critical point[10]

$$\lim_{n \to \infty} P_n[\psi_n] = P_\infty[\psi_\infty] .$$
(7.11)

Equation (7.11) states that the well-normalized large-scale block variables have a limiting probability distribution when $n \to \infty$.

7.1. Block Variables

Equation (7.11) seems to be rather natural. In order to understand its deep meaning, it is important to explain how it could fail. The first possibility is that the P_n could oscillate when n goes to infinity. If that happens, Eq. (7.11) must be modified; some other minor changes are needed, but this should not destroy the soundness of the approach. If we study simple ferromagnetic Ising systems, we can safely assume from physical arguments that no oscillations are present.[11] The real danger is that some of the ψ_n might *escape to infinity*, so that the limit $n \to \infty$ becomes rather tricky; a typical example of something that should not happen is the following:

$$P_n[\varphi_n] = \prod_{k_n} \tilde{P}_n(\varphi_n(k))$$

$$\tilde{P}_n(\varphi_n(k)) = \left(1 - \frac{1}{L_n^D}\right)\delta(\varphi_n(k)) + \frac{1}{L_n^D}\delta(\varphi_n^2(k) - L_n^D) \,. \quad (7.12)$$

In this case $\omega = \frac{1}{2}$; if the limit in Eq. (7.11) is taken in a naive way (pointlike convergence), we get

$$P_\infty(\psi_\infty(k)) = \delta(\psi_\infty(k))$$

$$\langle \psi_\infty^2(k) \rangle = \langle \psi_\infty^4(k) \rangle = \langle \psi_\infty^6(k) \rangle = \cdots = 0 \,, \quad (7.13)$$

whereas

$$\langle \psi_n^2(k) \rangle = 1 \,, \quad \langle \psi_n^4(k) \rangle = L_n \,, \quad \langle \psi_n^6(k) \rangle = L_n^2 \,. \quad (7.14)$$

In other words, we cannot find a probability distribution $P_\infty(\psi_\infty)$ such that

$$\langle \psi_n^q(k) \rangle \simeq \int dP_\infty(\psi_\infty(k))\psi_n^q(k) \neq 0 \quad (7.15)$$

for any choice of ω. If something of this kind happened, the whole approach would be in rather bad shape. Although a fully rigorous proof does not yet exist, the possibility of such a disaster is rather unlikely in the ferromagnetic Ising model because of the existence of the following rigorous inequality:[12]

$$\langle \varphi_n^4 \rangle \leq 3\langle \varphi_n^2 \rangle^2$$

$$\langle \psi_n^4 \rangle \leq 3 \quad \text{if } \langle \psi_n^2 \rangle = 1 \,. \quad (7.16)$$

The same inequality also holds in the Landau-Ginsburg model with nearest-neighbor coupling, the equality being realized for $g = 0$. Indeed, if we look at the probability distribution of a single block spin variable, Eq. (7.16) is strong enough to forbid pathologies like Eq. (7.12).[13]

The definition (7.1) of the block variables is not the only one possible; we could also write

$$\varphi_n(k) = \sum_i \left(\sigma_i \exp\left[-\frac{d^2(i, k_n)}{L_n^2} \right] \right), \tag{7.17}$$

where $d(i, k_n)$ is the distance of the point i from the center of the block k_n; the advantage of the original definition (7.1) is the possibility of using the recursive relation (7.2). Other recursive definitions are also possible, as we shall see later. It is generally believed that the existence of a limiting probability distribution [Eq. (7.11)] does not depend on the way that the block variables are defined (at least for "reasonable" choices).

A pictorial way to interpret Eq. (7.11) is as follows: We consider a two-dimensional system in which up spins are black and down spins are white. Let us take a television picture of the system from larger and larger distances. When the distance is small, we see individual spins; when the distance becomes large, we can only see the average of many spins [the resolution of the TV set acts like Eq. (7.17)]. The minimal size of the region we can resolve increases linearly with the distance. In order to avoid a uniform gray distribution, we must increase the contrast. The scaling hypothesis, Eq. (7.11) tells us that after an appropriate adjustment of the contrast, by looking at a picture taken from a very large distance, we cannot decide from which distance it has actually been taken if the distance is much greater than the lattice spacing. In other words, the probability distribution of the spins is invariant under scale transformations (after rescaling of the spins) in the large-distance region. Indeed, it can be shown (see subsection 7.4) that a sufficient (and practically necessary) condition for the validity of Eq. (7.11) for $\frac{1}{2} < \omega_c < 1$ is

$$\langle \sigma(lx_1) \cdots \sigma(lx_r) \rangle \xrightarrow[l \to \infty]{} l^{-rD(1-\omega_c)} \cdot f(x_1, \ldots, x_r) \equiv f(lx_1, \ldots, lx_r). \tag{7.18}$$

If P_∞ is a Gaussian probability distribution, only the two-point correlation function will be nonzero. It is easy to check that in the Gaussian model of the previous chapters

$$\omega_c = \frac{1}{2} + \frac{1}{D}. \tag{7.19}$$

When $D \leq 2$ we find $\omega_c \geq 1$; we already know that in this range of D the Gaussian model at the critical point does not make sense because of infinite correlation functions.

7.2. The recursion relation

The definition [Eqs. (7.2) and (7.10)] of the block variables is recursive, i.e., the probability distribution $P_{n+1}[\psi_{n+1}]$ is obtained directly from the probability distribution $P_n[\psi_n]$. We can describe this fact by writing

$$P_{n+1} = R(P_n)$$

$$\psi_{n+1}(k) = 2^{-D\omega_n} \sum_{i_n \in k_{n+1}} \psi_n(i_n),$$

(7.20)

where R is a nonlinear operator acting on probability measures. The asymptotic probability P_∞ satisfies the fixed-point condition:

$$P_\infty = R(P_\infty).$$

(7.21)

The ψ variables satisfy the normalization $\langle \psi_n^2 \rangle = 1$, and the value of ω_n will in general depend on n.

The basic idea of the renormalization group approach is to study the transformation R; as we shall see, we can connect in a simple way the singularities of the free energy (critical temperature, critical exponent) to the properties of R, where R remains regular at the transition.

We hope that in computing the action of R we can consider correlated fluctuations only on a scale of a few lattice spacings (i.e., of order L_n in the original lattice). The study of R should be possible also in a finite, not very large system; in this way we would bypass the difficulty pointed out at the beginning of this chapter.

It is often convenient to introduce the effective Hamiltonians for the block variables[14]

$$P_n[\psi_n] \propto \exp\{-H_n[\psi_n]\}.$$

(7.22)

Equation (7.20) induces a transformation on the space of all possible Hamiltonians

$$H_{n+1} = R_H[H_n].$$

(7.23)

If the original Hamiltonian $H_0 = \beta H$ is short range, it is believed (and it will be argued in Section 7.5) that H_n also remains essentially short range,[15] that is, the coefficients of terms of the type $(\psi_k - \psi_{k+l})^2$ are exponentially small when $l \to \infty$, even at the critical point where long-range correlations are present. The space of essentially short-range Hamiltonians is not so huge as the space of all possible probability distributions; it should be possible to use this fact both for a rigorous construction of the operator R_H and for finding simpler approximate transformations, e.g., by projecting Eq. (7.23) onto the "small" space of finite-range Hamiltonians. It is crucial that the transformation R_H always remain regular, even at the critical point. The present theory of critical phenomena assumes that all the singularities of the free energy and the onset of long-range correlations are generated by a regular R transformation.

7.3. Some mathematical preliminaries

Before obtaining the physically interesting results, let us study some general properties of Eq. (7.23). We assume that there is a (Banach?) space of Hamiltonians for which Eq. (7.23) makes sense. In order to visualize the situation, it is convenient to discuss the case in which we can define a function H_λ such that

$$H_\lambda|_{\lambda=n} = H_n$$
$$\frac{dH_\lambda}{d\lambda} = T(H_\lambda) , \qquad (7.24)$$

where T is a nonlinear operator (formally, $\exp T = R$). Equation (7.24) has the graphic advantage of describing a flow in a multidimensional space, the parameter λ playing the role of time. If H_f is a fixed point of the transformation (7.23), $R(H_f) = H_f$, we have

$$T(H_f) = 0 . \qquad (7.25)$$

Any H satisfying Eq. (7.25) is said to be a fixed point of the renormalization group transformation. Equation (7.11) tells us that, when the "time" λ goes to infinity, H_λ goes to a fixed point. For a given fixed point H_f, we can define an attraction basin (B_f), i.e., the set of all points H of the space such that

$$\lim_{\lambda \to \infty} H_\lambda = H_f \quad \text{if} \quad H_\lambda|_{\lambda=0} = H . \qquad (7.26)$$

Different points within the same basin have the same large "time" limit.

Equations (7.11) and (7.26) tell us that the whole space can be partitioned into attraction basins of different fixed points. A fixed point is said to be attractive if all the points nearby belong to its basin of attraction; it is said to be repulsive if its basin of attraction contains only one point (the fixed point itself!). There are also intermediate possibilities: The fixed point is attractive when approached from one direc-

7.3. Some Mathematical Preliminaries

tion and it is repulsive from the other directions (typical examples are shown in Fig. 7.1). If the r_a are the set of parameters on which the Hamiltonian depends (e.g., $H_\lambda^{[r]}[\psi] = \Sigma_a r_a^{(\lambda)} g_a[\psi]$, the g's being a "complete set" of functions of the block variables ψ), Eq. (7.24) can also be written as[16]

$$\frac{d}{d\lambda} r_a^{(\lambda)} = b_a(r). \qquad (7.27)$$

The fixed-point condition is now

$$b_a(r^f) = 0. \qquad (7.28)$$

The number of parameters r_a is infinite in the general case, but it is finite in most of the approximations. In order to study the mathematical properties of Eqs. (7.24)–(7.27) it is convenient to suppose that the number of parameters r_a is finite (let us say M) and that the true transformation T is gentle enough that there are essentially no differences between the finite- and the infinite-dimensional cases.[17] Neglecting pathologies, we can now classify the fixed point according to the dimension d_B (or the co-dimension $c_B = M - d_B$) of the basin of attraction (e.g., if the fixed point is attractive $c_B = 0$ and $d_B = M$). Now near the fixed point the transformation (7.24) may be linearized. If $H_\lambda = H_f + \delta H_\lambda$ and δH_λ is small, we have

$$\frac{d}{d\lambda} \delta H_\lambda = T_L \delta H_\lambda + 0(\delta H_\lambda^2), \qquad (7.29)$$

where T_L is now a linear operator. Correspondingly Eq. (7.27) becomes

$$\frac{d}{d\lambda} \delta r_a(\lambda) = \sum_c^M \tilde{b}_{ac} \delta r_c(\lambda) + 0(\delta r(\lambda)^2),$$

$$r_a(\lambda) = r_a^f + \delta r_a(\lambda), \qquad \tilde{b}_{ac} = \left.\frac{\partial b_a}{\partial r_c}\right|_{r_a = r_a^f}. \qquad (7.30)$$

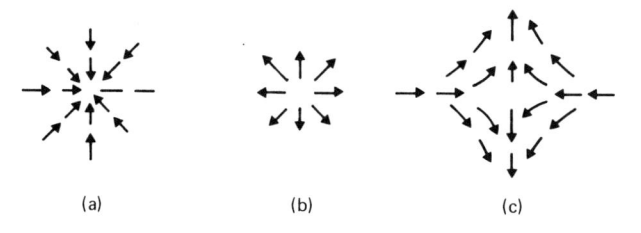

Figure 7.1. Examples of fixed points: (a) attractive; (b) repulsive; and (c) mixed.

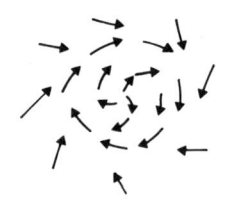

Figure 7.2. A flow producing a limiting cycle.

A well-known theorem states that if all the real parts of the eigenvalues of T_L (or $\tilde{b}$) are negative, the fixed point is attractive, if there are n eigenvalues with positive real part, $c_B = n$.[18]

In the typical situation that we shall study, there are two attractive fixed points, the low- and the high-temperature ones: the two basins of attraction are separated by a surface of co-dimension one (the so-called critical surface). (For simplicity we assume that there are no limiting cycles; see Fig. 7.2.) By continuity arguments, if a trajectory starts on this surface, it will remain there forever. Equation (7.11) implies that on this surface there must be at least one fixed point (H_c) which is attractive on the surface, but repulsive from outside, i.e., $\tilde{b}$ has one unstable (positive real part) eigenvalue. A two-dimensional example is shown in Fig. 7.3.

Usually the initial Hamiltonian at $\lambda = 0$ depends on β; in this way we have a whole family of trajectories $H(\lambda, \beta)$,

$$H(\lambda, \beta)|_{\lambda=0} = H_0(\beta) \equiv \beta H,$$
$$\frac{\partial H}{\partial \lambda} = T(H). \tag{7.31}$$

If $H_0(\beta)$ belongs to two different attraction basins for small and large values of β, respectively, there must be a value of β (β_c) such that $H_0(\beta)$ crosses the critical surface. If ω_R and ω_A are the two largest eigenvalues[19] of T_L at H_c ($\omega_R > 0 > \omega_A$), the corresponding eigenvalues being H_R and

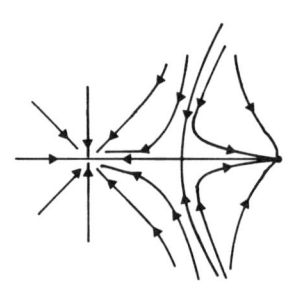

Figure 7.3. A flow with two attractive fixed points separated by a critical surface (dashed line) on which there is a partially attractive fixed point.

7.4. The Scaling Laws

H_A, we have for large λ and small $|\beta - \beta_c|$:

$$H(\lambda, \beta) \simeq H_c + H_A \exp(-\lambda|\omega_A|) + (\beta - \beta_c)\exp(\lambda\omega_R)H_R ;$$

if $\lambda^{\omega_R}|\beta - \beta_c| \ll 1$

$$H(\lambda, \beta) \approx H_{\pm}\left(\lambda + \frac{\ln(|\beta - \beta_c|)}{\omega_R}\right) + 0(|\beta - \beta_c|^{|\omega_A|/\omega_R}) \qquad (7.32)$$

if $\lambda + \dfrac{\ln(|\beta - \beta_c|)}{\omega_R} \gg 0$,

where H_+ and H_- are the two escape trajectories for $\beta < \beta_c$ and $\beta > \beta_c$, respectively, depending on the nonlinear terms neglected in Eq. (7.29); more precisely, $H_{\pm}(t) = H_0 \pm \exp(t\omega_R)H_R + 0[\exp(2t\omega_R)]$. If we do not start exactly from β_c but only close to β_c, the trajectory will stop near H_c for a long "time" proportional to $(1/\omega_R)\ln(1/|\beta - \beta_c|)$; for $\lambda \ll -\ln|\beta - \beta_c|/\omega_R$ the probability distribution of the block variables will be very similar to that at the critical point, while for $\lambda \gg -\ln|\beta - \beta_c|/\omega_R$ the probability distribution of the block variables will be very close to the low- (or high-) temperature fixed point. In this last region the connected correlation functions of the block variables will be essentially zero.

This picture is not essentially changed if there are two unstable eigenvalues of T_L^c: The critical surface now has co-dimension two, and two parameters must be tuned (e.g., temperature and magnetic field, or temperature and pressure) in H_0 in order to reach the critical point.[20]

7.4. The scaling laws

As we have seen, the asymptotic behavior of the block spin variables depends in a nonanalytic way on the temperature near the critical point [cf. Eq. (7.32)]; this implies that the correlation length must go to infinity as

$$\xi \sim |\beta_c - \beta|^{-1/\omega_R} \Rightarrow \nu = \frac{1}{\omega_R} . \qquad (7.33)$$

Indeed, the block variables will remain correlated as long as $\lambda + \ln(|\beta - \beta_c|)/\omega_R < 0$ and become uncorrelated (or with zero connected correlation) when $\ln(1/|\beta - \beta_c|)/\omega_R \ll \lambda$. The crossover between the two regimes is at $\lambda \simeq \ln(1/|\beta - \beta_c|)/\omega_R$, which corresponds to boxes of side $L_n \simeq |\beta - \beta_c|^{-1/\omega_R}$. We thus obtain eq. (7.33).

What happens to the free energy? Let us compute the partition function of a box of side 2^K for large K. We first notice that Eq. (7.22) fixes $H_n[\psi_n]$ apart from an additive constant; let us fix it in a conventional way, e.g., $H_n[\psi]|_{\psi=0} = 0$.

In this way we have

$$P_n[\psi] = \frac{1}{Z_n} \exp[-H_n[\psi]]$$

$$Z_n = \int d[\psi] \exp[-H_n[\psi]] .$$
(7.34)

We can define a *free-energy* density (the factor $1/\beta$ is omitted) for the block variables of order n ($n \leq K$):

$$f_n = -2^{-D(K-n)} \ln Z_n = -2^{-D(K-n)} \ln[P_n(\psi)|_{\psi=0}]$$

$$f_0 \equiv f ,$$
(7.35)

where the number of block variables is $2^{D(K-n)}$. Now we have

$$\exp[-2^{D(K-n-1)} f_{n+1}] = \exp[-2^{D(K-n)} f_n] \int_{\psi_{n+1}=0} d[\psi_n] \exp[-H_n[\psi_n]]$$

$$\equiv \exp[-2^{D(K-n)} (f_n + g[H_n])] ,$$
(7.36)

where the integral over the $2^{D(K-n)} \psi_n$ is under the constraint that the ψ_{n+1} be zero. The integral in Eq. (7.36) is a special case of the one needed to define the $H_{n+1}[\psi_{n+1}]$; it is thus reasonable to assume that $g[H]$ is a regular functional at the critical point as well. Equation (7.36) can be written in the recursive form

$$f_n = g[H_n] - 2^{-D} f_{n+1} ,$$
(7.37)

which implies

$$f \equiv f_0 = \sum_0^\infty 2^{-Dn} g[H_n] ;$$
(7.38)

the damping factor 2^{-Dn} goes to zero fast enough to imply good convergence in the sum. If we substitute the discrete index n with the continuous variables λ (or $L = \exp \lambda$), we get

$$f = \int_0^\infty d\lambda \exp(-D\lambda) g[H_\lambda]$$

$$= \int_1^\infty \frac{dL}{L} L^{-D} g[H(L)]$$
(7.39)

$$H(L) = H_\lambda|_{\lambda = \ln L} .$$

7.4. The Scaling Laws

If we substitute Eq. (7.32) in (7.39) it is easy to see that the leading singularity of the free energy is given by

$$f_{\text{sing}} \propto \xi^{-D} \propto |\beta - \beta_c|^{D/\omega_R} = |\beta - \beta_c|^{D\nu}. \tag{7.40}$$

Equation (7.40) can be intuitively understood: Equation (7.32) implies the existence of two different regimes for blocks of side L smaller than or larger than ξ. The number of variables needed to describe what happens in the crossover region is a fraction ξ^{-D} of the total number of variables. Its contribution to the free energy is thus proportional to ξ^{-D}. Equation (7.40) implies that the specific heat is given by

$$C_{\text{sing}} \sim |\beta - \beta_c|^{D\nu - 2} \Rightarrow \alpha = 2 - D\nu. \tag{7.41}$$

The same argument implies that at the critical point in a box of size L, $f(L) - f(\infty) \sim L^{-D}$. It is easy to see that Eq. (7.18) generalizes to

$$\langle \sigma(x_1)\sigma(x_2) \rangle \simeq \frac{1}{(x_1 - x_2)^{2D(1-\omega_c)}} f\left(\frac{x_1 - x_2}{\xi}\right) \quad |x_1 - x_2| \gg 1 \tag{7.42}$$

$$f(0) \neq 0, \quad f(z) \sim \exp -z \quad \text{for } z \to \infty,$$

which implies that

$$\eta = D(1 - 2\omega_c) + 2, \quad \chi = \int d^D x_2 \langle \sigma(x_1)\sigma(x_2) \rangle \propto \xi^{2-\eta/\nu} \Rightarrow \gamma = 2\nu - \eta. \tag{7.43}$$

More generally we have

$$\langle \sigma(x_1) \cdots \sigma(x_q) \rangle \simeq \frac{1}{(x_1 - x_2)^{qD(1-\omega_c)}} C\left(\frac{x_1 - x_2}{\xi}, \frac{x_1 - x_3}{\xi}, \ldots, \frac{x_1 - x_q}{\xi}\right) \tag{7.44}$$

when all distances $|x_i - x_j|$ are large. If we introduce an energy density

$$e_i = \sum_k J_{ik} \sigma_i \sigma_k, \tag{7.45}$$

similar arguments tell us that

$$\langle e(x)e(y) \rangle_c \sim \frac{1}{|x-y|^{2\omega_R}} f_e\left(\frac{x_1 - x_2}{\xi}\right)$$

$$\langle \sigma(x)\sigma(y)e(z) \rangle_c \sim \frac{1}{|x-y|^{\omega_R + 2D(1-\omega_c)}} C_e\left(\frac{x-y}{\xi}, \frac{y-z}{\xi}\right), \tag{7.46}$$

and so on.

Up to now we have considered the case of zero magetic field: If h is small enough to be treated as a perturbation, the magnetic field acting on the block variable increases as $L^{D\omega_c}$. In other words, there is another positive eigenvalue corresponding to the magnetic field ($\omega_H = D\omega_c$). Using arguments similar to the foregoing, we get[21]

$$\beta = \frac{\nu}{2}(D-2) + \eta/2, \qquad \delta = \frac{2D}{D-2+\eta} - 1, \qquad \gamma = (2-\eta)\nu. \qquad (7.47)$$

All the critical exponents can be computed starting from two inputs (ω_R and ω_c, or η and ν). In the following we shall see how to compute them.

Before closing this section we must show how Eqs. (7.18), (7.42), (7.44), and (7.46) are related to the scaling laws for the block variables. Let us begin by deriving Eq. (7.18) for the two-spin correlation from Eqs. (7.10) and (7.11). These equations imply that [cf. Eq. (7.3)]

$$\langle \psi_\infty^2 \rangle = \lim_{n \to \infty} \langle \psi_n^2 \rangle$$

$$\psi_n^2 = L_n^{-2D\omega} \sum_{i \in k_n} \sum_{j \in k_n} \langle \sigma_i \sigma_j \rangle$$

$$\simeq L_n^{-2D\omega} \sum_{i \in k_n} \sum_j \theta(dL_n - |i-j|)\langle \sigma_i \sigma_j \rangle$$

$$= L_n^{-(2D\omega - D)} \sum_j \theta(dL_n - |i-j|)\langle \sigma_i \sigma_j \rangle, \qquad d = 0(1). \qquad (7.48)$$

In deriving Eq. (7.48), we have substituted for the sum over two points of a very large box the sum over the first point inside the same box, while the second point is constrained to stay at a distance $0(L_n)$ (the side of the box). The correctness of this substitution may be checked *a posteriori* by ascertaining how much the result depends on d [in the region $d = 0(1)$].

If the correlation functions go to zero exponentially with the distance, the sum over j depends neither on L_n nor on d (if L_n is sufficiently large), and it is equal to the magnetic susceptibility: in this case ω is $\frac{1}{2}$.

If ω is larger than $\frac{1}{2}$, the sum over j must diverge when L_n goes to infinity:

$$\sum_j \theta(L_n - |i-j|)\langle \sigma_i \sigma_j \rangle \sim L_n^{D(2\omega - 1)}. \qquad (7.49)$$

If we neglect the possibility of oscillations in the correlation functions, we see that Eq. (7.49) may be satisfied only if

$$\langle \sigma_i \sigma_j \rangle \sim |i-j|^{2D(1-\omega)}. \qquad (7.50)$$

In the same way we can see that Eq. (7.18) implies that

7.4. The Scaling Laws

$$\langle \psi_n^4 \rangle = \frac{1}{L^{4D\omega}} \sum_{i_1,i_2,i_3,i_4} \langle \sigma_{i_1}\sigma_{i_2}\sigma_{i_3}\sigma_{i_4} \rangle$$

$$\simeq \frac{1}{L^{4D\omega}} \int_0^{L_n} d^D x_1 \, d^D x_2 \, d^D x_3 \, d^D x_4 \, f(x_1, x_2, x_3, x_4)$$

$$= \int_0^1 d^D y_1 \, d^D y_2 \, d^D y_3 \, d^D y_4 \, f(y_1, y_2, y_3, y_4), \quad (7.51)$$

in agreement with Eqs. (7.10) and (7.11). In deriving Eq. (7.51) we have implicitly assumed that the function $f(x, \ldots, x_4)$ defined in Eq. (7.18) is an integrable function. Now a careful analysis shows that although the function $f(x_1, \ldots, x_4)$ has singularities at coinciding points, it is still integrable: The proof contains some tricky points and will not be reported here. (In the simplest form one could use the Griffith and Lebowitz inequalities; see pp. 45 and 115.)

Let us now examine the implications of Eq. (7.42) for the expectation values of the φ_n^2. Using the same arguments as before, we get

$$\langle \varphi_n^2 \rangle \simeq \int_0^{L_n} d^D x_1 \, d^D x_2 \, \frac{f((x_1-x_2)/\xi)}{(x_1-x_2)^{2D(1-\omega_c)}}$$

$$= \int_0^{L_n/\xi} d^D y_1 \, d^D y_2 \, \frac{f(y_1-y_2)}{(y_1-y_2)^{2D(1-\omega_c)}} \cdot \xi^{2D\omega_c}$$

$$= L_n^{2D\omega_c} r\left(\frac{\xi}{L_n}\right). \quad (7.52)$$

Now, according to the definitions, we have

$$2^{D\omega_n} = \langle \varphi_{n+1}^2 \rangle / \langle \varphi_n^2 \rangle, \quad (7.53)$$

where the ω_n depend on the Hamiltonian H_n. If the Hamiltonian H_n is close to the critical fixed point, obviously $\omega_n \approx \omega_c$; if H_n is close to the high-temperature fixed point, $\omega_n \sim \frac{1}{2}$.

Equation (7.52) implies that

$$\omega_n = f\left(\frac{\xi}{L_n}\right) \quad \text{or} \quad \omega(\lambda) = g(\lambda - \ln \xi), \quad (7.54)$$

where we have set $2^n = \lambda$ in order to use the same notation as in the preceding section. This result is in perfect agreement with Eq. (7.32), which implies that

$$\omega_\lambda = g\left(\lambda + \frac{\ln(|\beta - \beta_c|)}{\omega_R}\right), \quad (7.55)$$

if we recall that

$$\xi = |\beta - \beta_c|^{-\omega_R}. \tag{7.56}$$

Similar but lengthy arguments can be offered to justify Eqs. (7.44) and (7.46).

7.5. Approximations

As usual we begin our investigation by studying what happens in the Gaussian model, because in this model everything can be explicitly computed.[22] The block variables have a Gaussian distribution: We have only to compute their correlation function,

$$G_n(k_n - l_n) = \langle \varphi_{k_n} \varphi_{l_n} \rangle. \tag{7.57}$$

If we introduce the Fourier transform $\tilde{G}_n$ of G_n defined by

$$\tilde{G}_n(p) = \sum_{k_n} \exp(ik_n p) G_n(k_n), \tag{7.58}$$

Eq. (7.2) takes a very simple form,

$$\tilde{G}_{n+1}(p) = \sum_{\vec{\rho}=0,1} \left\{ \prod_\nu^D [1 + \cos(p_\nu + \pi \rho_\nu)] \tilde{G}_n\left(\frac{\vec{p}}{2} + \pi\vec{\rho}\right) \right\}, \tag{7.59}$$

as can be explicitly checked. The sum over $\vec{\rho}$ goes over all the 2^D vectors of components 0 or 1; for example, if $D = 1$ Eq. (7.59) reduces to

$$\tilde{G}_{n+1}(p) = \left(1 + \cos\frac{p}{2}\right)\tilde{G}_n\left(\frac{p}{2}\right) + \left(1 + \cos\left(\frac{p}{2} + \pi\right)\right)\tilde{G}_n\left(\frac{p}{2} + \pi\right). \tag{7.60}$$

Although in principle we could simply apply Eq. (7.60) n times to get the $\tilde{G}_n$, it is more convenient to derive the asymptotic behavior of the $\tilde{G}_n$ by qualitative considerations, using the information on the model that we have already gathered.

We know that there are two possibilities, either the correlation is exponentially damped at large distances, or we stay at at the critical point and it decays like $1/r^{(D-2)}$ at large distances. If the correlations are short range, the previous arguments (cf. p. 113) tell us that when n goes to infinity the φ_n become uncorrelated variables; their correlation function is a delta function on the lattice, and it is a constant in momentum space,

7.5. Approximations

i.e.,

$$\tilde{G}_n(p) \simeq 2^{Dn}\tilde{G}_1(0). \tag{7.61}$$

However, if we stay at the critical point, we have approximately (if $k_n \neq 1_n$)

$$G_n(k_n - l_n) = \sum_{k \in k_n} \sum_{l \in l_n} G_1(k - l) \simeq \frac{2^{2Dn}}{[2^n(k_n - l_n)]^{D-2}}. \tag{7.62}$$

Indeed, when the box becomes very large, most of the points of the two boxes are at sufficiently large distance to apply the asymptotic expression $(G_1(r) \sim (r)^{(-D+2)})$ for the correlation. The factor 2^{2Dn} is the number of terms and $2^n(k_n - l_n)$ is the average distance between two points of the two boxes. Equation (7.62) becomes more and more accurate as $k_n - l_n$ goes to infinity. We finally conclude that

$$G_n(k) \approx 2^{(D+2)n} G_\infty(k), \tag{7.63}$$

where $G_\infty(k)$ behaves like $k^{-(D-2)}$ at large k. The previously defined exponent ω_c (see p. 114) is given by $(D+2)/2D = 1/2 + 1/D$, in agreement with Eq. (7.19).

After a long computation (which we omit), we find that if at the critical point $\tilde{G}_\infty(p) \sim 1/p^2$ (as we have already assumed),

$$\tilde{G}_\infty(p) = \sum_j \frac{1}{(p + 2\pi j)^2} \prod_{\nu}^{D} \frac{(1 - \cos p_\nu)}{(p_\nu + 2\pi j_\nu)^2}, \tag{7.64}$$

where the sum over j runs over the whole D-dimensional lattice. It is relatively easy to check that $\tilde{G}_\infty$ is left invariant by the transformation (7.59). The sum over j is convergent in any dimension; its presence is essential to make the function $\tilde{G}_\infty(p)$ periodic in p with period 2π. The singular term at $p = 0$ receives a contribution only from $j = 0$. Apart from this pole, $\tilde{G}_\infty(p)$ is an analytic function; $[G_\infty(p)]^{-1}$ will also be an analytic function with possibly some poles in the complex plane. We can thus write

$$[G_\infty(p)]^{-1} = \sum_k \cos(kp) J_k : \quad |J_k| \sim \exp(-\alpha k) \quad k \to \infty, \tag{7.65}$$

where α is the imaginary part of the nearest zero to the real axis of $G_\infty(p)$. Equation (7.65) is very interesting because the probability distribution for the ψ_∞ variables is given by

$$\exp[-H_\infty[\psi_\infty]]$$

$$H_\infty[\psi_\infty] = \frac{1}{(2\pi)^D} \frac{1}{2} \int_B dp |\psi_\infty(p)|^2 [G_\infty(p)]^{-1} \quad (7.66)$$

$$= \frac{1}{2} \sum_i \sum_k \psi_\infty(i) \psi_\infty(i+k) J_k .$$

The exponential decrease of the J_k implies that H_∞ may be well approximated by a Hamiltonian belonging to the space $\mathcal{H}^{(r)}$, i.e., the space of Hamiltonians of range r, where only variables of distance equal to or less than r are directly coupled;[23] the error involved in this approximation decreases exponentially as one increases the allowed range r.

This result is very important and it will be the basis of many approximations. Let us now attempt to derive it in a qualitative way.

The origin of the exponential damping can be seen in the following: the block variable Hamiltonian H_n can be written as

$$\exp\{-H_n[\psi_n]\} \propto \int d\psi_{n-1} \exp\{-H_{n-1}[\psi_{n-1}]\}$$

$$\cdot \prod_{k_n} \delta\left(\sum_{k_{n-1} \in k_n} \frac{\psi_{n-1}^{(k_{n-1})}}{2^{D\omega}} - \psi_n^{(k_n)} \right). \quad (7.67)$$

The integration over some of the variables ψ_{n-1} is trivial to perform because of the δ functions, and we are left with a fraction $1 - (\frac{1}{2})^D$ of nontrivial integrations to do. We call the variables associated with these integrations $\tilde{\psi}_n$. In two dimensions, if the variables ψ_{n-1} associated with the four blocks of Fig. 7.4 are denoted by ψ^1, ψ^2, ψ^3, and ψ^4, we have

$$\psi_n = \frac{\psi^1 + \psi^2 + \psi^3 + \psi^4}{2^{2\omega}}$$

$$\tilde{\psi}^1 = \psi^1 - \psi^2, \quad \tilde{\psi}^2 = \psi^3 - \psi^4 \quad (7.68)$$

$$\tilde{\psi}^3 = \psi^1 + \psi^2 - \psi^3 - \psi^4 .$$

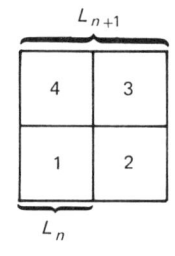

Figure 7.4. Four blocks of side L_n forming one block of side L_{n+1} in two dimensions.

7.5. Approximations

Equation (7.67) now becomes

$$\exp\{-H_n[\psi_n]\} = \int d\tilde{\psi}_{n-1} \exp\{-H[\psi_n, \tilde{\psi}_{n-1}]\} \,. \tag{7.69}$$

For the soundness of the renormalization group approach it is crucial that the $\tilde{\psi}_{n-1}$ variables have short-range correlations at fixed ψ_n.[24] If this is the case the integration over the $\tilde{\psi}_{n-1}$ variables cannot induce any long-range correlation in $H_n[\psi_n]$, and the transformation R will be regular. A careful analysis tells us that in the Gaussian model this is precisely what happens, which is why H^∞ for the Gaussian model is short range. We can argue that this is a general phenomenon. By fixing the ψ_n we have frozen some of the degrees of freedom: the $\tilde{\psi}_{n-1}$ are less coupled than the ψ_{n-1}, and for small ψ_n we expect that the integral in Eq. (7.69) will become singular at a temperature lower than the critical temperature of the whole system. As a consequence of this crucial assumption, the integration over the $\tilde{\psi}_{n-1}$ variables is "easy": it involves a number of degrees of freedom that do not go to infinity at the critical point. We can thus project the transformation R onto the space $\mathcal{H}^{(r)}$ and do the integrations in Eq. (7.69) in a box of size $L \gg r$. The error should be exponentially small both in L and in r.

This approximation is difficult to implement systematically: the space $\mathcal{H}^{(r)}$ is still very large and analytical computations in a finite but large box are not easy. A popular approach consists in using these ideas together with the high-temperature expansion. The high-T expansion is a powerful tool for estimating the integral over the $\tilde{\psi}$ variables, their correlation length being finite. Let us briefly see how it can be done. If we start with the nearest-neighbor Hamiltonian $H_0 = (-\beta/2) \Sigma_{i,k} J_{ik} \sigma_i \sigma_k$, after one interaction we have

$$H_1 = \sum_{l}^{\infty} \beta^l \tilde{H}_{1,l} \,, \tag{7.70}$$

where the $H_{1,l}$ connects variables at maximal distance l (the distance being defined as $\Sigma_\nu^D |i_\nu|$). We can now approximate H_1 with $H_1^{(r)} = \Sigma_l^r \beta^l H_{1,l}$ ($H_1^{(r)} \in \mathcal{H}^{(r)}$), this amounts to writing

$$H_1^{(r)} = H_1^{(r)}(\varepsilon)\big|_{\varepsilon=1}$$
$$H_1^r(\varepsilon) = \sum_{r'}^{r} \tilde{H}_{1,r'} \varepsilon^{r'} \,, \tag{7.71}$$

where $\tilde{H}_{1,r}$ connects variables at distance r. We can now compute $H_2(\varepsilon)$ from $H_1(\varepsilon)$ by formally expanding $H_2(\varepsilon)$ as function of ε and neglecting the terms of order ε^{r+1}; we set $\varepsilon = 1$ at the end. In this way we always remain in $\mathcal{H}^{(r)}$, and the whole computation can be performed using the standard techniques of the high-temperature expansion. Further simplifications are obtained by defining the block variables in such a way that they can only take the values ± 1 (in this way $\mathcal{H}^{(r)}$ becomes a finite-dimensional space). One defines ψ_n using Eq. (7.2) and sets[25]

$$\psi_n = \text{sign}(\varphi_n) \quad \text{if } |\varphi_n| \neq 0$$
$$\psi_n = r \quad \text{if } |\varphi_n| = 0, \quad (7.72)$$

where r is a random number taking the values -1 and 1 with equal probability. By use of this or similar techniques, quite spectacular results have been obtained, especially in two dimensions.

A rather simple but different computation is the following[26]: We start with a 4-by-4 Ising system with nearest-neighbor interaction; after one step of the renormalization group we have a 2-by-2 Ising system [the block variables are defined according to Eq. (7.72)]; the Hamiltonian for the block variables, in the absence of any external magnetic field, will have only nearest-neighbor interaction β_1, next-nearest-neighbor interaction (on the diagonal) β_2, and a four-spin interaction β_3. No other Hamiltonians are possible. If we start with a Hamiltonian containing all three of these interactions, the final Hamiltonian for the block variables will have the same form; in this way we obtain $\beta_i^{n+1} = b_i(\beta^n)$, $i = 1, 2, 3$.

The functions b_i can be computed by considering explicitly the $2^{16} = 65{,}536$ configurations of the system, which can be done quite easily on any computer. Apart from the zero and the high-temperature fixed point, one finds, if one starts with a purely ferromagnetic nearest-neighbor interaction, a third fixed point, the critical β being $\simeq 0.43$, in good agreement with the exact result. The critical exponent ν is rather accurate: $\nu = 0.937$ (ν is exactly 1). The critical couplings obtained at the nontrivial fixed point are

$$\beta_1^\infty = 0.307, \quad \beta_2^\infty = 0.084, \quad \beta_3^\infty = -0.004.$$

The fact that the nearest coupling is the largest, and the others are quite small, confirms the conjecture that the fixed-point Hamiltonian is essentially short range. The internal energy evaluated with this method is quite good: it is only a few percent off the correct value.

Different schemes with different definitions of the block variable can be found in the literature; normally in two dimensions the results are of the same quality, although there are no fixed rules for finding the most efficient scheme.

This last approach (exact enumeration of all configurations of a small system) is useful only for discrete spins and systems that are not too large (the number of spins may be of the order 10–20); extension of this approach to high dimensions presents some difficulties. The same computation in three dimensions would involve 64 spins (2^{64} configurations!). At the moment the most fashionable procedure is that of the Monte Carlo renormalization group, where the sum over the configurations is taken approximately by using the powerful Monte Carlo method described in the last chapter of this book.[27]

In this panoramic view of the different approaches we have ignored, for reasons of space, the approach based on the transfer matrix (see Chapter 12), which in many cases is rather simple and efficient.

7.6. The Migdal-Kadanoff approximation

The spirit of the Migdal-Kadanoff (MK) approximation[28] is very similar to the mean-field theory: one studies a trial Hamiltonian (H_T) such that the problem is soluble and uses a variational technique to choose H_T in order to minimize the difference $H - H_T$. Rigorous upper and lower bounds to the free energy are a byproduct of this approach. However, the space of trial Hamiltonians we consider here is much larger than in the mean-field theory, and nonclassical critical exponents are obtained. In the MK approach the renormalization group is formulated in a slightly different way from the previous sections: for each block of length L we pick one spin (e.g., the one at the lower left corner), and we sum over the other spins. In other words, the block spin variable is a single spin. This procedure (which is called decimation) can also be performed recursively for the Hamiltonians:

$$H_{n+1} = R_D(H_n). \qquad (7.73)$$

For the original Ising model, the decimation approach is not so sound: i.e., the single spins do not look like very good block variables.[29] In spite of this difficulty we can choose H_T such that it can be solved analytically by decimation.

We start from the usual convexity inequality; if

$$f(\varepsilon) = -\frac{1}{\beta V} \ln \sum_{\{\sigma\}} \exp[-(\varepsilon H_I(\sigma) + (1-\varepsilon) H_T(\sigma)]$$

$$\left.\frac{\partial f}{\partial \varepsilon}\right|_{\varepsilon=1} = \langle H_T - H_I \rangle_{H_I} = 0, \qquad (7.74)$$

we have

$$f_T \equiv f(0) \geq f(1) \equiv f_I \geq f(0) + \left.\frac{\partial f}{\partial \varepsilon}\right|_{\varepsilon=0}$$

$$f(\varepsilon) - f(1) = 0((1-\varepsilon)^2).$$
(7.75)

If $H_T - H_I$ can be considered small in some sense, $f(1) - f(0)$ is a quantity of second order in this difference, so that we can hope to be much smaller.[30]

If by adding $(1-\varepsilon)(H_T - H_I)$ to the original Ising Hamiltonian we increase the coupling between some of the spins and we decrease the coupling between some other spins at the same relative distance, the translational invariance of the Ising model implies that $\langle H_T - H_I \rangle_{H_I} = 0$.

Let us consider a very simple example, the Ising model on a triangular lattice. While the Ising Hamiltonian is given by $\Sigma_{(i,k)} \beta\sigma_i\sigma_k$, the first trial Hamiltonian we consider is given by $\Sigma_{(i,k)\in S} 2\beta\sigma_i\sigma_k$, where the set S contains half of the links, as shown in Fig. 7.5. We get

$$H_I - H_T = \sum_{(i,k)\notin S} \beta\sigma_i\sigma_k - \sum_{(i,k)\in S} \beta\sigma_i\sigma_k.$$
(7.76)

Now $\langle \sigma_i\sigma_k \rangle_{H_I}$ depends only on $i-k$, and it cannot depend on whether (i,k) belongs to S; $\langle H_I - H_T \rangle_{H_I}$ is therefore equal to zero.

Using the relation

$$\sum_{\mu=\pm 1} \exp(\beta(\sigma_1+\sigma_2)\mu) = \sum_{\mu\pm 1} \cosh^2\beta(1+\tanh\beta\sigma_1\mu)(1+\tanh\beta\sigma_2\mu)$$

$$= 2\cosh^2\beta[1+\tanh^2\beta\sigma_1\sigma_2]$$

$$= \frac{2\cosh^2\beta}{(1-\tanh^2\beta)^{1/2}} \exp[\mathrm{arctanh}(\tanh^2(\beta))\sigma_1\sigma_2],$$
(7.77)

we find that after decimation the Hamiltonian for the surviving spin is still

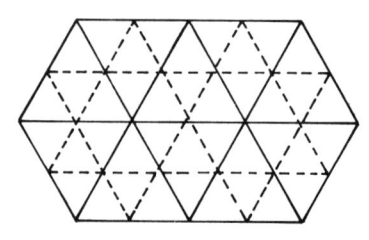

Figure 7.5. A triangular lattice: The links belonging to S are solid lines, the others are dashed lines.

nearest neighbor. We can now repeat the same operation (bond moving) at this level, decimate, and produce a new nearest-neighbor coupling Hamiltonian at the next level: this operation can be performed recursively, and the final trial Hamiltonian will contain bond moving at all levels. The free-energy density $f(0)$ is thus given by

$$\beta_0 = \beta \qquad \beta_{n+1} = b(\beta_n) = \text{arctanh tanh}^2(2\beta_n)$$

$$f = -\frac{1}{\beta} \sum_n^\infty \frac{1}{4^n} \frac{3}{4} \ln\left[\frac{2\cosh^2(2\beta_n)}{1-\tanh^2(2\beta_n)}\right]. \tag{7.78}$$

In the two limits of low and high temperatures we have

$$\beta_{n+1} \sim 2\beta_n \Rightarrow \beta_n \sim 2^n\beta \qquad f \approx -3 \quad \text{for } \beta \to \infty$$

$$\beta_{n+1} \sim 4\beta_n^2 \Rightarrow \beta_n \sim \frac{(4\beta)^{2^n}}{4} \qquad f \approx -\ln 2/\beta \quad \text{for } \beta \to 0 \tag{7.79}$$

In this case the fixed points are given by the solution of the equation $b(\beta) = \beta$.

There are two stable fixed points at $\beta = 0$ and $\beta = \infty$, separated by one unstable fixed point at $\beta = \beta_c$, where $\beta_c = b(\beta_c)$ ($\beta_c \simeq 0.3047$). The value of the critical temperature is 10% off the exact value, which is 0.2747. The error is larger for the value of $\nu = [(db/d\beta)|_{\beta=\beta_c}]^{-1} = 1.3383$, the exact value being $\nu = 1$ as in the square lattice.

Rather simple results are also obtained in the presence of a magnetic field; the zero-temperature fixed point now becomes unstable.

It is clear that the MK approximation is not very precise, but it gives a qualitatively correct picture in spite of its simplicity. Systematic improvements are possible if one studies what happens for ε near zero:[31] One discovers that there are ε-dependent critical $\beta[\beta_c(\varepsilon)]$ and a critical exponent $[\nu(\varepsilon)]$ which can be expanded in powers of ε [e.g., $\beta_c(\varepsilon) = 0.3047 - 0.0976\varepsilon + 0.0501\varepsilon^2$]. In this way better results can be obtained.[32]

It would be interesting to see what happens if the bond moving is done in such a way that the usual block spin approach becomes exact. Unfortunately, no results in this direction are yet available.

7.7. The hierarchical model

The hierarchical model is constructed in such a way that its solution can be easily found by using the block spin transformation.[33] Its Hamiltonian can be written as

$$H_H = \sum_m^\infty \beta_m \sum_{l,j \in S_m} \sigma_l \sigma_j, \tag{7.80}$$

where the set of pairs S_m is defined as follows: l and j belong to the same block of side L_{m+1}, but they belong to two different blocks of side L_m. It may be illuminating to write

$$H_H = \sum_{m}^{\infty} \left[\beta_m \sum_{(l_m, k_m)} \varphi_m(l_m) \varphi_m(k_m) \right]$$

$$= \sum_{m}^{\infty} \left[\beta_m \sum_{k_{m+1}} (\varphi_{m+1}(k_m))^2 \right], \qquad (7.81)$$

where the φ's are the usual block variables and the sum runs over all the pairs of block coordinates of level n such that l_m and k_m belong to the same block of size L_{m+1}.

The renormalization group transformation Eq. (7.7) becomes very simple for this strange Hamiltonian. It is useful to consider the Hamiltonian H_n in which the sum over m goes from 0 to n; for this Hamiltonian the variables φ_n are uncorrelated because there is no interblock interaction for $m > n$. Their probability distribution is therefore factorized. If we denote by $dP_n(\varphi_n) = P_n(\varphi_n) \, d\varphi_n$ the single block probability, the P_n satisfy a very simple recursion equation, which we write explicitly in one and two dimensions (see Fig. 7.7):

$$P_{n+1}(\varphi_{n+1}) \propto \int dP_n(\varphi_n^1) \, dP(\varphi_n^2) \delta(\varphi_{n+1} - \varphi_n^1 - \varphi_n^2) \exp(-\beta_n \varphi_{n+1}^2) \quad d=1$$

$$P_{n+1}(\varphi_{n+1}) \propto \int dP_n(\varphi_n^1) \, dP_n(\varphi_n^2) \, dP_n(\varphi_n^3) \, dP_n(\varphi_n^4) \qquad (7.82)$$

$$\times \delta(\varphi_{n+1} - \varphi_n^1 - \varphi_n^2 - \varphi_n^3 - \varphi_4^n) \exp(-\beta_n \varphi_{n+1}^2), \quad d=2$$

where the proportionality constant is chosen in such a way as to normalize the $P_{n+1}(\varphi_{n+1}) (\int d\varphi_{n+1} P_{n+1}(\varphi_{n+1}) = 1)$. In order to have similar recursion relations in any dimension and to unify the mathematical study of Eq. (7.82), it is convenient to modify the model slightly. Let us see what to do in two dimensions, the extension to higher dimensions being trivial.

If the original term in the block Hamiltonian (see Fig. 7.5) was

$$\beta_n(\varphi_n^{(1)} + \varphi_n^{(2)} + \varphi_n^{(3)} + \varphi_n^{(4)})^2 \equiv \beta_n(\varphi_{n+1})^2, \qquad (7.83)$$

in its place we can write

$$\tilde{\beta}_{2n}[(\tilde{\varphi}_{2n+1}^{(1)})^2 + (\tilde{\varphi}_{2n+1}^{(2)})^2] + \tilde{\beta}_{2n+1}(\tilde{\varphi}_{2n+2})^2$$

$$\tilde{\varphi}_{2n+1}^{(1)} = \varphi_n^{(1)} + \varphi_n^{(2)} \qquad \tilde{\varphi}_{2n+1}^{(2)} = \varphi_n^{(3)} + \varphi_n^{(4)}. \qquad (7.84)$$

7.7. The Hierarchical Model

If $\tilde{\beta}_{2n} = 0$ we recover the original Hamiltonian but we now have the freedom of setting $\tilde{\beta}_{2n} \neq 0$. We now have in any dimension

$$P_{n+1}(\tilde{\varphi}_{n+1}) \propto \int dP_n(\tilde{\varphi}_n^1) \, dP(\tilde{\varphi}_n^2)$$
$$\delta(\tilde{\varphi}_{n+1} - \tilde{\varphi}_n^1 - \tilde{\varphi}_n^2) \exp(-\tilde{\beta}_n(\varphi_{n+1})^2),$$
(7.85)

where the $\tilde{\varphi}_{Dn}$ are the usual block variables (φ_n) of the n level. Clearly the solution of the model depends on the choice of the $\tilde{\beta}_n$. The simplest choice is to take $\tilde{\beta}_n = \beta/2^{n\tilde{\omega}}$. Let us consider the case $\frac{1}{2} < \tilde{\omega} < 1$. A mathematically rigorous analysis shows that for this value of $\tilde{\omega}$, if we restrict ourselves to symmetric distributions, there are three fixed points: the high-temperature fixed point, $\omega = \frac{1}{2}$, $P_\infty(\psi_\infty) = [1/(2\pi)^{1/2}] \times \exp(-(\psi_\infty)^2/2)$; the low-temperature fixed point, $\omega = 1$, $P_\infty(\psi_\infty) = \delta(\psi_\infty - 1)$; and the critical one with $\omega = \tilde{\omega}$, the shape of $P_\infty(\psi_\infty)$ depending on $\tilde{\omega}$. The linearized transformation near the critical fixed points has only one positive eigenvalue in the space of even $P(\psi)$.

If $\tilde{\omega} < \frac{3}{4}$, $P_\infty(\psi_\infty)$ is Gaussian, when $\tilde{\omega} = \frac{3}{4} + \varepsilon$ with positive ε, the old Gaussian fixed point becomes seriously unstable (having two positive eigenvalues), but there is a critical fixed point nearby (with only one positive eigenvalue) which can be computed in powers series of ε: $P_\infty(\psi_\infty) \propto \exp(-\psi_\infty^2/2 - \varepsilon\psi_\infty^4 + \cdots)$.

The critical exponent ν has the classical value for $\tilde{\omega} < \frac{3}{4}$; for $\tilde{\omega} > \frac{3}{4}$ it can be expanded as a function of ε, $\nu = \frac{1}{2} + \frac{4}{3}\varepsilon$; and when $\tilde{\omega} \to 1$ it becomes infinite. The hierarchical model is quite important; it is the first model in which the renormalization group transformation can be set under full mathematical control, although it acts on an infinite-dimensional space.

Now how can this model be connected to the Ising model? The simplest possibility[34] consists in computing $\tilde{\omega}$ using the mean-field approximation ($\tilde{\omega} = 1/2 + 1/D$) and getting the value of ν. In this way we find the classical results for the exponents for $D > 4$; for $D = 4 - \varepsilon'$ the exponent ν can be expanded in powers of $\varepsilon' = (\varepsilon/16)(\nu = \frac{1}{2} + \frac{1}{12}\varepsilon)$ (cp. Section 9.4); for $D = 3$ we get $\nu \simeq 0.62$.

This procedure may be justified in three dimensions, where η is small ($\eta_3 \simeq 0.03$), but not in two dimensions, where η is not small ($\eta_2 = 0.25$; $\eta = D + 2 - 2D\omega_c$).

In principle it should be possible to study the Hamiltonian $H(\varepsilon) = H_H(1 - \varepsilon) + H_I\varepsilon$, at least for small ε, and to find a criterion to fix $\tilde{\omega}$ (i.e., to compute η). This is still an open problem. The same equations of the hierarchical model were derived more than ten years ago by Wilson using a different approach that would be too long to report here. Thus a systematic way to improve the approximation has not yet been obtained.

However, the main result of the hierarchical model is qualitative: in fewer than four dimensions there are two fixed points with nontrivial ω, the Gaussian fixed point (unstable) and the non-Gaussian one (stable). When the dimension goes to four, the non-Gaussian fixed point coalesces with the Gaussian one. The probability distribution $P_\infty(\psi)$ and the critical exponents can be computed as series expansions of $\varepsilon = 4 - D$. In dimensions greater than four the Gaussian model fixed point is the relevant one, and the Gaussian model (i.e., the mean-field approximation) is qualitatively correct near the transition. In the next chapter we shall see how the expansion in powers of $4 - D$ can be systematically computed for the critical exponents.

Notes for Chapter 7

1. The perturbative expansion of the Landau-Ginsburg model around the Gaussian limit ($g = 0$) seems at a first sight to be more promising; unfortunately, we have seen that perturbation theory cannot be used directly near the critical point as soon $D < 4$. The probability distribution of the large-scale fluctuations (for $g \neq 0$) is too different from that of the Gaussian model. However, by using the ideas introduced in this chapter, one will be able to transform the perturbative expansion in powers of g into a powerful tool in the next chapter.
2. We are interested in the singular terms (nonanalytic) of the thermodynamic functions. No singularities are present in a finite box. They can only come from the corrections due to the finite volume [if $f_\infty(\beta)$ is singular and $f_c(\beta)$ is regular, $f_\infty(\beta)$ and $f_\infty(\beta) - f_c(\beta)$ have the same singularities].
3. Introductions to the application of the renormalization group to critical phenomena can be found in G. Toulouse and P. Pfeuty, *Introduction au Groupe de Renormalization et à ses Applications*, Presses Universitaires, Grenoble (1975); S. K. Ma, *Modern Theory of Critical Phenomena*, Benjamin, New York (1976); D. Amit, *Field Theory, the Renormalization Group, and Critical Phenomena*, McGraw-Hill, New York (1978).
4. Block variables were introduced by L. P. Kadanoff, *Ann. Phys.* (N.Y.) **2**, 263 (1966); see also L. P. Kadanoff et al., *Rev. Mod. Phys.* **39**, 395 (1967).

5. Usually k denotes the coordinates of the box in units of L_n. Sometimes we shall use the notation k_n to stress that we indicate a box of side L_n; when no ambiguity is possible, the subscript n will be omitted.
6. The recursivity of the definition of the block variables will play a crucial role: everything becomes much simpler if we concentrate our attention on the recursive transformation, as it was discovered by Wilson in his fundamental papers, *Phys. Rev. B 4*, 3174, 3184 (1971).
7. This may happen only if we do not stay at the critical point.
8. This statement is precise only if $\langle \sigma \rangle \equiv m = 0$; only the connected correlation functions are zero in the general case.
9. In most cases (possible exceptions may be zero-temperature antiferromagnetic materials), ω must satisfy the bounds $\frac{1}{2} \leq \omega \leq 1$.
10. For a discussion of this point see G. Gallavotti, G. Jona Lasinio, *Commun. Math. Phys. 41*, 301 (1975) and *Advances in Physics 27*, 913 (1978) (and references therein).
11. In more complex systems like Ising systems with competing ferromagnetic and antiferromagnetic interactions, or at commensurate–incommensurate transitions, oscillations may be present. This possibility is under current investigation.
12. J. L. Lebowitz, *Commun. Math. Phys. 35*, 87 (1974).
13. A sketch of a possible rigorous proof is the following: one starts by proving some correlation inequalities like Eq. (7.16). These inequalities should imply that the allowed space of probability measures for the well normalized block-variables is a compact space in an appropriate topology; that would be enough: it would then be possible to extract a subsequence (n_s) such that $\lim_{s \to \infty} P_{n_s}[\psi] \equiv P_\infty[\psi_\infty]$ where P is a nontrivial probability distribution. This kind of proof would not exclude oscillations, i.e., the existence of different subsequences that do not have any limit (or have another limit).
14. It is convenient to consider the temperature as one of the parameters characterizing the probability distribution; for the initial case, it is convenient to write $P_0 \propto \exp[-\beta H] \equiv \exp[-H_0]$.
15. A review of the rigorous results obtained on this very important question can be found in K. Gawedzki and A. Kupainen, *Cargèse Summer School 1983*, ed. by P. K. Mitter et al., Plenum Press, New York (1984); for a more heuristic approach see C. Bachas, *Nucl. Phys. B 235* [FS11], 172 (1984).
16. For a general discussion of the long-time behavior of the solution of Eq. (7.26) see, for example, L. Pontriaguine, *Equations Differentielles Ordinaries*, Mir, Moscow (1969).
17. More complicated asymptotic behavior (like the limiting cycle shown in Fig. 7.2 or "turbulent" (aperiodic) behavior) are excluded by hypothesis [Eq. (7.11)].

18. We must also consider the nonlinear terms in order to study what happens when the real part of the eigenvalues is exactly equal to zero.
19. For simplicity we assume ω_R and ω_A to be real.
20. The dimension of the generic intersection of two sets A and B belonging to the space S is given by $\dim(A) + \dim(B) - \dim(S) = \dim(A) - \text{codim}(B)$: If a negative dimension is obtained, no generic intersection is present.
21. These scaling laws are derived under the hypothesis that $g(H_c) \neq 0$; if this hypothesis fails, as happens if P_∞ is also Gaussian (e.g., for $D > 4$), some of the scaling laws do not hold anymore, and they are replaced by inequalities, some of which can also be rigorously proven.
22. T. L. Bell and K. G. Wilson, *Phys. Rev. B 11*, 3431 (1975); we could also take as a starting point the $0(N)$ symmetric model in the limit N going to infinity; in this case exact results can be obtained: see S. K. Ma, *Rev. Mod. Phys. 45*, 589 (1973); *J. Math. Phys. 15*, 1886 (1974); A. Guha, M. Okawa, and J. B. Zuber, *Nucl. Phys. B 240* [FS12], 566 (1984).
23. For example, the J's that couple block fields at distances one, two, and three are, respectively, in two dimensions proportional to 1, -0.2187, 0.0492; see A. Guha *et al.*, art. cit.
24. Of course the long-range correlations of the $\tilde{\psi}_{n+1}$ induce a long-range correlation on the $\tilde{\psi}_n$, if we integrate them.
25. If the number of spins in the block is even, Eq. (7.55) may be ambiguous ($\tilde{\varphi}_n$ may be zero). The solution consists in defining ψ_n to be ± 1 with 50% probability in this case: $P(\psi_n = 1) = P(\tilde{\varphi}_n > 0) + \frac{1}{2} P(\tilde{\varphi} = 0)$.
26. M. Nauenberg and B. Nienhuis, *Phys. Rev. 33*, 344 (1974).
27. The most recent Monte Carlo renormalization group computations are those of D. J. Wallace, A. S. Pawley, R. Swendson, G. and K. G. Wilson, *Phys. Rev. B 29*, 403 (1986) (they estimate for the Ising model in three dimensions $\beta_i = 0.22654 \pm 0.00006$, $\nu = 0.631 \pm 0.002$, $\eta = 0.031 \pm 0.004$); D. J. Callaway and R. Petronzio, *Nucl. Phys. B 240* [FS12], 577 (1984); R. Swendson, *Phys. Rev. Lett. 52*, 1165 (1984). The last paper is particularly interesting because the fixed-point Hamiltonian is directly estimated for the two-dimensional and the three-dimensional Ising models at the critical point; the apparent exponential decrease of the coupling with the distance is explicitly verified.
28. A. A. Migdal, *Sov. Phys. JEPT 69*, 810; L457 (1975); L. P. Kadanoff, *Ann. Phys.* (N.Y.) *100*, 359 (1976); and *Rev. Mod. Phys. 49*, 267 (1977).
29. The correlation function of the decimated variables is given by

$\langle \sigma_{i_n} \sigma_{k_n} \rangle = \delta_{i_n, k_n} + O(L_n^{-(D-2+\eta)})$. A nontrivial result is obtained simply by removing the δ function by fiat and rescaling the σ's. The remaining σ's over which we have summed are likely to have a rather long-range correlation (much larger than L_n when $n \to \infty$), as can be checked in the Gaussian model. In this approach the theoretical basis for the regularity of R_D at the critical point is not clear.

30. We can guess, but we cannot prove, that

$$f(0) + \frac{1}{2} \left.\frac{\partial f}{\partial \varepsilon}\right|_{\varepsilon=0} > f(1).$$

31. We could also consider more sophisticated bond moving operations. See for example S. Caracciolo, G. Parisi, and N. Sourlas, *Nucl. Phys. B 205* [FS5], 345 (1982).
32. G. Martinelli and G. Parisi, *Nucl. Phys. B 180* [FS2], 483 (1980); see also S. Caracciolo, *Nucl. Phys. B 180* [FS2], 405 (1981).
33. A review of the rigorous results obtained for the hierarchical model can be found in P. Collet and J. P. Eckmann, Renormalization Group Analysis of the Hierarchical Model in Statistical Mechanics, Lecture Notes in Physics No. 74, Springer, Berlin (1978), where the results (which we state in this section) are derived with many details; in particular, Fig. 7.6 is taken from their book.
34. There are many alternative ways to derive the hierarchical model as an approximation to the Ising model: e.g., in the original approach the hierarchical model was obtained by considering an approximate renormalization group transformation in momentum space [K. Wilson, *Phys. Rev. B 4*, 3174, 3184 (1971)]. Recent efforts in the same spirit as the original paper by Wilson are described in K. E. Newman and E. K. Riedel, *Phys. Rev. B 25*, 264 (1982).

CHAPTER 8

Perturbative Evaluation of the Critical Exponents

8.1. The basic definitions

The methods described in the previous chapter go under the name of the "real space renormalization group" approach because most of the arguments are carried out in configuration space. A different method, the "field theoretical approach," is based on the perturbative expansion for the Landau-Ginsburg model described in Chapters 5 and 6.[1] Using this method it is possible to construct the expansion for the critical exponents in powers of $\varepsilon = 4 - D$ in a systematic way. By pushing the computation further, it is possible to obtain the coefficient of ε^k with arbitrary k; at present the coefficients of the ε expansion (up to ε^4) have been evaluated;[2] in this book we shall describe the computation up to second order in ε. This method can also be used to compute the exponents directly in fixed dimensions without having to change the dimensions of the space in the middle of the computation.[3]

Let us see which are the basic ingredients of this approach. We have already seen that the naive perturbative expansion breaks down at the transition point; however, using the information we have already gathered in the preceding chapter on the structure of second-order phase transitions, it is possible (as we shall see) to reshuffle the perturbative expansion in such a way that it will also make sense near the critical point. It is clear that this reshuffling may be particularly useful near four dimensions, where the fixed-point probability distribution is not far from being Gaussian. We would be very happy if we could write down directly the perturbative expansions for the fixed-point probability distribution treating the deviations from a pure Gaussian law as a perturbation.

We already know that the naive perturbative expansion around the Gaussian model in powers of the coupling constant g breaks down near the critical point because the effective dimensionless coupling constant

8.1. The Basic Definitions

$u = g\xi^{4-D}$ goes to infinity with ξ. We can now use the scaling laws of the previous chapter to bypass this difficulty. We first observe that the two-spin correlation function can be written in momentum space as

$$G(p) = \frac{z_1}{p^2 + m^2 + O(p^4)} \qquad (8.1)$$

where:[4]

$$\begin{aligned} z_1 &= 1 & m^2 &= \xi^{-2}, & &\text{if } g = 0 \\ z_1 &\approx \xi^{-\eta} & m^2 &\simeq \xi^{-2}, & &\text{if } \xi \to \infty \text{ at } g \neq 0. \end{aligned} \qquad (8.2)$$

If we introduce the so-called renormalized field φ_R defined as $\varphi_R = \varphi/z_1^{1/2}$, it is evident that the renormalized two-point correlation function (i.e., the correlation function of two φ_R fields) must satisfy the normalization condition

$$G_R(p) = \frac{1}{p^2 + m^2 + O(p^4)}. \qquad (8.3)$$

The scaling law implies that the correlation functions of the renormalized field behave in a very simple way when m goes to zero at fixed p/m in momentum space (or at fixed mx in configuration space). For an N-point function we have only an overall factor that is proportional to $1/m^{N(1+D/2)-D}$ or to $m^{N(1-D/2)}$ in momentum or in configuration space, respectively. If everything is measured in units of m or ξ^{-1} the correlation functions of the renormalized field have a well-defined limit when we go to the critical point where g/m^{4-D} is infinite.

It is particularly useful to consider the four-point connected amputated correlation function of the renormalized field, evaluated at zero external momentum, which for simplicity we shall denote as Γ_4^R.

We can thus define

$$\lambda = m^{D-4} z_1^2 \Gamma_4 \equiv m^{D-4} \Gamma_4^R. \qquad (8.4)$$

We have

$$\begin{aligned} \lambda &= \frac{g}{m^{4-D}} + O\left(\left(\frac{g}{m^{4-D}}\right)^2\right) & &\text{for } \frac{g}{m^{4-D}} \sim 0 \\ \lambda &\to \lambda_c & &\text{for } \frac{g}{m^{4-D}} \to \infty. \end{aligned} \qquad (8.5)$$

The finiteness of λ at the critical point (i.e., $\lambda_c < \infty$) is a direct consequence of Eq. (7.44); moreover, λ_c depends on the probability distribution at the fixed point, and consequently it must go to zero when the probability distribution becomes Gaussian, as should happen near four dimensions according to the discussion in the preceding chapter. It is reasonable to conjecture (and we shall verify later) that $\lambda_c \to 0$ when $D \to 4$. It is also quite possible that λ_c is not extremely large directly in three dimensions.

It is clear that a perturbative expansion in powers of λ may be extremely useful for computing what happens at the critical point, i.e., at $\lambda = \lambda_c$. Equation (8.4) implies that λ is proportional to the dimensionless coupling constant u at small u (i.e., when m^2 is large) and remains finite when $u \to \infty$ (from now on we set $u = g/m^{4-D}$). This fact suggests that we transform the perturbative expansion in powers of u into a perturbative expansion in powers of λ, bypassing in this way the problem of using a perturbative expansion at very large couplings. If λ_c is small enough, this program may be successful; if λ_c is too large, it may be difficult to use the perturbative expansion at $\lambda = \lambda_c$. The critical exponents may be computed by starting from the following scaling laws:

$$m^2 \propto (\mu - \mu_c)^{2\nu} \Rightarrow z_2(\lambda) \equiv \frac{d\mu}{dm^2} \sim m^{2C_2},$$

$$C_2 = \frac{1}{2\nu} - 1, \qquad (8.6)$$

$$z_1(\lambda) \equiv \left(\frac{G(p)^2}{dG/dp^2} \right) \bigg|_{p^2=0} \propto m^{2C_1}, \qquad C_1 = \frac{\eta}{2},$$

which suggests that we define the following functions:

$$C_1(\lambda) \equiv m^2 \frac{\partial}{\partial m^2} \bigg|_{g \text{ fixed}} \ln[z_1(\lambda)],$$

$$C_2(\lambda) \equiv m^2 \frac{\partial}{\partial m^2} \bigg|_{g \text{ fixed}} \ln[z_2(\lambda)], \qquad (8.7)$$

$$C_1(\lambda_c) = C_1 \qquad C_2(\lambda_c) = C_2.$$

It is crucial that both the z's and the C's be dimensionless functions so that they can depend only on the dimensionless parameter λ (or u). How can we compute λ_c? We define the function

$$b(\lambda) = m^2 \frac{\partial}{\partial m^2} \bigg|_{g \text{ fixed}} \lambda = (D-4)u \frac{\partial}{\partial u} \bigg|_{m^2 \text{ fixed}} \lambda. \qquad (8.8)$$

Now λ_c is fixed by the condition $b(\lambda_c) = 0$. The situation is illustrated in Fig. 8.1.

8.1. The Basic Definitions

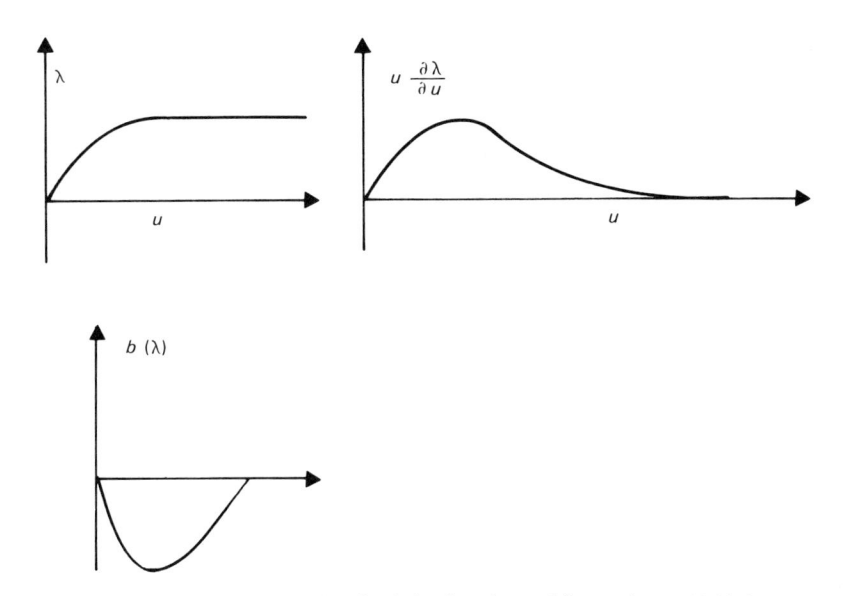

Figure 8.1. Qualitative sketch of the functions $\lambda(u)$, $u\partial\lambda/\partial u$, and $b(\lambda)$.

Equations (8.7) and (8.8) can be easily integrated. Using the boundary conditions (8.3)–(8.5), we get

$$u(\lambda) = \lambda \exp\left\{\int_0^\lambda \left(\frac{D-4}{2b(\lambda')} - \frac{1}{\lambda'}\right) d\lambda'\right\}$$

$$z_1(\lambda) = \exp\left\{\int_0^\lambda \frac{C_1(\lambda')}{b(\lambda')} d\lambda'\right\} \quad (8.9)$$

$$z_2(\lambda) = \exp\left\{\int_0^\lambda \frac{C_2(\lambda')}{b(\lambda')} d\lambda'\right\}.$$

More precisely, in the region of $\lambda \sim \lambda_c$ (i.e., $u \to \infty$) we have

$$u = E\left(\frac{\lambda_c - \lambda}{\lambda_c}\right)^{(-4+D)/2b'} \Rightarrow \lambda = \lambda_c\left[1 - \left(\frac{u}{E}\right)^{-2b'/(4-D)}\right]$$

$$b' = \frac{d}{d\lambda} b(\lambda)\big|_{\lambda = \lambda_c} \quad (8.10)$$

$$E = \lambda_c \exp\left\{\int_0^{\lambda_c}\left[-\frac{4-D}{2b(\lambda')} - \frac{1}{\lambda'} - \frac{4-D}{2b'(\lambda_c - \lambda')}\right] d\lambda'\right\}.$$

Equation (8.10) tells us that when $\xi \to \infty$ the dependence of λ on the value of g vanishes as $m^{-2b'}$; this is in agreement with Eq. (7.32): near the critical point there is residual dependence of the trajectories on the Hamiltonian for the block spin variables, which vanishes as ξ^{ω_A}. We finally get

$$-\omega_A = 2b', \tag{8.11}$$

where ω_A is defined on p. 120 of the last chapter.

In this way we have also obtained the exponent characterizing the scaling corrections. Indeed, we expect that

$$\chi = \frac{1}{(T-T_c)^\gamma} [a_0 + a_1(T-T_c)^{t_1} + a_2(T-T_c)^{t_2} + \cdots] \tag{8.12}$$

where $t_i = -\omega_i^A/\omega_R$, ω_i^A are the various attractive eigenvalues (supposedly real) of the linearized renormalization group transformation.[5] It is possible to estimate even the subleading exponents (t_i, $i>1$), but we postpone to Chapter 11 the discussion of this tricky matter (see Section 11.3).

The practical implementation of the program is rather straightforward. If ultraviolet divergences were neglected (they are present when $\Lambda \to \infty$ in $D \geq 2$), we should compute only m^2 and λ as functions of μ and g as a power series of g, and order by order find μ and u as functions of m and λ.

In the presence of ultraviolet divergences we have already seen that the perturbative expansion cannot be carried out at fixed μ, but it should be done at fixed μ_R; on the other hand, $(\partial\mu/\partial m^2)|_{g \text{ fixed}}$ or similar quantities are free of ultraviolet divergences, the divergent part of the counterterms being independent of μ_R (and, consequently m). We have in our hands all the tools we need to begin the evaluation of the exponents.

This will be done in the next section.

8.2. A simple example

Let us compute the critical exponents using the formalism of the preceding section at the lowest order in perturbation theory (i.e., only one-loop diagrams are considered). The only diagrams relevant are shown in Fig. 8.2.

Figure 8.2. One-loop diagrams for the self-energy (a) and the vertex (b).

8.2. A Simple Example

At this order the self-energy diagram is momentum independent, so that we get

$$G(p) = \frac{1}{p^2 + m^2} + O(g^2)$$

$$\Gamma_4(0) = g - g^2 \frac{3}{2} \frac{1}{(2\pi)^D} \int \frac{d^D q}{(q^2 + m^2)^2}$$

$$= \left\{ u - \frac{3}{2} u^2 \frac{\Gamma(2 - D/2)}{(4\pi)^{D/2}} \right\} m^{2(2-D/2)} \qquad (8.13)$$

$$m^2 = \mu^2 - \frac{1}{2} \frac{g}{(2\pi)^D} \int \frac{d^D q}{q^2 + m^2}.$$

We then obtain

$$\lambda = u - \frac{3}{2} \frac{\Gamma(2 - D/2)}{(4\pi)^{D/2}} u^2$$

$$\frac{\partial m^2}{\partial \mu} = 1 - \frac{1}{2} \frac{\Gamma(2 - D/2)}{(4\pi)^{D/2}} u, \qquad (8.14)$$

which imply

$$u = \lambda + \frac{3A\lambda^2}{(4-D)} + O(\lambda^3) \qquad A = \frac{\Gamma(3 - D/2)}{(4\pi)^{D/2}}$$

$$b(\lambda) = -\frac{4-D}{2} \lambda + \frac{3}{2} A\lambda^2 + O(\lambda^3) \qquad (8.15)$$

$$C_2(\lambda) = \frac{1}{4} A\lambda + O(\lambda^2) \qquad C_1(\lambda) = O(\lambda^2).$$

If terms of higher order in λ are neglected, we get

$$\lambda_c = \frac{4-D}{3A} \equiv \lambda_c^{(1)}$$

$$C_1 = 0 \qquad D < 4 \qquad (8.16)$$

$$C_2 = \frac{4-D}{6}.$$

The critical exponents are

$$\eta_D = 0 \equiv 2C_1$$

$$\nu_D = \frac{1}{2-(4-D)/3} \equiv \frac{1}{2-2C_2}, \quad D < 4. \tag{8.17}$$

We obtain $\nu_3 = -0.6$ and $\nu_2 = 0.75$, which should be compared with $\nu_3 = 0.629$, $\nu_2 = 1$, $\eta_3 \simeq 0.033$, and $\eta_2 = 0.25$. It is evident that the three-dimensional results are much more accurate than the two-dimensional ones. Indeed, we see that λ_c goes to zero when $D \to 4$. In the next chapter we shall see that this fact remains true at all orders in perturbation theory. This should not be a surprise. Indeed, λ_c just tells us how far the stable fixed point is from the Gaussian fixed point, which starts to be unstable as soon as $D < 4$.[6]

We can generalize these results to the case of N-component spins, the only difference being the multiplicities of the diagrams in Fig. 8.2 $((N+8)/6$ and $(N+2)/6$, respectively).[7] We finally obtain

$$\lambda_c = \frac{3(4-D)}{(N+8)A} \equiv \lambda_c^{(1)}, \quad \eta_D = 0$$

$$\nu_D = \frac{1}{2 - \left(\frac{N+2}{N+8}\right)(4-D)}. \tag{8.18}$$

For $N \to \infty$ we recover the exact results $\nu_D = 1/(D-2)$; indeed, an explicit computation shows that higher-order diagrams in units of $\lambda/\lambda_c^{(1)}$ are suppressed by factors $1/N$.

8.3. A more sophisticated computation

The computations of the previous section can be extended to higher orders. At two loops the only diagrams are shown in Fig. 8.3. Their

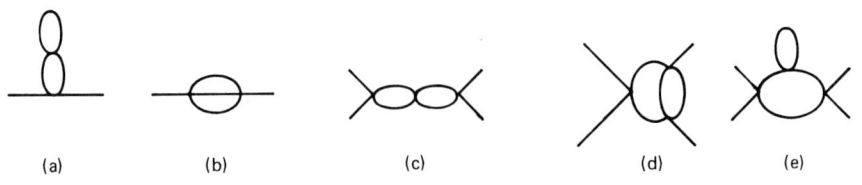

Figure 8.3. Two-loop diagrams for the self-energy (a), (b) and the vertex (c), (d), (e). Diagrams (a) and (e) are eliminated in the expansion at fixed m^2.

8.3. A More Sophisticated Computation

evaluation is easy (the techniques needed are presented in Sec. 8.5). The final result is

$$b(\lambda) = -\frac{4-D}{2} \lambda \left\{ 1 - r + r^2(4-D) \right.$$
$$\left. \times \left[\frac{10N+44}{(N+8)^2} f(D) - \frac{N+2}{(N+8)^2} h(D) \right] \right\}$$

$$C_1(\lambda) = (4-D)^2 r^2 \frac{N+2}{4(N+8)^2} h(D)$$

$$C_2(\lambda) = \frac{(N+2)}{2(N+8)} (4-D) \left\{ r - (4-D)r^2 \frac{3f(D) - h(D)/2}{N+8} \right\} \quad (8.19)$$

$$r = \frac{\lambda}{\lambda_c^{(1)}} = \frac{\lambda A(N+8)}{3(4-D)}$$

$$h(4) = 1, \quad h(3) \simeq 0.59, \quad h(2) \simeq 0.46;$$

$$f(4) = 1, \quad f(3) = \tfrac{2}{3}, \quad f(2) \simeq 0.56.$$

We see that when $D \to 4$ or $N \to \infty$ the higher-order terms are negligible. For $N = 1$ we find

$$-\frac{2b(\lambda)}{\lambda(4-D)} = 1 - r + 0.41 r^2, \quad 2C_2(\lambda) = \frac{1}{3} r - 0.069 r^2,$$

$$C_1(\lambda) \simeq 0.011 r^2 \quad \text{for } D = 3$$

$$-\frac{2b(\lambda)}{\lambda(4-D)} = 1 - r + 0.71 r^2 \quad (8.20)$$

$$2C_2(\lambda) \simeq \frac{2}{3} r - 0.25 r^2 \quad C_1(\lambda) = 0.034 r^2, \quad \text{for } D = 2.$$

The vanishing of the two-loop corrections when N goes to infinity is due to the fact that the model in this limit is soluble (see Chapter 6) and that the one-loop diagrams already give the correct answer, as can be seen by looking at the explicit formulae. This result could also be obtained by using the fully renormalized expansion of Section 8.5, where all diagrams with more than one loop vanish when N goes to infinity.

The results of the one-loop approximation are correct when D goes to 4 due to the structure of the functions $b(\lambda)$, $C_1(\lambda)$ and $C_2(\lambda)$, which have a well defined series expansion in powers of λ in four dimensions, as can be seen from Eq. (8.19) at the two-loop level. Moreover, $b(\lambda) = -\tfrac{1}{2}(4 - D)\lambda + C\lambda^2 + 0(\lambda^3)$ $(C > 0)$, and therefore $b(\lambda)$ must have a zero of order $4-D$ for D close to 4. The behavior of the theory near four dimensions will be amply discussed in the next chapter, where we shall see that this structure will persist, as well, if higher-loop contributions are included.

In the real case (e.g., $D = 3$), the terms of order r^2 are not negligible and give substantial corrections; in such cases it is desirable to use some refined method for decreasing the errors involved in truncating the Taylor expansion (in general, the best method depends on the function one is considering). In the present case a comparison with the results of a many-loop computation shows that the fastest way to evaluate the corrections induced by the two-loops terms is to consider $r(\tilde{\varepsilon})$ as the solution of the equation

$$-\frac{2b(\lambda)}{\lambda(4-D)} = 1 - \tilde{\varepsilon}.$$

We get, for example, for $N = 1$ in $D = 3$ and $D = 2$, respectively,

$$r(\tilde{\varepsilon}) = \tilde{\varepsilon} + 0.41\tilde{\varepsilon}^2$$

$$\tilde{C}_2(\tilde{\varepsilon}) = C_2[r(\tilde{\varepsilon})] = \frac{\tilde{\varepsilon}}{6}[1 + 0.19\tilde{\varepsilon}] \qquad \text{for } D = 3$$

$$r(\tilde{\varepsilon}) = \tilde{\varepsilon} + 0.71\tilde{\varepsilon}^2$$

$$\tilde{C}_2(\tilde{\varepsilon}) = C_2[r(\tilde{\varepsilon})] = \frac{1}{3}\tilde{\varepsilon}[1 + 0.34\tilde{\varepsilon}] \qquad \text{for } D = 2.$$

(8.21)

If we set $\varepsilon = 1$ in Eq. (8.21) we get

$$r_c = 1.41$$

$$C_2 = 0.198 \Rightarrow \nu_3 = 0.623 \quad \text{for } D = 3$$

$$r_c = 1.71$$

$$C_2 \simeq 0.50 \Rightarrow \nu_2 \simeq 1.01 \qquad \text{for } D = 2.$$

(8.22)

We see that the two-loop results are more accurate than the one-loop, but at this stage we cannot estimate the error to be attached to such a computation. Fortunately, a few years after the computation of the two-loops results, a quite sophisticated 6-loop computation (1142 diagrams) was carried out for the b and C functions in $D = 3$, the coefficients being estimated with an accuracy of 10^{-8}.[8]

We have in $D = 3$

8.3. A More Sophisticated Computation

$$b(\lambda) = \sum_{1}^{\infty} b_k \lambda^k = -\frac{4-D}{2} \lambda \left[\sum_{1}^{\infty} \tilde{b}_k r^{k-1} \right]$$

$$\simeq -\frac{4-D}{2} \lambda [1 - r + 0.41r^2 - 0.35r^3 + 0.33r^4$$

$$- 0.49r^5 + 0.75r^6 + 0(r^7)].$$

(8.23)

The asymptotic behavior of the b_k for large k, can be computed using similar techniques to those in Chapter 5.[9] One finds in $D = 3$

$$\tilde{b}_k \to -c(-R)^k \Gamma\left(k + \frac{9}{2}\right)\left[1 + 0\left(\frac{1}{k}\right)\right]$$

(8.24)

$$c \simeq 0.03996, \quad R \simeq 0.147742.$$

The asymptotic formula (8.24) is good enough even at relatively small k: by comparing Eq. (8.34) with the exact results we see that at $k = 6$ and $k = 7$ Eq. (8.24) is off by only a few percent. From all this information one finally estimates, using the Borel technique for summing the non-convergent expansion (8.23)[10]

$$\nu_3 = 0.630 \pm 0.015,$$

$$\eta_3 = 0.031 \pm 0.004,$$

(8.25)

$$\lambda_c = 1.416 \pm 0.005.$$

These numbers are in good agreement with the results coming from the high-temperature expansion. For example, an analysis done for the series for the bcc lattice gives

$$\nu_3 = 0.629 \pm 0.003,$$

$$\eta_3 = 0.031 \pm 0.005.$$

(8.26)

Unfortunately, in two dimensions the fixed point is at high values of λ, and only four-loop diagrams have been computed. One finds

$$\eta_2 = 0.13 \pm 0.07,$$

$$\nu_2 = 0.97 \pm 0.08,$$

(8.27)

$$\lambda_c = 1.85 \pm 0.10.$$

Equation (8.27) is qualitatively correct, but the results are far from being as good as in $D = 3$.[11] At least in three dimensions the problem of evaluating the critical exponents is essentially solved: if simple computations are done, approximate results are obtained; if we perform more refined computations, we get more and more precise results.

8.4. On the scaling laws

In the previous sections we have looked at what happens to dimensionless quantities when g/m^{4-D} goes to infinity. Let us study here what happens to the correlation functions as a function of the momenta. In this case a new scale is present, and new dimensionless quantities can be formed. The simplest possibility is to consider what happens when $u \to \infty$ at fixed p^2/m. The quantities that have a good limit in this situation are the so-called renormalized correlation functions, e.g.,

$$G_2^R(p) = z_1^{-1}(u) G_2(p),$$
$$G_4^R(p) = z_1^{-2}(u) G_4(p), \qquad (8.28)$$
$$\Gamma_4^R(p) = z_1^{2}(u) \Gamma_4(p),$$

where we denote by p the set of all momenta on which the correlation functions depend. Indeed, the scaling laws tell us that the renormalized correlation functions have a limit when $g \to \infty$ at fixed p and m.[12] The computation of the G^R's at $\lambda = \lambda_c$ is easy: We simply develop the G^R's as functions of λ, and we evaluate the perturbative series (eventually resummed à la Borel) at $\lambda = \lambda_c$.

This method, however, is not so efficient when p/m goes to infinity. As soon as $D < 4$ we have

$$\lim_{t \to \infty} t^2 G_2(t) = 1,$$
$$\lim_{t \to \infty} \Gamma_4(t) = g. \qquad (8.29)$$

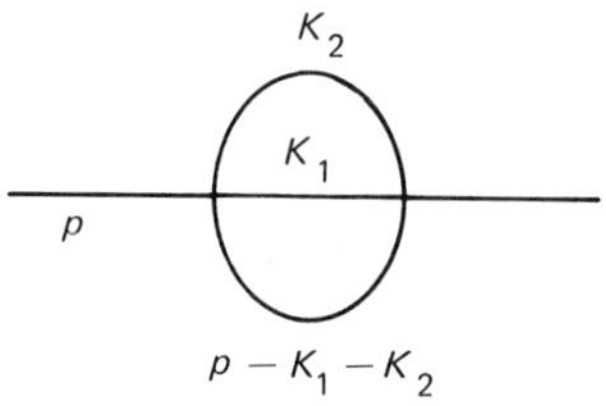

Figure 8.4. The same diagram as in Figure 8.3(b) with the momentum flow used in its evaluation.

8.4. On the Scaling Laws

Indeed, all loop diagrams vanish when $t \to \infty$, so we remain with the zero-loop result. Equation (8.29) implies that

$$\lim_{t\to\infty} t^2 G^R(t) = z_1^{-1}$$
$$\lim_{t\to\infty} \Gamma_4^R(t) = z_1^2 g . \qquad (8.30)$$

The r.h.s. of Eq. (8.30) diverges when $\lambda \to \lambda_c$. The scaling implies that precisely at $g = \infty$ (i.e., at $\lambda = \lambda_c$)

$$t^2 G_2^R(t) \sim t^\eta$$
$$\Gamma_4^R(t) \sim t^{4-D-2\eta} . \qquad (8.31)$$

It is clear that the simple truncated perturbative expansion in λ, evaluated at $\lambda = \lambda_c$, cannot reproduce the behavior displayed by Eqs. (8.30) and (8.31).

In the perturbative expansion for the renormalized correlation function, all the definitions (i.e., $G_2^R = 1/(p^2 + m^2 + O(p^4))$, $\Gamma_4^R(0) = \lambda m^{4-D}$) are at small external momenta, and we can only hope that the correlation functions at small momenta are smooth functions of λ. In order to bypass this difficulty, we have to write the correlation functions at large momenta as functions of some other correlation functions at small momenta. This can be done by generalizing Eqs. (8.6) and (8.7). Let us simply discuss the case of a two-field correlation function. By dimensional analysis we get

$$m^2 \frac{\partial}{\partial m^2}\bigg|_{g,p} G_2(p) = -2p \frac{\partial}{\partial p}\bigg|_{m^2,g} G_2(p) + \frac{D-4}{2} g \frac{\partial}{\partial g}\bigg|_{m^2,p} G_2(p)$$
$$- G_2(p) \qquad (8.32)$$
$$\Leftrightarrow G_2(p, m^2, g) = m^{-2} \tilde{G}_2\left(\frac{p^2}{m^2}, \frac{g}{m^{4-D}}\right),$$

where G_2 is considered a function of m^2 and g. By multiplying Eq. (8.32) by z_1^{-1} and using the relation

$$g \frac{\partial}{\partial g}\bigg|_{p^2,m^2} G_2^R(p) = z_1^{-1} g \frac{\partial}{\partial g}\bigg|_{p,m^2} G_2(p) + g \frac{\partial z_1^{-1}}{\partial g}\bigg|_{m^2} G_2(p)$$
$$\qquad (8.33)$$
$$= z_1^{-1} g \frac{\partial}{\partial g}\bigg|_{p,m^2} G_2(p) - \frac{2}{4-D} m^2 \frac{\partial z_1^{-1}}{\partial m^2}\bigg|_{g} G_2(p),$$

we obtain

$$z_1^{-1} m^2 \frac{\partial}{\partial m^2}\bigg|_{g,p} G_2(p) = \left[-p^2 \frac{\partial}{\partial p^2}\bigg|_{m^2,g} + \frac{D-4}{2} g \frac{\partial}{\partial g}\bigg|_{p,m^2} + C_1(g) \right] G_2^R(p). \tag{8.34}$$

We can now go to the variables λ and m^2 by using the chain rule:

$$g \frac{\partial}{\partial g}\bigg|_{m^2} = g \frac{\partial \lambda}{\partial g}\bigg|_{m^2} \frac{\partial}{\partial \lambda}\bigg|_{m^2} = \frac{2}{D-4} b(\lambda) \frac{\partial}{\partial \lambda}\bigg|_{m^2}. \tag{8.35}$$

We finally get the so-called Callan-Symanzik equation,[13]

$$\Delta G_2^R(p) \equiv z_1^{-1} m^2 \frac{\partial}{\partial m^2}\bigg|_{p^2,g} G_2(p)$$

$$= \left\{ -p^2 \frac{\partial}{\partial p^2}\bigg|_{\lambda,m^2} + b(\lambda) \frac{\partial}{\partial \lambda}\bigg|_{m^2,p^2} - [1 - C_1(\lambda)] \right\} G_2^R(p). \tag{8.36}$$

Equation (8.36) can be used to compute G_2^R as a function of ΔG_2^R. This possibility is particularly interesting because $\Delta G_2^R / G_2^R = m^2 [\partial G_2(p)/\partial m^2] / G_2(p)$ is supposed to vanish when $p \to \infty$. This can be checked in perturbation theory, where the large-momentum behavior of p is m independent (see the example in the next section). It is also evident that if the correlation functions exist (as they should) at the critical point $m^2 \to 0$, with p^2 and g fixed, $\lim_{m^2 \to 0} \partial \ln G_2 / \partial \ln m^2$ must be zero.[14]

If we stay exactly at λ_c, we see that when p^2 is large $G_2^R(p)$ satisfies the simplified differential equation

$$\left[-p^2 \frac{\partial}{\partial p^2} - (1 - C_1) \right] G_2^R(p^2) = 0, \tag{8.37}$$

implying $G_R^2(p) = 1/p^{(2-2C_1)}$.

A similar equation can be written for the other correlation functions [e.g., $\Delta \Gamma_4^R = [-2p(\partial/\partial p) + b(\lambda)(\partial/\partial \lambda) + (4 - D)/2 - 2C_1(\lambda)] \Gamma_4^R(p)$]. We get in this way correct asymptotic behavior at large momenta at $\lambda = \lambda_c$. This analysis may be carried further by integrating Eq. (8.36), using the method of the characteristic curves. We introduce an "effective running" coupling constant $\lambda_r(\lambda, t)$ satisfying the equation

$$t \frac{\partial}{\partial t} \lambda_r(\lambda, t) = -\frac{1}{2} b[\lambda_r(\lambda, t)], \tag{8.38}$$

with the boundary condition $\lambda_r(\lambda, 1) = \lambda$.

8.4. On the Scaling Laws

Equation (8.38) may be solved by introducing the function $\rho(\lambda) = [u(\lambda)]^{1/(4-D)}$ and its functional inverse ρ^{-1} (i.e., $\rho^{-1}(\rho(x)) = \rho(\rho^{-1}(x)) = x$):

$$\lambda_r(\lambda, t) = \rho^{-1}(\rho(\lambda)/t). \quad (8.39)$$

In different limits the function ρ behaves as

$$\begin{aligned} \rho(\lambda) &\to \lambda^{1/(4-D)} && \text{if } \lambda \to 0 \\ \rho(\lambda) &\to (\lambda_c - \lambda)^{-1/(2b')} && \text{if } \lambda \to \lambda_c \\ \rho^{-1}(x) &\to x^{(4-D)} && \text{if } x \to 0 \\ \rho^{-1}(x) &\to \lambda_c - x^{-2b'} && \text{if } x \to \infty. \end{aligned} \quad (8.40)$$

Equations (8.9) and (8.10) imply that

$$\lambda_r(\lambda_c, t) = \lambda_c$$

$$\lambda_r(\lambda, t) \xrightarrow[t \to \infty]{} u(\lambda)/t^{(4-D)} = \frac{g}{(mt)^{4-D}}, \quad \lambda < \lambda_c. \quad (8.41)$$

The solution of Eq. (8.36) can be written as

$$G_2^R(tp, \lambda) = \frac{-2}{t^2} \int_0^t \frac{dx}{x} x^2 \frac{z_1(\lambda_r(\lambda, t/x))}{z_1(\lambda)} \cdot \Delta G_2^R\!\left(xp, \lambda_r\!\left(\lambda, \frac{t}{x}\right)\right). \quad (8.42)$$

We see that when $t \to \infty$, the integral is dominated by the region of finite x, because $\Delta G_R(xp)$ goes to zero fast enough for large x, so that the integral in Eq. (8.42) is dominated by ΔG_2^R evaluated for coupling constants near $\lambda_r(\lambda, t)$. At the critical point $\lambda = \lambda_c$ [Eq. (8.43)] becomes

$$G_2^R(tp, \lambda_c) = -2/t^{2-2C_1} \int_0^t \frac{dx}{x} x^{2-2C_1} \Delta G_2^R(xp, \lambda_c)$$

$$\xrightarrow[t \to \infty]{} \frac{A}{t^{2-2C_1}} \quad (8.43)$$

$$A = -2 \int_0^\infty \frac{dx}{x} x^{2-2C_1} \Delta G_2^R(xp, \lambda_c).$$

When $t \to \infty$ the integral in Eq. (8.43) can be extended to infinity, thus producing a convergent result.

Equation (8.42) can be studied in different limits—e.g., it can be used to recover the correlation functions at the critical point $m^2 = 0$, $g \neq 0$, which we call $G_2^c(p \mid g)$, from the function $\Delta G_2^R(p, \lambda)$ evaluated at $m^2 = g^{2/(4-D)}$. After some algebra we get

$$G_2^c(tp \mid g) = \frac{-2}{t^2} \int_0^\infty \frac{dx}{x} x^2 z_1\left[\lambda_r\left(\tilde{\lambda}, \frac{t}{x}\right)\right] \cdot \Delta G_2^R\left(tp, \lambda_r\left(\tilde{\lambda}, \frac{t}{x}\right)\right) \quad (8.44)$$

where $\tilde{\lambda}$ satisfies the condition $\rho(\tilde{\lambda}) = 1$.[15] In this case we have the simplification $\lambda_r(\tilde{\lambda}, t/x) = \rho^{-1}(x/t)$. Here, moreover, when $t \to \infty$ or $t \to 0$, the integral is dominated by coupling constants near $\lambda_r(\tilde{\lambda}, t)$. Different values of λ_r are relevant in different regions of momentum space, as can be seen by Eq. (8.41).[16]

While in the large-momentum region we recover the perturbative expansion, in the low-momentum region we find $G_2^c(p) \sim p^{-2}(p/g^{4-D})^\eta$ in perfect agreement with the scaling law [Eq. (7.18)]. Similar results can also be obtained for the higher-order correlation functions.

At this stage we can safely use the perturbative expansion for ΔG_2^R in powers of λ, which is reliable only at small momenta, to compute G_2 at all momenta, by using Eq. (8.44). Equations (8.42)–(8.44) look very similar to Eqs. (7.24) and (7.32). This similarity is not only formal. We have a sequence of different theories, labeled by g. Equations (8.42)–(8.44) tell us that if we rescale the momenta by a factor t, the behavior in this momentum region is controlled by the theory with coupling constant equal to $\lambda_r(\tilde{\lambda}, t)$.

In this respect the value of $\lambda_c = \lim_{t \to 0} \lambda_r(\lambda, t)$, is called an infrared stable fixed point, whereas $\lambda = 0$ ($\lim_{t \to \infty} \lambda_r(\lambda, t) = 0$ if $\lambda < \lambda_c$) is called the ultraviolet stable fixed point. These quantities are relevant for studying the theory in the small- and large-momentum regions, respectively.[17]

8.5. The fully renormalized expansion

In this section we show how to compute directly the functions $b(\lambda)$, $C_1(\lambda)$, $C_2(\lambda)$ by using a perturbative expansion in powers of λ and bypassing the g expansion. This approach will be crucial in the next chapter in order to construct the ε expansion ($\varepsilon = 4 - D$).

In order to implement this program we must go through two steps: (a) we must construct the perturbative expansion in powers of λ for the correlation functions of the renormalized field φ_R (this expansion is sometimes called the fully renormalized perturbative expansion); (b) we must express the functions $b(\lambda)$, $C_1(\lambda)$, and $C_2(\lambda)$ in terms of the correlations functions of the renormalized field φ_R low-momentum region. These correlations functions can be computed using the techniques described in (a).

8.5. The Fully Renormalized Expansion

Let us start by considering point (a). Our aim is to modify the rules for evaluating a diagram in order to obtain directly the desired expansion. Indeed, we have already seen that the expansion at fixed z_1/m^2 can be obtained by substituting in all self-energy diagrams $\Sigma(p)$ their subtracted value $(\tilde{\Sigma}(p) \equiv \Sigma(p) - \Sigma(0))$: this construction guarantees the correct normalization of the propagator and corresponds to adding a term $\frac{1}{2}\Sigma(0)\varphi^2$ in the Hamiltonian density (see p. 85).

By the same token, when we construct the perturbative expansion in powers of g for the normalized correlation functions at fixed m^2, we must substitute for $\Sigma(p)$ the so-called renormalized self-energy:

$$\Sigma_R(p) \equiv \Sigma(p) - \Sigma(0) - p^2 \frac{d}{dp^2} \Sigma(p^2)\bigg|_{p^2=0}. \quad (8.45)$$

It is also evident that

$$G_R(p) \equiv 1/(p^2 + m^2 + \Sigma_R(p)) \quad (8.46)$$

satisfies the normalization condition [Eq. (8.3)] and that the subtraction of a term $p^2(d/dp^2)\Sigma(p^2)|_{p^2=0}$ in the propagator can be considered as the effect of changing the coefficient of the term $\frac{1}{2}(\partial_\nu\varphi)^2$ in the Hamiltonian. Indeed, the coefficient of the kinetic term, when the Hamiltonian is expressed in terms of the renormalized field, is no longer equal to $\frac{1}{2}$:

$$(\partial_\nu\varphi)^2 = z_1(\partial_\nu\varphi_R)^2. \quad (8.47)$$

In the same way we can obtain the fully renormalized expansion in powers of λ. If we use the foregoing rules, a diagram (with loops) gives a contribution $\Gamma(p)$ to the connected amputated four-point function. We declare that the contribution of such a diagram is its subtracted value:

$$\Gamma_R(p) = \Gamma(p) - \Gamma(0). \quad (8.48)$$

This rule is correct if there are no subdiagrams with four external legs. If such subdiagrams are present, the rule must first be applied recursively to these subdiagrams, and later to the full diagram, without paying attention to the number of external legs of the full diagram.

It should by now be clear that, if we use the appropriate subtraction scheme for the self-energy and the vertex, we obtain the expansion for the correlation functions of φ_R at fixed λ, by introducing automatically the correct counterterms in the Hamiltonian, without any need of an explicit separate evaluation of the corresponding counterterms. We shall see that this approach will be very important in the next chapter at four dimensions, where the counterterms are divergent (i.e., of order g/ε to some power near four dimensions), whereas no divergences will be present in the fully renormalized expansion.

Step (b) is relatively simpler. The starting point will be the Callan-Symanzik equation of the previous subsection.

Inserting the normalization conditions

$$[G_2^R(p)]^{-1} = p^2 + m^2 + 0(p^4)$$
$$\Gamma_4^R(p) = \lambda m^{4-D} \tag{8.49}$$

in the r.h.s. of the Callan-Symanzik equation, we get

$$\Delta [G_2^R(p)]^{-1}|_{p^2=0} = [1 - C_1(\lambda)]m^2$$

$$\Delta \frac{d}{dp^2}[G_2^R(p)]^{-1}|_{p^2=0} = -C_1(\lambda) \tag{8.50}$$

$$m^{D-4}\Delta\Gamma_4^R(p)|_{p=0} = \frac{4-D}{2}\lambda + b(\lambda) - 2\lambda C_1(\lambda).$$

The last two equations give us the functions $b(\lambda)$ and $C_1(\lambda)$, and the first equation can be used to fix the normalization of Δ. Indeed, we know that Δ is proportional to $d/d\mu$ and that

$$\frac{d}{d\mu}\langle \varphi(x_1)\varphi(x_2)\cdots\varphi(x_N)\rangle_c = \frac{1}{2}\int dy \langle \varphi^2(y)\varphi(x_1)\varphi(x_2)\cdots\varphi(x_N)\rangle_c. \tag{8.51}$$

Therefore

$$\Delta G_N^R(p_1\cdots p_N) = z_1^{-N/2}\Delta G_N(p_1\cdots p_N)$$
$$= FG_{N,\varphi^2}(p_1\cdots p_N), \tag{8.52}$$

where $G_{N,\varphi^2}(p_1\cdots p_N)$ is the correlation function with an extra factor φ^2 at momentum 0, the factor F being fixed by the condition (8.50).

To simplify the notation we introduce the so-called renormalized φ_R and φ_R^2 fields [see Eq. (8.9)]

$$\varphi_R = z_1^{-1/2}\varphi, \qquad \varphi_R^2 = z_2^{-1}\varphi^2. \tag{8.53}$$

As long as $z_1 \neq z_2$, $\varphi_R^2 \neq (\varphi_R)^2$. The notation may sometimes be confusing, but it is a widely accepted one. The renormalized correlation functions are the correlation functions of the renormalized field.

It is easy to see that[18]

8.5. The Fully Renormalized Expansion

$$-\frac{1}{2}[G^R_{2,\varphi^2}(p)]|_{p=0} = m^{-4}. \tag{8.54}$$

Consequently Δ is given by the insertion of the factor $-\frac{1}{2}[1-C_1(\lambda)]\int d^D y\, \varphi_R^2(y)$. We finally obtain[19]

$$\Delta G^R_N = -\frac{m^2}{2}[1 - C_1(\lambda)]G^R_{N,\varphi^2}. \tag{8.55}$$

The repetition of the arguments of the previous sections tells us that

$$-\frac{1}{2}\Delta[G^R_{2,\varphi^2}(p)|_{p=0}m^4] = C_2(\lambda) - C_1(\lambda). \tag{8.56}$$

Let us see how all this works in practice. The insertion of $\frac{1}{2}\varphi^2$ corresponds to replacing a propagator $1/(p^2 + m^2)$ with its derivative with respect to m^2 (i.e., $-1/(p^2 + m^2)^2$; this operation can be symbolically denoted by adding a cross to the propagator.

The relevant diagrams for $\Delta\Gamma_4$ at one and two loops and for C_1 at two loops are shown in Fig. 8.5. These diagrams refer to the g expansion. If we recall the diagrammatical expansion of Γ_4 as a function of g at one loop, we see that the two-loop diagram for $\Delta\Gamma_4$ simplifies to those shown in Fig. 8.6.[20]

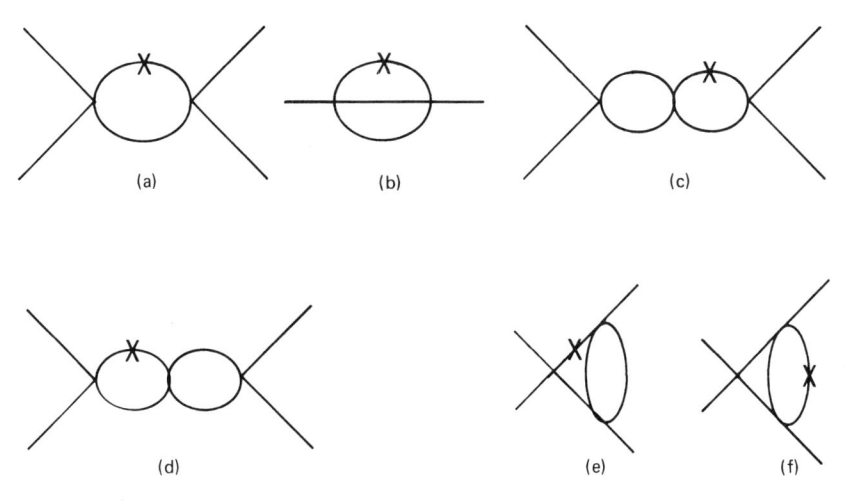

Figure 8.5. The nonvanishing diagrams for $\Delta\Gamma_4$ and $d/dp^2\,\Delta G^{-1}(p)$ in the expansion at fixed m^2; we have not repeated diagrams that differ only in the way we put the crosses on symmetric lines (e.g., Figure 8.5(b) has multiplicity $\frac{1}{2} = 3 \times \frac{1}{6}$ and not $\frac{1}{6}$).

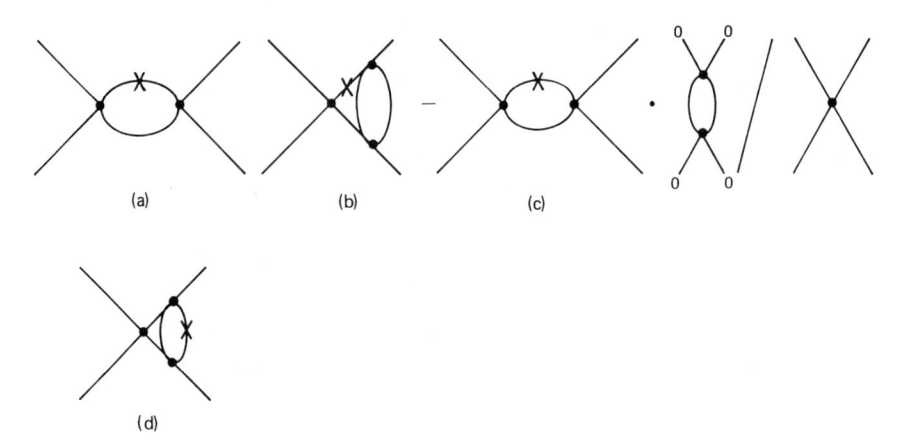

Figure 8.6. Diagrams for the expansion of $\Delta\Gamma_4$ in powers of λ (λ is denoted by a black dot) at the order λ^3. Diagram (c) is the product of the diagram of (a) with the one-loop contribution to the Γ_4 at zero momenta. For practical purposes it is convenient to evaluate (b), (c), and (d) together. The zero in (c) reminds us that the diagram is evaluated at zero external momenta.

We finally find that

$$C_1(\lambda) = \frac{\lambda^2 m^{2(4-D)}}{6(4\pi)^{D/2}} \frac{d}{dp^2} m^2 \frac{d}{dm^2} \int_0^1 da\, db\, dc\, \Gamma(3-D)$$

$$\times \frac{\delta(a+b+c-1)}{(ab+bc+ca)^{D/2}} \left[\frac{p^2 abc}{ab+bc+ca} + m^2\right]^{D-3}$$

$$= \frac{\lambda^2}{6(4\pi)^{D/2}} \Gamma(5-D) \int_0^1 \frac{\delta(a+b+c-1)}{(ab+bc+ca)^{D/2+1}} abc$$

$$m^{D-4}\Delta\Gamma_4^R(0) = m^{D-2} \frac{d}{dm^2} \left\{\frac{3}{2(2\pi)^D} g^2 \int \frac{d^D k}{(k^2+m_2)^2}\right.$$

$$- \frac{3g^3}{(2\pi)^{2D}} \int \frac{dk_1\, dk_2}{(k_1^2+m^2)^2(k_2^2+m^2)((k_1+k_2)^2+m^2)}\bigg\}$$

$$- \frac{3g^3}{(2\pi)^D} \int \frac{d^D k_1}{(k_1^2+m^2)} \frac{m^{D-2}d}{dm^2} \int \frac{d^D k_2}{(k_2^2+m^2)^2}$$

$$= -\frac{3}{2(4\pi)^{D/2}} \lambda^2 \Gamma\left(\frac{6-D}{2}\right) + \frac{3\lambda^3}{(4\pi)^D} \Gamma(5-D) \int_0^1 a\, da\, db\, dc$$

$$\times \delta(a+b+c-1)\left[\frac{1}{(ab+bc+ca)^{D/2}} - \frac{1}{a^{D/2}} \frac{1}{(b+c)^{D/2}}\right],$$

(8.57)

8.5. The Fully Renormalized Expansion

where we have used the expression:

$$3m^{2(D-4)} \frac{g^3}{(4\pi)^D} \Gamma(4-D) \int_0^1 a\,da\,db\,dc\, \frac{\delta(a+b+c-1)}{(ab+bc+ca)^{D/2}} \quad (8.58)$$

for the diagram of Fig. 8.3(d) at zero external momentum.

The number of diagrams to be computed in this approach is smaller than in the previous case; moreover, after terms of different signs are combined in the same integral, the final result is finite when $D = 4$. This fact will be further elaborated upon in the next chapter.

The reader may find it useful to summarize the logic of the approach described in this chapter. We recall that our theory depends on two parameters m and g. We are interested in studying the limit $m \to 0$ of the correlation functions at a fixed external momentum p (which may eventually be zero).

Our first step was to study the theory at fixed p/m in the limit m going to zero; in this case the theory is parametrized by the other dimensionless quantity, i.e., $u = g/m^{4-D}$, which in fewer than four dimensions becomes infinite when m goes to zero. It is convenient to parametrize the theory in terms of its correlation functions. This has been done by introducing $\lambda = \Gamma_4^R/m^{4-D}$, which can be proven to remain finite at the phase transition point (i.e., $m = 0$), where it becomes equal to λ_c. Especially near four dimensions, λ_c is quite small, and the expansion in powers of λ is quite efficient for obtaining the correlation functions at $\lambda = \lambda_c$ at fixed p/m.

It is crucial to understand that if we measure everything in units of m, the large-momentum behavior of the theory changes drastically when λ becomes equal to λ_c, so that the perturbative expansion in powers of λ cannot be used directly to compute what happens in the region of large p/m, and more specifically at fixed p when m goes to zero.

The usefulness of this approach is, however, based on two facts: (a) the Callan-Symanzik equation makes it possible for us to compute the correlation functions in the high-momentum region, using as input the correlation functions in the low-momentum region. The identification of λ_c is an easy task. (b) The previous remarks imply that the λ expansion is not uniformly convergent for all values of p, especially in the large p/m region. As long as the equations of motion (see p. 164) give the correlation functions in the low-momentum region as integrals over the correlation functions at all momenta, the high-momentum region included, we may fear that the nonuniformity of convergence of the λ expansion may propagate from the high-momentum region to the low-momentum region. Although this reflection is perfectly correct, we are partially protected against this effect: indeed, we have expressed the low-momentum correlation functions using as parameters the low-momentum correlation functions themselves evaluated at some fixed external momenta [i.e., $G(0)$, $p^2(d/dp^2)G(p)|_{p^2=0}$, $\Gamma_4(0)$]. An explicit computation (see below, note 5) shows that the final effect is a weak cut at $\lambda = \lambda_c$ of the form $(\lambda_c - \lambda)^\delta$ with $\delta \gg 1$.

We have studied the theory at fixed p/m in order to obtain information on the theory at fixed p in the limit m going to zero. This detour was necessary because infrared divergences appear in the perturbative expansion at fixed p in the limit m going to zero, as soon as the order of the perturbative expansion is greater than $2/\varepsilon$, as we have already noticed in Chapter 6. A careful study of Eq. (8.45) shows that the final result is not C^∞ in the coupling constant g, and terms proportional to $g^{2/\varepsilon}$ appear naturally [in three dimensions we can get terms proportional to $g^3 \ln(g)$ or something similar].[21]

When the dimension goes to four ($\varepsilon = 4 - D \to 0$), these problems fade because the order in g at which they appear goes to infinity as $2/\varepsilon$. As we shall see in the next chapter, an ε expansion for the critical properties can be constructed working directly in the massless theory, but as soon as ε is different from zero, we need to consider the massive ($m \neq 0$) theory and to follow the approach described in this chapter.

Notes for Chapter 8

1. In this approach most of the computations are performed in momentum space.
2. D. I. Kazakov, O. V. Tarasov, and A. A. Vladimirov, *Sov. Phys. JEPT* **50**, 516 (1979).
3. The fixed-dimensions approach has been introduced in this context in G. Parisi, *Cargèse Lectures Notes* (1973), published with some modifications as *J. Stat. Phys.* **23**, 49 (1980). In this chapter we follow this paper quite closely; see also G. Parisi, *Nuovo Cimento A* **21**, 179 (1974).
4. At $g = 0$, $m^2 = \mu$. Here m is not the magnetization; the notation comes from relativistic field theory, where m is the mass of the particle associated with the field, as we shall see in Chapter 15.
5. The existence of these extra terms near $T = T_c$ explains why Eqs. (8.7) and (8.8) are a much more efficient way to find λ_c than defining $M(\lambda) = \lambda/u$ and searching for a zero of $M(\lambda)$. Equation (8.10) tells us that, unless $(4 - D)/2b'$ is an integer (which is not the general case), $M(\lambda)$ has a cut at $\lambda = \lambda_c$ proportional to $(\lambda_c - \lambda)^{(4-D)/2b'}$: the perturbative expansion in λ will be poorly convergent near $\lambda = \lambda_c$.

Indeed, a perturbative approach may work only if the functions we need to compute are as smooth as possible. On the other hand the existence of other irregular terms in $\lambda(u)$ (i.e., $\lambda(u) = \lambda - a_1 u^{-t_1}\omega_R + a_2 u^{-t_2}\omega_R + \cdots$ for $u \to \infty$) tells us that a residual singularity of the form $(\lambda - \lambda_c)^{t_2/t_1}$ is expected to be present in $b(\lambda)$. Fortunately there are indications that there is a strong enough gap between the first and the second attractive eigenvalues so that the cut at $\lambda = \lambda_c$ has a high enough power to be neglected, i.e., t_2/t_1 is relatively large (see Section 11.3).

6. An easy computation tells us that the Kurtosis $(\langle \varphi_n^4 \rangle - 3\langle \varphi_n^2 \rangle^2)/\langle \varphi_n^2 \rangle^2$ of a block spin variable of side $L_n \gg \xi$ is given by $-\lambda_c (\xi/L_n)^D$. This formula permits a numerical determination of λ_c, using the Monte Carlo techniques described in the last chapter of this book; one finds that λ_c/λ_c^1 is 1.42 ± 0.07 and 1.80 ± 0.06 in three and two dimensions, respectively (F. Cooper, B. Freedman, and D. Preston, *Nucl. Phys. B 210* [FS6], 210 (1982).

7. In order to compute the N-dependent multiplicity it may be convenient first to compute the multiplicity at $N = 1$, and later to use the graphical representation (see subsection 6.3) for the vertex to extract the dependence on N, using the value at $N = 1$ as normalization.

8. G. A. Baker, B. G. Nickel, M. S. Green, and D. I. Meiron, *Phys. Rev. Lett. 36*, 1351 (1976); G. A. Baker, B. G. Nickel, and D. I. Meiron, *Phys. Rev. B 17*, 1365 (1978); B. G. Nickel, *J. Math. Phys. 19*, 542 (1978).

9. E. Brézin and G. Parisi, *J. Stat. Phys. 19*, 269 (1979).

10. J. C. Le Guillou and J. Zinn-Justin, *Phys. Rev. Lett. 39*, 95 (1977).

11. Using the exact expressions for the correlation functions of the Ising model, it is possible to evaluate λ_c analytically. Unfortunately this interesting computation has not yet been done.

12. This last statement contains most of the information coming from the scaling law.

13. This equation is the backbone of our approach; it has been proven in K. Symanzik, *Commun. Math. Phys. 18*, 222 (1970) [see also *Commun. Math. Phys. 23*, 61 (1971)] and in J. Callan, *Phys. Rev. D 1*, 1541 (1971). Here we have essentially followed Symanzik's proof, which is much simpler.

14. If $m^2 \to 0$ with fixed p^2 and g, $u \to \infty$, $\lambda \to \lambda_c$, and $p^2/m^2 \to \infty$.

15. A different value of m^2 could also be used; the condition $\rho(\tilde{\lambda}) = 1$ becomes now $\rho(\tilde{\lambda}) = g/m^{4-D}$.

16. For this reason $\lambda_r(\lambda, x)$ is called the effective running coupling constant.

17. The necessary conditions for $\lambda_U(\lambda_I)$ being an ultraviolet or infrared stable fixed point, are $b(\lambda_I) = 0$ [$b(\lambda_U) = 0$] and $b'(\lambda_I) \geq 0$ [$b'(\lambda_U) \leq 0$], respectively. If $b(\lambda) = b'(\lambda) = 0$ we have to consider higher derivatives.

18. If no indication is given of the momentum carried by the φ_R^2 field, the momentum is assumed to be automatically zero.
19. At the two-loop level the factor $1 - C_1(\lambda)$ can be nearly everywhere replaced by 1, in the evaluation of Δ, $C_1(\lambda)$ being of order λ^2.
20. Indeed, by substituting in diagram 8.6(a) the expression of λ as a function of g, we generate new terms, which cancel diagram 8.6 and reproduce the missing diagrams [8.5 and its symmetric partner 8.5(d); see Fig. 8.7].

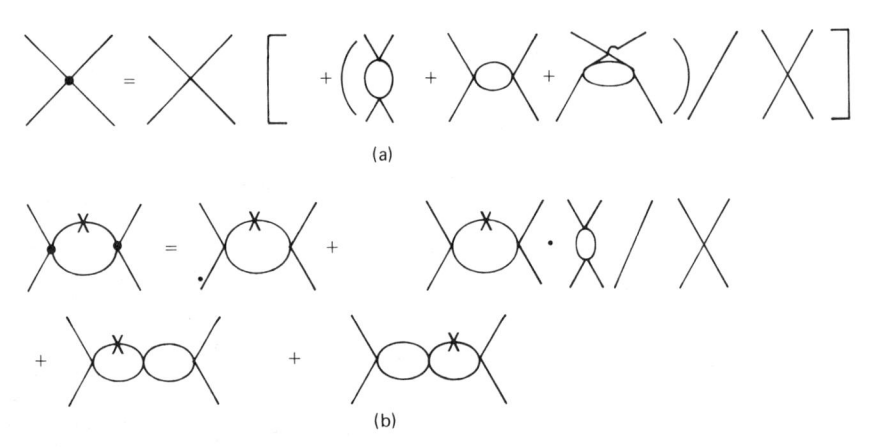

Figure 8.7. $\Gamma_4(0)$ as function of g is shown in (a). The value of (a) in terms of the g expansion at order g^3 is shown in (b). See footnote 20.

21. A careful discussion of the analyticity properties in g is contained in K. Symanzik, *Lett. Nuovo Cimento* **8**, 771 (1973) and "Cargèse Lecture Notes 1973," unpublished; see also G. Parisi, *Nucl. Phys. B* **254**, 58 (1985), and references therein.

CHAPTER 9

Near Four Dimensions

9.1. Noninteger dimensions

In the previous chapter we gathered evidence for the stability of the Gaussian fixed point for $D > 4$ and for the existence of a stable fixed point that bifurcates from it as soon as $D < 4$. In particular, the one-loop $b(\lambda)$ function is [neglecting multiplicative factors of $O(1)$]

$$2b(\lambda) = -(4-D)\lambda + \lambda^2 . \tag{9.1}$$

The infrared stable fixed point λ_c (i.e., the value of the renormalized coupling when $m^2 \to 0$) is given by

$$\begin{aligned} \lambda_c &= 4 - D, & D < 4, \\ \lambda_c &= 0, & D \geq 4. \end{aligned} \tag{9.2}$$

Equation (9.2) suggests that it may be possible, in the same way as in the hierarchical model, to study what happens for small $\varepsilon = 4 - D$ and to build an expansion[1] for critical exponents in powers of ε. Unfortunately, to carry on this program we must work in a space of noninteger dimensions, which is not a clearly defined mathematical entity.

Different points of view are possible: noninteger dimensional spaces are just a useful trick without any physical mathematical meaning; or noninteger dimensional spaces are a natural extension of integer dimensional spaces and they will sooner or later receive the place they deserve in mathematical textbooks.[2] Although it is not clear at present what kind of animal a noninteger dimensional space is, it is straightforward to compute the various perturbative expansions in such a space. The coefficients of the high-temperature expansion are polynomials[3] in D, and we can just evaluate them at whatever value of D we want.[4]

In the normal perturbative expansion of the Landau-Ginsburg model, a diagram with ℓ loops can be written as

$$\int d\alpha_i \frac{1}{[4\pi\mathcal{D}(\alpha)]^{D/2}} \exp\left[-\frac{\mathcal{N}(\alpha, p)}{\mathcal{D}(\alpha)} + m^2 \Sigma \alpha_i\right], \qquad (9.3)$$

where there is an α_i associated with each internal line, $\mathcal{D}(\alpha)$ and $\mathcal{N}(\alpha, p)$ are polynomials of order l and $l+1$, respectively, in the α's, $\mathcal{N}(\alpha, p)$ depends linearly on the invariants $P_{ik} = p_i \cdot p_k$, and p_i and p_k are the external momenta.[5]

Equation (9.3) has been derived for integer dimensions, but its generalization to noninteger dimensions does not present serious difficulties, $\mathcal{D}(\alpha)$ being always non-negative. The high- and low-temperature expansions are convergent, and the perturbative expansion is Borel summable (at least in integer dimensions), so we can hope that they can be used to define the theory. A direct definition is more problematic. One would be very happy if it could be proven that there is only one smooth function, which interpolates between integer dimensions.[6]

Another possibility consists in using the rotational invariance of the theory. The correlation functions $\langle \varphi(x_1)\varphi(x_2)\cdots\varphi(x_N)\rangle$ are functions of the invariants $X_{ik} = x_i x_k$. We can therefore consider them as functions of the X_{ik} without reference to the original space. Now there are identities that connect correlation functions of different orders (the so-called equations of motion); they can be derived from the identity[7]

$$0 = \int d[\varphi] \frac{\delta}{\delta\varphi(y)} [\varphi(x_1)\cdots\varphi(x_N)] \exp[-H[\varphi]]$$

$$= \int d[\varphi] \left\{ \sum_i^N \left[\prod_{k\neq i}^N \varphi(x_k)\right] \cdot \delta(y - x_i) - \prod_k^N \varphi(x_k) \frac{\delta H}{\delta\varphi(y)} \right\} \exp[-H[\varphi]]$$

$$\propto \left\langle \sum_i^N \left[\prod_{k\neq i}^N \varphi(x_k)\right] \delta(y - x_i) - \prod_k^N \varphi(x_k) \frac{\delta H}{\delta\varphi(y)} \right\rangle. \qquad (9.4)$$

The equations (9.4) in the Landau-Ginsburg model connect correlation functions of order $N+1$ with correlation functions of order $N+3$ and $N-1$; they form a set of coupled differential equations for the correlation functions, e.g.,

$$(-\Delta_y + m^2)\langle \varphi(x)\varphi(y)\rangle + \frac{g}{3!}\langle \varphi^3(y)\varphi(x)\rangle = \delta(x - y),$$

$$(-\Delta_y + m^2)\langle \varphi(x_1)\varphi(x_2)\varphi(x_3)\varphi(y)\rangle + \frac{g}{3!}\langle \varphi^3(y)\varphi(x_1)\varphi(x_2)\varphi(x_3)\rangle$$

$$= \delta(x_1 - y)\langle \varphi(x_2)\varphi(x_3)\rangle + \delta(x_2 - y)\langle \varphi(x_1)\varphi(x_3)\rangle \qquad (9.5)$$

$$+ \delta(x_3 - y)\langle \varphi(x_1)\varphi(x_2)\rangle$$

and so on. Similar equations hold also for the higher-order correlation functions.

It is believed that the "equations of motion" (9.4) have only one solution (the correct one) if appropriate conditions are imposed. Perhaps it will be possible to transform Eq. (9.4) into equations for the correlation functions as functions of the X_{ik} and to prove the existence and uniqueness theorem for noninteger dimensions as well. It seems rather doubtful, however, that the final correlation functions could be interpreted as expectation values of a probability measure when D is not an integer.

Irrespective of the success of this rigorous construction, we can safely use noninteger dimensional spaces as useful tools for computing the properties of theories defined in integer dimensions. In this spirit we proceed now to construct the $4 - \varepsilon$ expansion. The first step will be the study of the renormalized perturbative expansion at $D = 4$.

9.2. At four dimensions

In the preceding chapter we claimed that the perturbative expansion for the renormalized correlation functions in powers of the coupling constant λ is well defined (no ultraviolet divergences are present) in four dimensions. The full proof of this fact can be found in many books on field theory; here we limit ourselves to a brief description of the various ingredients of the proof as far as the correlation functions of the field φ_R are concerned.[8]

The basic ingredient is the BHP theorem. In simplified form it states that[9] if a diagram contains no divergent subdiagrams ($\mathscr{D} < 0$ for each subdiagram) [cf. Eq. (5.35)] and the diagram itself is not divergent ($\mathscr{D} < 0$), the diagram is finite when $\Lambda \to \infty$ and $\mu \neq 0$.[10] If a diagram contains no divergent subdiagrams, but the diagram itself is divergent,[11] the diagram is finite in configuration space, but the divergences arise in the Fourier transform.[12] Its renormalized value is defined by

$$V_R(p) = V(p) - \sum_{n\,0}^{N(\mathscr{D})} \frac{1}{n!} \left(\frac{\partial}{\partial p}\right)^n V(p) \cdot p^n , \qquad (9.6)$$

where $V(p)$ is the value of the diagram (which diverges when $\Lambda \to \infty$), $N(\mathscr{D})$ is the largest integer less than or equal to $\mathscr{D}$, and by $(\partial/\partial p)^n$ we denote all possible derivatives of order n with respect to the external momenta p. In other words, we subtract from the diagram its Taylor expansion truncated at the order $N(\mathscr{D})$ around the point $p = 0$.[13] Now if a diagram is divergent and contains divergent subdiagrams, the renormalized value of the diagram is defined by applying Eq. (9.6) to the whole diagram and to each of the subdiagrams with $\mathscr{D} \geq 0$.

The BHP theorem, which generalizes the considerations of subsection 5.5, states that the renormalized value of a diagram is finite when $\Lambda \to \infty$. The proof is quite delicate and involves careful combinatorics; it will not be reproduced here.

Now the essence of the BHP theorem is that any possible ultraviolet divergences may be removed from the theory by adding to the Hamiltonian g-dependent counterterms (which diverge when $\Lambda \to \infty$); moreover, such counterterms are polynomial in φ and in its derivatives. Indeed, the value of a diagram can be changed by adding a counterterm in the Hamiltonian, as was done in Sec. 8.5 for the expansion in the renormalized coupling. Let us give an example. Consider the diagram in Fig. 9.1 for the 6-point amputated one-line irreducible correlation function. The value of the diagram, neglecting the δ functions, is given by

$$V(q_1, q_2, q_3) = \frac{g^3}{(2\pi)^D} \int \frac{d^D k}{[(k+q_1)^2 + m^2][(k+q_2)^2 + m^2][k^2 + m^2]}$$

$$q_1 = p_1 + p_2, \qquad q_2 = p_3 + p_4, \tag{9.7}$$

$$q_3 = p_5 + p_6 = -q_1 - q_2.$$

The diagram has a degree of divergence $\mathscr{D} = D - 6$: For $6 \leq D < 8$ its normalized part is defined as

$$V_R(q_1, q_2, q_3) = \frac{g^3}{(2\pi)^D} \int d^D k$$

$$\times \left[\frac{1}{[(k+q_1)^2 + m^2][(k+q_2)^2 + m^2][k^2 + m^2]} - \frac{1}{(k^2 + m^2)^3} \right]. \tag{9.8}$$

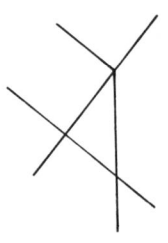

Figure 9.1. The first nontrivial diagram for the 6-point amputated one-line irreducible correlation function computed in the text Eq. (9.7).

9.2. At Four Dimensions

The extra term we have subtracted can be considered to have originated in a term $(A/6!) \int d^D x \, \varphi^6(x)$ in the Hamiltonian with

$$A = \frac{g^3}{(2\pi)^D} \int d^D k \, 1/(k^2 + m^2)^3 . \tag{9.9}$$

If this prescription is used, the theory will be defined perturbatively in any dimensions; however, the Hamiltonian we must use may be very complicated and contains every possible functional dependence on p. We want to consider here theories in which the Hamiltonian in the presence of the counterterm has the same form as the original Hamiltonian[14] (such theories are called renormalizable). If $D > 4$, Eq. (5.36) for $\mathcal{D}$ in the φ^4 theory $[\mathcal{D} = (D-4)L + 4 - E]$ tells us that, going to a high enough order in g, all diagrams are divergent: an infinite number of counterterms is needed, and the theory is not renormalizable.[15] If $D = 4$, the only divergent diagrams are the self-energy ($\mathcal{D} = 2$) and the 4-point vertex ($\mathcal{D} = 0$). The degree of divergence of these diagrams is independent of the order in g, and all powers of the coupling constant g will be present in the counterterms; the counterterms cannot be computed in a closed form. A simpler situation holds in fewer than four dimensions: only a finite number of diagrams is divergent (the degree of divergence decreases with the order in g), and the counterterms needed to remove the divergences are a polynomial in g less than or equal to $2/(4-D)$. In this last case the theory is said to be super-renormalizable.

In the super-renormalizable case a knowledge of the needed counterterms makes the construction of rigorous theorems (like the Borel summability of the perturbative expansion) relatively easy. On the other hand, for renormalizable theories, lack of control over the counterterms beyond the perturbation theory makes life very hard for those who try to construct rigorous results. As we shall see later the very existence of the non-Gaussian theory has been questioned in the infinite cutoff limit.

The general rule is that a theory is renormalizable only if the coupling constant is dimensionless; the allowed counterterms are proportional to

$$\int d^D x \, O_a(x) f_a(g, \Lambda) \tag{9.10}$$

where $O_a(x)$ is a local operator[16] of dimensions $d_a \leq D$, and

$$\begin{aligned} f_a(g, \Lambda) &\sim \ln^n \Lambda \qquad \Lambda \to \infty \text{ if } d_a = D \\ f_a(g, \Lambda) &\sim \Lambda^{D-d_a} \qquad \Lambda \to \infty \text{ if } d_a < D , \end{aligned} \tag{9.11}$$

as can be seen by dimensional considerations (the Hamiltonian, being the argument of an exponential, must be dimensionless).

In the $D = 4$ Landau-Ginsburg model, the only operators of dimensions d fewer than or equal to 4 are[17] φ^2, φ^4, and $(\partial_\nu \varphi)^2$.

The BHP renormalized perturbative expansion can thus be constructed at $D = 4$ by using (as in the preceding chapter) order-by-order in perturbation theory a renormalized propagator and a renormalized vertex such that

$$G_R(p) = \frac{1}{(p^2 + m^2 + O(p^4))} \qquad \Gamma_R(p)|_{p=0} = \lambda . \qquad (9.12)$$

This can be done by subtracting from the contribution coming from the loops[18] the value for the self-energy and for the vertex at zero external momentum; for the self-energy we must also subtract the term of the Taylor expansion in p^2 proportional to p^2, the renormalized self-energy being fixed by the condition

$$\Sigma_R(p) = O(p^4) . \qquad (9.13)$$

A careful combinatoric analysis shows that the prescription we have just described produces the correct expansion for the correlation functions of the renormalized fields in powers of the renormalized coupling constant;[19] a full proof of this statement and of the finiteness of the BHP renormalization can be found in many books on field theory. It can be proven that the existence (and finiteness) of the perturbative expansion for four-dimensional theory implies that the correlation functions satisfy (at least perturbatively) the Callan-Symanzik equation (see p. 152) in four dimensions, and the functions $b(\lambda)$ and $C(\lambda)$ of the preceding chapter have a finite perturbative expansion; this last result will be the basis of the expansion of the critical exponents around four dimensions (the so-called ε expansion).[20]

We stress that the existence of the perturbative expansion by no means implies the existence of the theory in four dimensions. Indeed, the counterterms we must add to the Hamiltonian in order to have a finite limit when the cutoff Λ (or the inverse of the lattice spacing a) goes to infinity, have a formal expansion in powers of the coupling constant λ and $\ln \Lambda$ (or $-\ln a$); this expansion clearly breaks down when $\ln \Lambda$ goes to infinity at fixed λ. A careful examination shows that we can write

$$g = \lambda \sum_{k,j}^{\infty} a_{k,j} (\lambda \ln \Lambda)^k \lambda^j \qquad (9.14)$$

(a similar formula holds also for z), and perturbation theory can therefore be trusted only in the region

$$\lambda \ln \Lambda < O(1) . \qquad (9.15)$$

9.3. The Massless Theory

The renormalized perturbative expansion succeeds in confining all the divergences in the counterterms, leaving us with a finite expansion for the renormalized correlation functions if expanded in powers of another renormalized correlation function (i.e., $\lambda = \Gamma_R^4(0)$). Unfortunately this is not enough to define a theory. The renormalized perturbative expansion is not convergent (as usual), and there are strong indications that for a wide class of theories (among them the $g\varphi^4$ interaction) the perturbative expansion cannot be summed à la Borel (see p. 87).[21] At present no one knows how to construct a procedure such that the resummed correlation functions can be interpreted as the moments of a positively defined probability distribution, i.e.,

$$\langle \varphi(x_1) \cdots \varphi(x_n) \rangle = \int d\mu[\varphi] \varphi(x_1) \cdots \varphi(x_n). \tag{9.16}$$

If an arbitrary resummation technique is used, there is no reason why the measure $d\mu[\varphi]$ (if it exists) should be positive.

The problem of the existence of a finite theory in the infinite cutoff limit is not extremely relevant for purely statistical mechanics (with applications to real systems), because in this framework the size of the atoms produces a natural cutoff. This problem is, however, of the greatest importance as far as the applications of this formalism to relativistic quantum field theory are concerned (see Chapter 15). Here no cutoff is apparently present: the role of the lattice is played by ordinary space-time, which is assumed to be continuous at all scales.[22] Although no rigorous results are present, arguments based on the renormalization group have been used to throw some light (as we shall see soon) on this long-debated question.

9.3. The massless theory

We have already seen that if we try to construct the perturbative expansion for the massless (critical) theory (i.e., $m^2 = 0$) in the presence of a cutoff, we find serious problems in fewer than four dimensions: higher-order diagrams are infrared divergent (i.e., the integrals are too singular at zero momentum). On the other hand, if we stay at four and higher dimensions, the diagrams that we have previously considered (cf. p. 106) produce no infrared divergences. A careful power counting argument shows that this result (i.e., the absence of infrared divergences in dimensions greater than or equal to four) is valid at all orders of the perturbative expansion.

The typical result of the perturbative expansion for the two-field correlation function in more than four dimensions ($\varepsilon < 0$) at smaller momenta will be

$$G_2^{-1}(p) = p^2 + g(\Lambda^{-\varepsilon}p^2 + p^{2-\varepsilon} + p^4\Lambda^{-\varepsilon} + \cdots)$$
$$+ g^2(\Lambda^{-2\varepsilon}p^2 + \Lambda^{-2\varepsilon+2}p^4 + \cdots + p^{2-\varepsilon}\Lambda^{-\varepsilon} + p^{2-2\varepsilon}) \quad (9.17)$$

$$D = 4 - \varepsilon,$$

where Λ is the cutoff. This expansion may be rearranged as

$$G_2^{-1}(p) = \sum_k^\infty [f_k^{(0)}(g\Lambda^{-\varepsilon})p^{2-k\varepsilon}\Lambda^{k\varepsilon} + f_{(k)}^{(1)}(g\Lambda^{-\varepsilon})p^{4-k\varepsilon}\Lambda^{-2+k\varepsilon} + \cdots],$$
$$(9.18)$$

where the functions $f_k^{(j)}$'s can be computed without difficulty in perturbation theory. In the small-momentum limit only the first term survives ($\varepsilon < 0$) and the Gaussian model is recovered. This result shows that the critical exponent η takes the classical value (i.e., $\eta = 0$) in more than four dimensions; similar arguments can be put forward for the other exponents.

In four dimensions the situation is somewhat more complex, and logarithmic corrections are present:[23]

$$G_2^{-1}(p) = p^2 \sum_k^\infty f_k^{(0)}(g)\left(\ln\frac{p}{\Lambda}\right)^k + \frac{p^4}{\Lambda^2}\sum_k^\infty f_k^{(1)}(g)\left[\ln\frac{p}{\Lambda}\right]^k + \cdots.$$
$$(9.19)$$

These corrections may be very relevant in the small-momentum region because by summing logarithms we can get any power we want. In order to study what happens in this region, it is convenient to consider the renormalized perturbative expansion as previously done. Unfortunately, it is not consistent to impose the conditions given by the equations directly at $m = 0$ because $\partial G^{-1}/\partial p^2$ becomes infinite in perturbation theory when p goes to zero. The normalization conditions must be imposed at a nonzero external momentum; for example, we can require that [24]

$$\Sigma(0) = 0 \Leftrightarrow G^{-1}(0) = 0$$

$$\Sigma(p^2)|_{p^2=M^2} = 0 \Leftrightarrow G^{-1}(p^2)|_{p^2=M^2} = M^{-2} \quad (9.20)$$

$$\Gamma_4(p)|_{p_i^2=M^2, p_i p_{j_{i\neq j}} = -\frac{1}{3}M^2} = \lambda.$$

The first condition implies that the theory remains massless, while the other two are substitutes for Eqs. (9.12) and (9.13): They give the

9.3. The Massless Theory

normalization of the propagator and define the renormalized coupling constant λ, respectively. With these normalizations the infinite cutoff limit can be taken in the perturbative expansion in the same way as for the massive theory; the typical result one obtains for the two-point correlation function is

$$G_2^{-1}(p) = p^2 \sum_{0}^{\infty} {}_k f_k(g)\left[\ln\left(\frac{p}{M}\right)\right]^k. \tag{9.21}$$

At first sight it appears that something strange has happened: the original theory depends on only one parameter, i.e., the coupling constant, while apparently the renormalized theory depends on two parameters, the coupling constant and the renormalization point M. The paradox disappears when we realize that the value of M only controls the convention we have followed in defining the renormalized correlation functions; in other words, if we use a different value of M (e.g., $\tilde{M}$) there must be a different value of the renormalized coupling constant (e.g., $\tilde{\lambda}(\lambda, M, \tilde{M}) \equiv \tilde{\lambda}(\lambda, \tilde{M}/M)$ such that the correlation functions evaluated at $\tilde{\lambda}(\lambda, \tilde{M}/M)$, $\tilde{M}$ are proportional to those evaluated at λ, M. This statement may be transcribed by saying that there are two functions, $\tilde{\lambda}(\lambda, \tilde{M}/M)$ and $z_1(\lambda, \tilde{M}/M)$, such that the correlation functions satisfy the equations[25]

$$z_1(\lambda, \tilde{M}/M)G_2(p \mid \tilde{\lambda}, \tilde{M}) = G_2(p \mid \lambda, M)$$

$$z_1^2(\lambda, \tilde{M}/M)G_4(p \mid \tilde{\lambda}, \tilde{M}) = G_4(p \mid \lambda, M) \tag{9.22}$$

$$z_1^{-2}(\lambda, \tilde{M}/M)\Gamma_4(p \mid \tilde{\lambda}, \tilde{M}) = \Gamma_4(p \mid \lambda, M) \ldots$$

The functions $\tilde{\lambda}$ and z_1 can be easily computed using the conditions stated in Eq. (9.20). We find

$$\tilde{M}^{-2} = G_2(\tilde{M} \mid \tilde{\lambda}, \tilde{M}) = z_1^{-1}\left(\lambda, \frac{\tilde{M}}{M}\right)G_2(\tilde{M} \mid \lambda, M)$$

$$\tilde{\lambda} = \Gamma_4(\tilde{M} \mid \tilde{\lambda}, \tilde{M}) = z_1^2\left(\lambda, \frac{M}{M}\right)\Gamma_4(\tilde{M} \mid \lambda, M). \tag{9.23}$$

These equations imply

$$\tilde{\lambda}\left(\lambda, \frac{\tilde{M}}{M}\right) = \Gamma_4(\tilde{M} \mid \lambda, M)[G(\tilde{M} \mid \lambda, M)\tilde{M}^2]^2$$

$$z_1\left(\lambda, \frac{\tilde{M}}{M}\right) = [G(\tilde{M} \mid \lambda, M)\tilde{M}^2]^{-1}. \tag{9.24}$$

It is convenient to study Eqs. (9.22) in differential form, i.e., by taking the derivative with respect to M at fixed λ and $\tilde{M}$; in this way we find that

$$\left[-n \frac{\partial}{\partial M} [\ln z_1(\lambda, \tilde{M}/M)] + z_1(\lambda, \tilde{M}/M)^{-n} \frac{\partial \tilde{\lambda}(\lambda, \tilde{M}/M)}{\partial M} \frac{\partial}{\partial \tilde{\lambda}} \right]$$
$$\times G_{2n}(p \mid \tilde{\lambda}, \tilde{M}) = \frac{\partial}{\partial M} G_{2n}(p \mid \lambda, M). \tag{9.25}$$

If we evaluate Eq. (9.25) at $\tilde{M} = M$ we get

$$M \frac{\partial}{\partial M} G_{2n}(p \mid \lambda, M) = \left[-nC(\lambda) + b(\lambda) \frac{\partial}{\partial \lambda} \right] G_{2n}(p \mid \lambda, M)$$

$$C(\lambda) = M \frac{\partial}{\partial M} z_1(\lambda, \tilde{M}/M)\big|_{\tilde{M}/M=1}, \tag{9.26}$$

$$b(\lambda) = M \frac{\partial}{\partial M} \tilde{\lambda}(\lambda, \tilde{M}/M)_{\tilde{M}/M=1}.$$

Now dimensional analysis tells us that the correlation functions are homogeneous functions of fixed degree in M and the momenta: the derivative with respect to M may be traded for a derivative with respect to the momenta. One finally finds the equation

$$r \frac{\partial}{\partial r} G_{2n}(rp \mid \lambda, M) = \left[+D - n(2 + D - C(\lambda)) + b(\lambda) \frac{\partial}{\partial \lambda} \right] G_{2n}[rp \mid \lambda, M] \tag{9.27}$$

$$r \frac{\partial}{\partial r} \Gamma_4(rp \mid \lambda, M) = \left[4 - D - 2C(\lambda) + b(\lambda) \frac{\partial}{\partial \lambda} \right] G_{2n}(rp \mid \lambda, M).$$

The reader will notice that Eq. (9.27) is essentially the Callan-Symanzik equation [cf. Eq. (8.36)] where the r.h.s. has been neglected; this is consistent with the statement (made in the preceding chapter) that the r.h.s. of the Callan-Symanzik equation is negligible with respect to each term of the l.h.s. in the regime where m^2 goes to zero.

The final result [which could also have been obtained directly from Eqs. (9.23) and (9.24)] can be derived by integrating Eq. (9.27) using the method of characteristic curves (see p. 152):

$$G_{2n}(rp \mid \lambda, M) = z_1(\lambda, r)^n r^{-n(2+D)+D} G_{2n}(p \mid \lambda^2, M)$$

$$z_1(\lambda, r) = \exp \int_1^{\ln r} dt \, C(\tilde{\lambda}(\lambda, t)) = \exp \int_\lambda^{\tilde{\lambda}(\lambda, r)} \frac{d\bar{\lambda}}{b(\bar{\lambda})} C(\bar{\lambda}) \tag{9.28}$$

$$\frac{\partial \tilde{\lambda}}{\partial \ln r} = b(\tilde{\lambda}) \Leftrightarrow \int_\lambda^{\tilde{\lambda}} d\bar{\lambda} [b(\bar{\lambda})]^{-1} = \ln r.$$

9.3. The Massless Theory

The function $\tilde{\lambda}(\lambda, r)$ is often called the effective running coupling constant because it is the appropriate coupling constant for constructing the perturbative expansion at the scale r. We thus arrive at the conclusion that there is essentially only one massless theory, i.e., theories with different renormalized coupling constants may be considered as the same theory observed at two different scales.

It is evident that the small- and large-momentum behaviors of the theory are controlled by the functions C and b; if we compute these functions to the first nontrivial order, we find that[26]

$$b(\lambda) = \frac{N+8}{6} \frac{\lambda^2}{(4\pi)^2}$$

$$C(\lambda) = \frac{N+2}{36} \frac{\lambda^2}{(4\pi)^4}.$$

(9.29)

The corresponding $\tilde{\lambda}$ and z_1 functions are

$$\tilde{\lambda}(\lambda, r) = \frac{\lambda}{\lambda - [(N+8)/6]\lambda^2/4\pi^2 \ln r},$$

$$z_1(\lambda, r) = \exp \frac{N+2}{6(N+8)} \left[\frac{\tilde{\lambda}(\lambda, r) - \lambda}{(4\pi)^2} \right].$$

(9.30)

We see that in this approximation $\tilde{\lambda}(\lambda, r)$ goes to zero when the momentum goes to zero, while it has a pole at large momentum; this last feature is clearly nonsense because it can be rigorously proven that the correlation functions must be bounded at nonzero momenta. Indeed, it is clear that the perturbative expansion can be trusted only in the region where the renormalized coupling constant is small (in this case in the small-momentum region), not where the coupling constant becomes large.

For the time being let us forget all the problems connected with behavior at large momenta and let us concentrate on the small-momentum region. The positivity of the one-loop contribution to $b(\lambda)$ implies that, at least for $\lambda \ll 1$, $b(\lambda)$ must be positive, and therefore the running coupling constant decreases when the momentum goes to zero. Therefore, if λ is not too large, by decreasing the momentum, we will always be driven to the small coupling region where the perturbative expansion can be used. We finally find

$$G_2(p) \sim \frac{1}{p^2} \left[1 + O\left(\frac{1}{\ln(p^2/M^2)}\right) \right].$$

(9.31)

The critical exponent η also takes the classical value in four dimensions. Similar results can be obtained for the exponents ν by studying the theory in the presence of a small mass. One finds

$$\chi(\mu - \mu_c) \sim \frac{1}{\mu - \mu_c} [-\ln(\mu - \mu_c)]^{1/3} . \qquad (9.32)$$

The exponent γ also takes its classical value ($\gamma = 1$), apart from a small logarithmic correction. Equation (9.32) is in agreement with the data coming from Monte Carlo simulations and the high-temperature expansion, although a precise determination of the power of the logarithm is difficult to obtain experimentally.[27]

These results easily generalize to an N-component theory; one finds that

$$\chi(\mu - \mu_c) \sim \frac{1}{\mu - \mu_c} [-\ln(\mu - \mu_c)]^{(N+2)/(N+8)} . \qquad (9.33)$$

The sign of the one-loop contribution to the β function was crucial for obtaining this result; with the opposite sign, we could not control the theory in the small-momentum region, while the perturbative expansion can be successfully used in the large-momentum region. It is common to distinguish between these two situations by saying that the theory is asymptotically free (in the ultraviolet) if $b(\lambda)$ is negative.[28] This distinction is very important, as the two classes of theories have quite different behavior.

9.4. The ε expansion

We have just seen that in four dimensions the renormalized perturbative expansion exists and is finite. In four dimensions we have (see Fig. 9.2)

$$2b(\lambda, \varepsilon) = -\varepsilon\lambda + \tilde{b}(\lambda, \varepsilon)$$

$$\tilde{b}(\lambda, \varepsilon) = \sum_{k}^{\infty} \lambda^k \tilde{b}_k(\varepsilon) , \qquad (9.34)$$

where the $\tilde{b}_k(\varepsilon)$ have a finite nonzero limit when $\varepsilon \to 0$. Similar results hold for the functions C_1 and C_2. The functions b and C can be computed in perturbation theory to all orders in the massive theory, while in the massless theory infrared divergences are present at the order λ^{n_c}, $n_c \simeq D/\varepsilon$ as an effect of the presence of nonanalytic terms of the form $\lambda^{4/\varepsilon}$. However, the order n_c at which this disaster occurs goes to infinity with $1/\varepsilon$, so for ε infinitesimal we can use the perturbative expansion in either

9.4. The ε Expansion

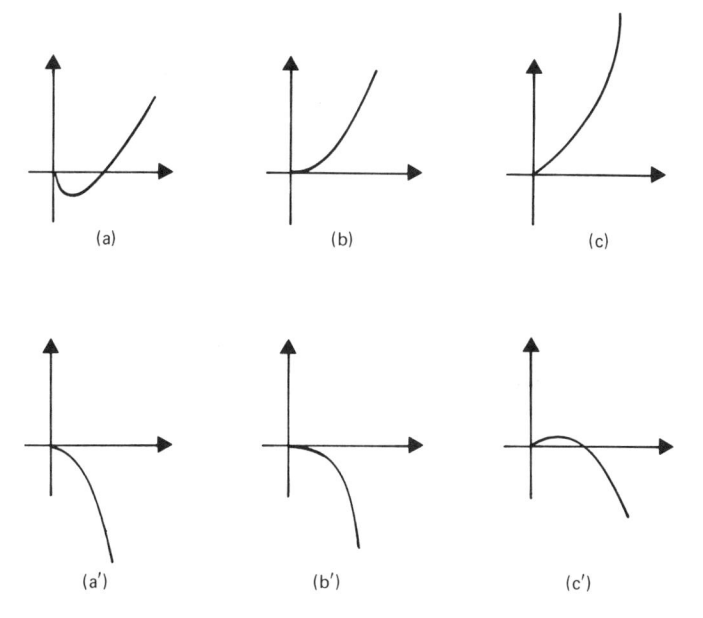

Figure 9.2. The $b(\lambda)$ function in $4 - \varepsilon$ (a, a'), 4 (b, b'), and $4 + \varepsilon$ (c, c') dimensions for positive b_2 (a), (b), (c) and negative b_2 (a'), (b'), (c'), respectively, in the region of small positive ε and λ.

the massive or the massless case. (For practical purposes the massless one is easier to compute and is the usual choice.)

We can now find $\lambda_c(\varepsilon)$ [i.e., the solution of the equation $b(\lambda)$] as a power series in ε, with an accuracy ε^l if the diagrams up to l loops are computed. By substituting $\lambda_c(\varepsilon)$ in the C functions one finds the critical exponents as functions of ε. For example, it is known that

$$\eta = \frac{\varepsilon^2}{2} \frac{N+2}{N+8} \tag{9.35}$$

$$\nu = \frac{1}{2} + \frac{N+2}{4(N+8)} \varepsilon + \frac{(N+2)(N^2 + 23N + 60)}{8(N+8)^3} \varepsilon^2, \tag{9.36}$$

as can be seen after some algebra[29] from Eq. (8.19). If we set $\varepsilon = 1$, for $N = 1$ we find

$$\nu_3 = 0.628 \qquad \eta_3 = 0.018.$$

It is known that the ε expansion is not convergent, but it is hoped that it is Borel summable. Using asymptotic estimates of the same type as Eq. (8.24), one gets[30]

$$\eta(\varepsilon) \simeq \sum_{k}^{\infty} \eta_k \varepsilon^k \; ; \qquad \nu(\varepsilon) \simeq \sum_{k}^{\infty} \nu_k \varepsilon^k$$

$$\eta_k \sim k^{7/2} k! (-A)^k C_\eta$$

$$\nu_k \sim k^{9/2} k! (-A)^k C_\nu \qquad (9.37)$$

$$A = \frac{1}{16\pi} .$$

The constants C have been computed only quite recently.[31] Estimates based on the Borel summation techniques using a four-loop computation of the η_k and ν_k up to $k = 4$, give for $D = 3$, $N = 1$

$$\eta_3 = 0.033 \pm 0.001 \qquad \nu_3 = 0.628 \pm 0.002 , \qquad (9.38)$$

in very good agreement with previous determinations.

9.5. The range of the renormalized coupling

An interesting problem is the determination of the possible values that the renormalized coupling constant may assume. In fewer than four dimensions, when the bare coupling constant g ranges from zero to infinity, the renormalized dimensionless coupling λ ranges from 0 to λ_c, which is the maximal allowed value of the renormalized coupling constant. As long as $\lambda_c \to 0$ when $D = 4$, the four-dimensional theory in the infinite cutoff limit exists only for $\lambda = 0$, i.e., free (or Gaussian) theory.

The implication seems to be that in four dimensions in the continuum limit a non-Gaussian φ^4 theory does not exist, in spite of the formal renormalized perturbative expansion. This conclusion, first suggested by the Russian school in the fifties,[32] seems to be physically compelling if the regularized theory is defined by using analytic continuation in the dimensions.[33] However, this negative result may only indicate that a different approach must be used to define a non-Gaussian theory in $D = 4$. It would be very important to arrive at a firm conclusion regarding the possibility or impossibility of defining such a theory. Although the majority (to which I belong) believes in the impossibility, the question is far from being settled in a satisfactory way. Here I shall try to present further arguments that point in the same direction; in the absence of a full understanding of the problem, the arguments will be more qualitative than usual.

We try now to reach the same conclusions working directly in four dimensions: we consider the theory in the presence of a cutoff Λ (the theory with cutoff is certainly well defined). The crucial question is the following: can we find a function $g(\Lambda^2/m^2)$ such that

9.5. The Range of the Renormalized Coupling

$$\lim_{\Lambda^2 \to \infty} \lambda\left(\frac{\Lambda^2}{m^2}, g\left(\frac{\Lambda^2}{m^2}\right)\right) \neq 0 ? \qquad (9.39)$$

Now $m^2(\partial/\partial m^2)\lambda|_{g_B} = -\Lambda^2(\partial/\partial \Lambda^2)|_{g_B}\lambda = b(\lambda)$ (g_B is dimensionless); if we use the one-loop expression for $b(\lambda) = \lambda^2$ [and for simplicity suppose that $\lambda(1, g) = g$], we get

$$\lambda\left(\frac{\Lambda^2}{m^2}, g\right) = \frac{g}{1 + g \ln \Lambda^2/m^2} < \frac{1}{\ln \Lambda^2/m^2} \quad \text{if } g \geq 0. \qquad (9.40)$$

On the other hand, if the function $b(\lambda)$ has a zero for $D = 4$ at $\lambda = \lambda_c$ (see Fig. 9.3), which does not seem to be indicated by the four-loop computation, it is possible to find a $g(\Lambda^2/m^2)$ such that Eq. (9.39) holds, if we start with a sufficiently large bare coupling constant (we neglect the dependence of $b(\lambda)$ on Λ^2/m^2 for simplicity). In other words, there are two possibilities: when $\Lambda \to \infty$ the domain of attraction of the origin contains (a) the whole real g_B axis; (b) only a part of the axis. If possibility (a) is satisfied, a bound similar to Eq. (9.40) is valid and only a free theory is obtained; if possibility (b) holds, a nonfree theory ($\lambda \neq 0$) can be obtained.

Numerical simulations and high-temperature analysis seem to indicate that possibility (a) is the correct one. Moreover, unless something quite peculiar happens, if a nontrivial fixed point exists in four dimensions, by continuity it should also exist in $4 + \varepsilon$ dimensions. Therefore a nonfree theory should also exist in more than four dimensions, but this is rather unlikely. One can prove[34] that in the nearest-neighbor interaction Landau-Ginsburg model on the lattice, $\lambda \leq C(am)^{D-4}$, $a \approx 1/\Lambda$ being the lattice spacing and C being a constant of order 1. Therefore if $\Lambda \approx a^{-1}$ goes to infinity, λ goes to zero if $D > 4$.

It seems possible that in the near future someone will prove that the bound equation (9.40) holds in $D = 4$ with a particular regularization

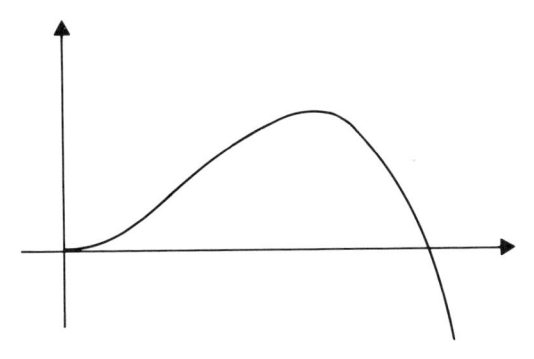

Figure 9.3. A possible shape of the $b(\lambda)$ function in four dimensions.

scheme; it would be much more difficult to prove this result for all possible schemes and to exclude completely the possibility of a non-Gaussian theory in $D = 4$.

The sign of the one-loop contribution to $b(\lambda)$ is crucial in the argument (as was stressed by Landau). If $b(\lambda) = -b_2\lambda^2$ with $b_2 > 0$, the one-loop approximation tells us

$$\lambda = \frac{g_B}{1 - g_B b_2 \ln \Lambda^2/m^2}, \qquad (9.41)$$

and it is sufficient to take $g_B \approx 1/(b_2 \ln \Lambda^2/m^2)$ to get a nontrivial result. The critical exponents could be computed here in $D = 4 + \varepsilon$ dimensions, ε being positive. This kind of theory is called asymptotically free (in the ultraviolet) because after the limit $\Lambda \to \infty$, it becomes more and more Gaussian on shorter and shorter scales.

It is believed that it should be possible to prove rigorously the existence of asymptotically free theories in the limit $\Lambda \to \infty$, but nobody has yet succeeded in this difficult task. From the point of view of relativistic quantum field theory (as will be seen in Chapter 15), we are interested in a theory defined on the continuum $\varphi(x)$ with $\Lambda = \infty$ and the lattice is only an auxiliary construction. The existence of a theory in the limit $\Lambda \to \infty$ is crucial for this kind of application.

Before ending this section let us reformulate the preceding arguments in slightly different language.[35] As we discussed in Chapter 5, from a physical point of view at the scale of the lattice, the limit $a \to 0$ at fixed m corresponds to the limit $m \to 0$ at fixed a, i.e., to a critical point, because the correlation length goes to infinity in lattice spacing units when $ma \to 0$.

In the Landau-Ginsburg model, for $D < 4$ there are two fixed points on the critical surface, the Gaussian and the nontrivial, the first having one unstable eigenvalue on the critical surface, the other being attractive. If we work at fixed g (i.e., $g = \infty$) we do not have any parameter to change, and when $ma \to 0$ the probability distribution for block variables of size $L_n \sim (ma)^{-1}$ is given by the nontrivial fixed point. If we send m and g to zero together (always in lattice spacing units) at fixed $u = g/m^{4-D}$, the distribution of the block variables will be practically Gaussian at short distances, and it will start to be non-Gaussian only at distances of order m^{-1}. In other words, we have to integrate an equation of the kind $d\lambda/dt = -b(\lambda)$, where the time t runs from 0 to $t_M = -\ln m^2 a^2$. Although the point $\lambda = 0$ is repulsive and the point $\lambda = \lambda_c$ is attractive, we can take the initial point $\lambda(0) = ga^{4-D}$ close enough to zero in a dependent way so that when $t_M \to \infty$, $\lambda(t_M)$ remains fixed at a value between 0 and t_M.

If the Gaussian fixed point is attractive and there are no other fixed points on the critical surface, which belongs entirely to the attraction basin of the Gaussian fixed point, all trajectories arrive in a finite "time" close to the Gaussian fixed point, so that in the limit $t_M \to \infty$, the large-scale block variable distribution is always Gaussian.

When the theory is renormalizable, the Gaussian fixed point is stable or unstable, depending on the sign of $b(\lambda)$ near $\lambda = 0$. If the theory is asymptotically free, it is sufficient to take $\lambda(0) \sim 1/t_M$ [to be compared to $\lambda(0) \sim \exp[-t_M(4 - D)/2]$ for the Landau-Ginsburg model in $D < 4$] to get a non-Gaussian limit when $t_M \to 0$. In the absence of other fixed points, a non-asymptotically free theory is always free when $t_M \to \infty$. An example of an asymptotically free theory will be shown in the next chapter.

Notes for Chapter 9

1. K. G. Wilson and M. E. Fisher, *Phys. Rev. Lett.* 28, 234 (1972); K. G. Wilson, *Phys. Rev. Lett.* 28, 548 (1972).
2. Mathematicians have nearly always been able to make rigorous what physicists have found to be useful (e.g., distributions).
3. We have already observed this phenomenon in Chapter 4; a few orders of the high-temperature expansion of the Ising model in generic dimensions can be found in R. Balian, J. M. Drouffe, and C. Itzykson, *Phys. Rev. D* 11, 2114 (1975).
4. Paradoxical results may be obtained: for example, the coefficient of K^4 in the high-temperature expansion of the Ising model is proportional to the number of squares per site $(D(D-1)/2)$, which becomes negative for $0 < D < 1$.
5. There are useful and simple graphic rules for evaluating $\mathcal{D}(\alpha)$ and $N(\alpha, p)$ without explicitly doing the Gaussian integrals (see p. 96).
6. For example, Carlson's theorem states that if there are two analytic functions $f_1(D)$ and $f_2(D)$ such that, for integer D, $f_1(D) = f_2(D)$ and both functions satisfy the bound $|f_i(D)| < \exp(c|D|)$, in the right complex plane with $c < \pi$, then $f_1(z) = f_2(z)$. See E. C. Titchmarch, *Theory of functions*, Oxford University Press (1939).
7. The integral in Eq. (9.4) is zero because the integrand is a total derivative.
8. There would be no additional difficulties in considering as well the correlation functions of φ_R^2.

9. See, for example, C. Itzykson and J. B. Zuber, *An Introduction to Field Theory*, McGraw-Hill, New York (1980); K. Hepp, *Théorie de la renormalization*, Springer, Berlin (1969); W. Zimmermann, *Elementary Particles and Quantum Field Theory*, ed. by S. Deser et al., MIT Press, Cambridge (1970); D. Amit, *Field Theory, The Renormalization Group, and Critical Phenomena*, McGraw-Hill, New York (1978); see also S. Coleman in *International School Ettore Maiorana*, Editrice Compositori, Bologna (1973).
10. This theorem deals with ultraviolet divergences, and it is convenient to exclude the possibility of infrared divergences by assuming $\mu \neq 0$. In the limit $\mu \to 0$ there is an analogous theorem on the structure of infrared divergences [G. Parisi, *Nucl. Phys. B 150*, 163 (1978)], which mimics the BHP theorem with the exchange of configuration space and momentum space.
11. All momentum integrations but the last one are convergent.
12. An example is the self-energy diagram of Fig. 8.3 or any function $f(x)$ such that $f(x) \sim 1/x^\alpha$ with $\alpha > D$.
13. There is nothing special about the point $p = 0$; any other point would be the same. Indeed, this second solution is the one adopted in the massless case, as we shall see in the next subsection.
14. In the Landau-Ginsburg model this is equivalent to requiring that the only possible counterterms be obtained by changing μ, g, or the $\frac{1}{2}$ in front of the kinetic term.
15. It is possible that nonrenormalizable theories may be properly defined by using a different approach, e.g., the $1/N$ expansion. See G. Parisi, *Nucl. Phys. B 100*, 368 (1975) and *Nucl. Phys. B 254*, 58 (1985).
16. This local operator is a polynomial in the field $\varphi(x)$ and in its derivatives computed in the same point. Though it may seem rather strange here (function of φ would be more appropriate), the word "operator" is chosen for historical reasons.
17. If we stay at $h = 0$, counterterms proportional to φ or φ^3 are excluded by symmetry arguments; moreover, a theorem due to Symanzik states that, if the correlation functions are ultraviolet finite at $h = 0$, they are also at $h \neq 0$ (apart from a possible multiplicative redefinition of h); see K. Symanzik, *Cargèse Lecture Notes* 1970, ed. by D. Bessis (1971).
18. Equation (9.6) applies only to diagrams with loops.
19. Indeed a counterterm proportional to $(\partial_\nu \varphi)^2$ corresponds to a redefinition of the field φ, the normalization of the kinetic term being a convention. If we apply in $D < 4$ the same subtraction scheme needed in the four-dimensional theory [Eq. (9.13)], we recover the perturbative expansion of the preceding chapter.

20. For a review of the ε expansion see E. Brézin, J. C. Le Guillou, and J. Zinn-Justin in *Phase Transitions and Critical Phenomena*, Vol. VI, C. Domb and M. S. Green, eds. (Academic Press, New York, 1976) and references therein.
21. G. Parisi, *Nuovo Cimento A* 21, 179 (1974); *Phys. Rep.* 49, 215 (1978) and references therein.
22. There are speculations that at distances smaller than 10^{-33} cm, where quantum gravity induces strong fluctuations in the metric, space-time resembles a foamlike structure and the radius of the bubbles of the foam may provide a natural cutoff; a discussion of this kind of speculation (which must remain at a very preliminary stage, given the difficulties in quantizing the gravity) would lead us very far from the subject of this book.
23. K. Symanzik, in *New Developments in Gauge theories*, ed. by G. 't Hooft *et al.*, Plenum Press, New York (1980) and references therein.
24. This choice of external momenta may seem rather peculiar. Many others are possible (as long as no combination of the external momenta is zero). This particular choice has the advantage of maximal symmetry—indeed, if we set $p_i p_j = A\delta_{ij} + B$ and we require that $p_i^2 = M^2$, momentum conservation (i.e., $p_1 + p_2 + p_3 + p_4 = 0$) implies $A = \tfrac{4}{3}M^2$, $B = -\tfrac{1}{3}M^2$, and hence Eq. (9.20).
25. These equations were introduced long ago [M. Gell-Mann and F. E. Low, *Phys. Rev.* 95, 1300 (1954)] and have been called "the renormalization group equations." Their possible relevance for the theory of second-order phase transitions was stressed only much later (C. Di Castro and G. Jona Lasinio, *Phys. Lett.* 29A, 322 (1969)). The approach set forth in Chapter 7 has been called the renormalization group approach in honor of these equations.
26. The functions $b(\lambda)$ at the order λ^3 and $c_i(\lambda)$ at the first nontrivial order coincide both for the Callan-Symanzik equation and for the renormalization group equation.
27. It is not simple to distinguish between a logarithm and a small power $\mu^{-\alpha} \sim 1 - \alpha \ln \mu + O(\alpha^2 \ln^2 \mu)$ when $\mu \to 0$, unless one controls the function over quite a large range of $\ln \mu$.
28. An example of an asymptotically free theory (in the ultraviolet) can be found in the next chapter. Asymptotically free field theories are very important for relativistic quantum field theory: it is currently believed that strong interactions are correctly described by an asymptotically free theory (see, for example, the review article of G. Altarelli, *Phys. Rep.* 81C, 1 (1982), and references therein).
29. An explicit computation of the critical exponents up to the order ε^3 has been carried out by E. Brézin, J. C. Le Guillou, J. Zinn-Justin, and B. G. Nickel, *Phys. Lett. A* 44, 227 (1973).

30. E. Brézin, J. C. Le Guillou, and J. Zinn-Justin, *Phys. Rev. D 15*, 1544 (1977).
31. A. J. McKane, D. J. Wallace, and O. F. de Alcontara Bonfim, *J. of Phys. A 17*, 1861 (1984).
32. L. Landau in *Niels Bohr and the Development of Physics*, McGraw-Hill, New York (1955); E. S. Fradkin, *JEPT 28*, 750 (1955).
33. Possible applications of these ideas to the phenomenology of high-energy physics (in particular to the electroweak interactions) have been proposed in L. Maiani, G. Parisi, and R. Petronzio, *Nucl. Phys. B 136*, 115 (1978).
34. M. Aizenman, *Phys. Rev. Lett. 47*, 1 (1981) and *Commun. Math. Phys. 86*, 1 (1982); J. Frolich, *Nucl. Phys. B 200* [*FS*4] (1982); the results of these papers are based on the method presented in Chapter 15 of this book.
35. Similar considerations can also be found in K. Wilson and J. Kogut, *Phys. Rep. 12*, 175 (1974).

CHAPTER 10

On Spontaneous Symmetry Breaking

10.1. Perturbation theory

In the previous chapters we have studied the Landau-Ginsburg model in the region $T > T_c$ ($\mu > \mu_c$) at $h = 0$; in this chapter we shall consider the case $T < T_c$ where spontaneous magnetization is present for $h \neq 0$.

It is convenient to study the model with $h(x) = h$, i.e., constant nonzero magnetic field. We recall that the Hamiltonian of the model is given by

$$\beta H = \int d^D x \left[\frac{1}{2} (\partial_\nu \varphi)^2 + \frac{1}{2} m^2 \varphi^2 + \frac{1}{4!} g \varphi^4 - h\varphi \right]. \tag{10.1}$$

The perturbative expansion of the previous chapter can be extended to the study of the Hamiltonian (10.1) by using the following procedure: we look for the constant field φ_0 that minimizes $H(\varphi(x) = \varphi_0)$ and we set

$$\begin{aligned} \varphi &= \varphi_0 + \tilde{\varphi} \\ H(\varphi) &\equiv H(\varphi_0) + \tilde{H}(\tilde{\varphi}) \,. \end{aligned} \tag{10.2}$$

Because φ_0 minimizes $H(\varphi)$, $\tilde{H}(\tilde{\varphi})$ contains no linear term in $\tilde{\varphi}$; it starts with a quadratic term. The cubic and quartic terms vanish when $g \to 0$, and the perturbative expansion in g can be constructed as before. Indeed, we find that φ_0 can be found as the absolute minimum of[1]

$$H(\varphi_0) = \frac{\mu \varphi_0^2}{2} + \frac{g}{4!} \varphi_0^4 - h\varphi_0 \,. \tag{10.3}$$

If $\mu > 0$ and $h = 0$, $\varphi_0 = 0$ and we recover the case previously studied. In the general case we find

$$H(\varphi) = \int d^Dx \left[H(\varphi_0) + \frac{1}{2}(\partial_\nu \tilde{\varphi})^2 + \frac{\tilde{\mu}}{2}\tilde{\varphi}^2 + \frac{1}{3!}g_3\tilde{\varphi}^3 + \frac{1}{4!}g_4\tilde{\varphi}^4 \right]$$

$$\mu\varphi_0 + \frac{g}{6}\varphi_0^3 = h \qquad (\varphi_0 h > 0) \tag{10.4}$$

$$\tilde{\mu} = \mu + \frac{1}{2}g\varphi_0^2 \qquad g_3 = g\varphi_0 \qquad g_4 = g.$$

When $g \to 0$ we have, for small h,

$$\varphi_0 \sim \frac{h}{\mu}, \qquad \mu > 0$$

$$\varphi_0 \sim \text{sign}(h)\sqrt{\frac{-6\mu}{g}} + O(h), \qquad \mu < 0. \tag{10.5}$$

In this case the distribution probability of $\tilde{\varphi}$ becomes Gaussian: To leading order in g we get

$$\langle \varphi \rangle = \varphi_0 \qquad f(h) \equiv -\frac{1}{V}\ln Z = H[\varphi_0]$$

$$(\tilde{\mu} - \Delta_x)\langle \tilde{\varphi}(x)\tilde{\varphi}(y)\rangle_C = \delta(x - y). \tag{10.6}$$

The two-field connected correlation function in momentum space is $\tilde{G}_0(p) = 1/(\tilde{\mu} + p^2) = \int d^Dx \exp(ipx) G_0(x)$, $G_0(x) = \langle \tilde{\varphi}(x)\tilde{\varphi}(0) \rangle$. In the limit $h \to 0^+$ for $\mu < 0$ we find

$$\tilde{\mu} = -2\mu \qquad \varphi_0 = \left(-\frac{6\mu}{g}\right)^{1/2} \qquad g_3 = \sqrt{-6\mu g}. \tag{10.7}$$

When $\mu < 0$, spontaneous symmetry breaking is present:

$$\lim_{h \to 0^+} \langle \varphi \rangle = -\lim_{h \to 0^-} \langle \varphi \rangle \neq 0$$

$$f(h) = f(0) - |h|\left(-\frac{6\mu}{g}\right)^{1/2} + O(h^2). \tag{10.8}$$

Let us now see how to set up the perturbative expansion. For simplicity let us suppose that we are working in a theory with a cutoff where no ultraviolet divergences are present.[2] The diagrams contributing to $\langle \varphi \rangle$ and $\langle \varphi\varphi \rangle_C$ are shown in Fig. 10.1; the final result is, at the one-loop

10.1. Perturbation Theory

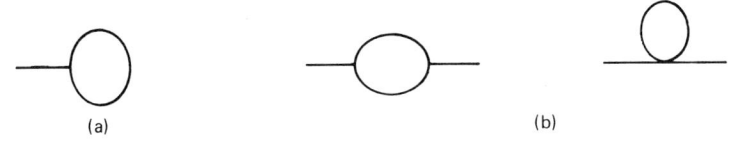

Figure 10.1. Diagrams contributing (a) to $\langle \varphi \rangle$ and (b) to $\langle \varphi\varphi \rangle_c$.

order,

$$\langle \varphi(0) \rangle = \varphi_0 - g_3 \frac{1}{3!} \int \langle \tilde{\varphi}(0)\tilde{\varphi}^3(x) \rangle_c \, d^D x$$

$$= \varphi_0 - g_3 \frac{1}{2} \int d^D x \langle \tilde{\varphi}(0)\tilde{\varphi}(x) \rangle_c \langle \tilde{\varphi}^2(x) \rangle_c$$

$$= \varphi_0 - \frac{1}{2} g_3 \int d^D x \, G_0(x) G_0(0) = \varphi_0 - \frac{1}{2} g_3/\tilde{\mu} \int \frac{d^D p}{(2\pi)^D} \frac{1}{p^2 + \tilde{\mu}}$$
(10.9)

$$\tilde{G}(p) = \frac{1}{p^2 + \tilde{\mu} - \Sigma(p)}$$

$$\Sigma(p) = \frac{1}{2} g_3^2 \int \frac{d^D k}{(2\pi)^D} \frac{1}{(p-k)^2 + \tilde{\mu}} \frac{1}{k^2 + \tilde{\mu}} - \frac{1}{2} g_4 \int \frac{d^D k}{(2\pi)^D} \frac{1}{k^2 + \tilde{\mu}}.$$

Similarly, the first correction to the free-energy density is given by the Gaussian integral over the field $\tilde{\varphi}$:

$$f(h) = H(\varphi_0) + \frac{1}{2} \int \frac{d^D k}{(2\pi)^D} \ln(k^2 + \tilde{\mu}) . \qquad (10.10)$$

Equation (10.10) can also be written as

$$f(h) = [\gamma(\varphi) - h\varphi]_{\varphi = \varphi_0}$$

$$\gamma(\varphi) = \frac{\mu}{2} \varphi^2 + \frac{g}{4!} \varphi^4 + \frac{1}{2} \int \frac{d^D k}{(2\pi)^D} \ln(k^2 + \tilde{\mu}(\varphi)) \qquad (10.11)$$

$$\tilde{\mu}(\varphi) = \mu + \frac{1}{2} g\varphi^2 ,$$

where $\gamma(\varphi)$ no longer depends on h.

Now it is important to note that although Eq. (10.10) has been derived here for $\langle \varphi \rangle = \varphi_0$, Eq. (10.9) for $\langle \varphi \rangle$ could have been obtained by setting $\langle \varphi \rangle = \bar{\varphi}$, where $\bar{\varphi}$ is the solution of the equation

$$\left. \frac{\partial \gamma}{\partial \varphi} \right|_{\varphi = \bar{\varphi}} = h , \qquad (10.12)$$

as can be easily verified. Now it is possible to generalize Eq. (10.12) to all orders and to find simple rules for computing $\gamma(\varphi)$.[3] Toward this end it is convenient to define the magnetic field as a function of the expectation value [i.e., $h(\varphi)$] in such a way that

$$\langle \varphi \rangle |_{h=h(\varphi)} = \varphi . \qquad (10.13)$$

In the most general case one can consider Eq. (10.12) for an x-dependent $\langle \varphi \rangle \equiv \varphi(x)$: h would also depend on x, and it would be a functional of φ; for simplicity here we study only the case in which neither h nor φ depends on x.

For later convenience we introduce a φ-dependent free-energy density $\tilde{f}(\varphi)$ and an auxiliary function $\gamma(\varphi)$ defined by[3]

$$\tilde{f}(\varphi) \equiv f(h(\varphi)) = \gamma(\varphi) - h\varphi . \qquad (10.14)$$

In other words, $\tilde{f}(\varphi)$ is the value of the free-energy density when $\langle \varphi \rangle = \varphi$. In the zero-loop approximation we obviously have

$$\gamma(\varphi) = \frac{\mu^2}{2} \varphi^2 + \frac{g\varphi^4}{4!} , \qquad (10.15)$$

while at one-loop order $\gamma(\varphi)$ is given in Eq. (10.11).

Intuitively the variational properties of the free energy suggest that, at a given h, $\langle \varphi \rangle$ is fixed by the condition

$$\langle \varphi \rangle = \varphi_m , \qquad (10.16)$$

where φ_m is the minimum of $\tilde{f}(\varphi)$. Obviously φ_m must satisfy the equation

$$\left. \frac{d\gamma}{d\varphi} \right|_{\varphi = \varphi_m} = h . \qquad (10.17)$$

The intuitive argument is indeed correct: using the known relation

$$\frac{df}{dh} = \langle \varphi \rangle \equiv \varphi , \qquad (10.18)$$

one easily derives Eq. (10.17):

$$\frac{d\gamma}{d\varphi} = \frac{d}{d\varphi} [f(h) + \varphi h] = \frac{df}{dh} \frac{dh}{d\varphi} + \varphi \frac{dh}{d\varphi} + h = h(\varphi) . \qquad (10.19)$$

It is also clear that, if Eq. (10.17) has many solutions, the one having the smallest value of $\tilde{f}$ will have minimal free energy and it will be the physical one.

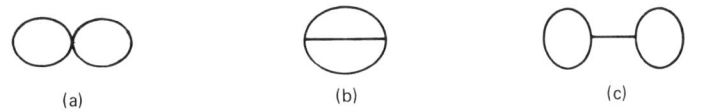

Figure 10.2. Two-loop contribution to the free energy; diagram (c) is one-line reducible and does not contribute to $\gamma(\varphi)$.

The transformation from an h-dependent free energy to a φ-dependent free energy (where $\varphi = -df/dh$) is a very common procedure in thermodynamics, which goes under the name of the Legendre transform;[4] it is also used for going from the Lagrangian to the Hamiltonian formalism.

There are graphical rules for computing $\gamma(\varphi)$ directly; it turns out that only one-line irreducible diagrams need be evaluated, greatly simplifying the computation. We shall not give the proof of this very useful result, neither shall we study further the properties of the function $\gamma(\varphi)$, as this would take too much space; in the rest of this chapter we shall consider the case $N \neq 1$, in which interesting new phenomena are present.

Before going on to look at $N \neq 1$, however, a few remarks are in order. The function $\gamma(\varphi)$ can be written as

$$\gamma(\varphi) = \sum_{0}^{\infty} \frac{1}{n!} \Gamma_n \varphi^n, \quad (10.20)$$

where $\Gamma_1 = 0$, $\Gamma_2 = [G_2(p)|_{p=0}]^{-1}$, and Γ_n is the sum of the amputated (i.e., without the contribution of external legs) one-particle-line irreducible diagrams. It is easy to convince ourselves of the correctness of Eq. (10.20) for small values of n (say, $n = 2, 3, 4$) by differentiating with respect to φ [Eq. (10.17)] $n-1$ times and reinterpreting the result in a graphical way (see Fig. 10.2). A full proof requires some tricky combinatorial analysis; in the simplest approach the proof is constructed by induction in n.

In order to compare the experimental data near the phase transition with the theory, it is particularly interesting to have a reliable result for the function $\gamma(\varphi)$ [or equivalently $h(\varphi)$] at the critical point. A few orders of the $4 - D$ expansion for $\gamma(\varphi)$ have been computed.[5] Unfortunately we lack a high-loop computation done directly at three dimensions.

10.2. No-go theorems

When T is greater than T_c, the Ising model ($N = 1$) and the Heisenberg model ($N = 3$) look very similar, the only difference being in the weights of the loop diagrams which are smooth functions of N. However, the theories with $N \neq 1$ have qualitatively new features as soon as $T < T_c$ at $h = 0$: spin waves are present, i.e., configurations whose energy differs from the ground state by an arbitrarily small amount. The $O(N)$ invariance of the theory implies that the spontaneous magnetization may point in an arbitrary direction on the N-dimensional sphere of radius φ_0 ($\sum_a^N \langle \varphi_a \rangle^2 = \varphi_0^2$). At $h = 0$ we have an infinite number of equilibrium, translational invariants, configurations which can be labeled by a continuous parameter (not only two as happens in the Ising model). We would now like to rotate the spins coherently in a large region of the space: the increase in energy, if the rotation angle is a slowly varying function of the position, will not be high. Consider, for example, the case $N = 2$. We set

$$\varphi_1(x) = \varphi_0 \cos \theta(x)$$
$$\varphi_2(x) = \varphi_0 \sin \theta(x) .$$
(10.21)

The energy for such a configuration will be (always at $h = 0$)

$$H[\theta] = H[\theta]\big|_{\theta(x)=0} + \frac{1}{2} \varphi_0^2 \int d^D x (\partial_\nu \theta)^2 .$$
(10.22)

θ enters the Hamiltonian only through its derivatives, so Eq. (10.22) may be very small if θ is slowly varying. In this approximation we find that the $\theta\theta$ correlation function is $1/p^2$ in momentum space, corresponding to a correlation

$$\langle \theta(x)\theta(0y) \rangle \sim \frac{1}{x^{D-2}} \quad x \to \infty \quad \text{if} \quad D > 2 .$$
(10.23)

We shall see in the next section that Eq. (10.23) is an exact statement.

We expect long-range correlations to be present when the $O(N)$ symmetry is spontaneously broken; moreover, something peculiar must happen when $D = 2$, because the Fourier transform of $1/p^2$ does not exist owing to infrared divergences.[6]

These qualitative observations are sharpened by the Mermin-Wigner theorem: there is a critical dimension D_c such that for any short-range[7] Hamiltonian at $T \neq 0$, no spontaneous symmetry breaking is possible, D_c being 1 for discrete symmetries ($N = 1$) while $D_c = 2$ for continuous symmetries ($N > 1$).

This can be proven in many different ways: for example, in the Heisenberg model with nearest-neighbor interaction it is possible to derive[8] the inequality for $D = 2$,

$$|\langle \varphi \rangle| < \frac{CT^{-1/2}}{\ln(-|h|)} ,$$
(10.24)

10.2. No-Go Theorems

where C is a constant. Equation (10.24) implies the absence of spontaneous magnetization in the limit $h = 0$ (at $T \neq 0$). Here we present a proof of these results for D_c, stressing the physical origin of the phenomenon. Although we skip some points, the proof can be made fully rigorous.[9] We already know that, if at $h = 0$ symmetric boundary conditions are used, the expectation value of the magnetization is zero. If a magnetic field is applied on the boundary (or, equivalently, the spins on the boundary are constrained to stay in one given direction), a nonzero magnetization in the center of the box is present (when the volume goes to infinity) if the symmetry is spontaneously broken. In order to prove the absence of spontaneous symmetry breaking, we need only show that the magnetization in the center of the box does not depend on what happens on the boundaries.[10] We begin the proof by making some preliminary observations, which hold in both cases ($N = 1$ and $N \neq 1$). We denote by φ_a^+ and φ_a^- two opposite values of φ, such that

$$\varphi_a^+ = -\varphi_a^-. \tag{10.25}$$

We consider a system of side L with $\varphi(x) = \varphi^+$ on the boundary. We now compare two configurations: In the first configuration, $\varphi = \varphi^+$ everywhere; in the second, $\varphi(x) = \varphi^-$ inside a box of side $L/2$ concentric with the first one. This condition does not completely fix the second configuration: let $\Delta E(L)$ be the minimum energy difference between the energies of the first and second types of configuration.

We claim that, if

$$\lim_{L \to \infty} \Delta E(L) \equiv \Delta E < \infty, \tag{10.26}$$

the spontaneous magnetization must be zero. The sufficiency of condition (10.26) can be shown first by restricting the sum over all possible configurations to a representative set of trial configurations. We consider concentric boxes of side $L_k = 2^k L_0$. Our trial configurations are constructed as follows: the φ field on the boundary of the kth box takes the value $\varphi^+ \sigma_k$, where $\sigma_k = \pm 1$ ($k = 0, \ldots, n-1$), $\sigma_n = 1$. Between the boundaries of two given boxes we have two possibilities:

(a) The field φ has the same value on both boundaries. In this case we set φ between the boundaries equal to φ on the boundaries.

(b) The field φ takes two different values on the boundaries. In this case φ in between is a smooth interpolation such that the contribution to the energy increase coming from that region is $\Delta E(L_k)$, which for large k or L_0 may be well approximated by ΔE.[11]

If only these configurations are considered, we recover a one-dimensional Ising chain with effective Hamiltonian

$$H_{\text{ef}} = \frac{-\Delta E}{2} \sum_{k\,0}^{n-1} \sigma_k \sigma_{k+1}, \qquad (10.27)$$

with the boundary condition[12] $\sigma_n = 1$. The expectation value of $\varphi(\varphi^c)$ in the central box is given by

$$\varphi_a^c = \langle \sigma_0 \rangle \varphi_a^+ . \qquad (10.28)$$

It is straightforward to compute $\langle \sigma_0 \rangle$ with the techniques described in Section 4.5: We get

$$\langle \sigma_0 \rangle = \left(\tanh \frac{\Delta E}{2}\right)^n = L^{\ln_2[\tanh(\Delta E/2)]} . \qquad (10.29)$$

As soon as $\Delta E < \infty$, $\ln_2[\tanh(\Delta E/2)] < 0$ and $\langle \sigma_0 \rangle$ vanishes, at least as a power of L.

If condition (10.26) is also satisfied when the internal box has side L and the external has side $L + L_0$, we could follow the same argument as before, considering concentric boxes of side nL_0 (not 2^n). In this case we would obtain an exponential decrease of φ^c with L ($\varphi^c \simeq \exp(-L)$). Forgetting technicalities, if we could reverse the sign of the spins in an arbitrarily large region by paying only a finite amount of energy, we could eliminate completely the effects of the boundary terms.[13] The proof should end by showing that the same results hold if we sum over all possible configurations; i.e., we have considered a representative set. At this stage the proof becomes more complicated and will be omitted.

It usually happens that this kind of argument underestimates rather than overestimates the fluctuations: Very often the correlation functions decay exponentially in cases where Eq. (10.29) predicts a power-law decay. In intuitive terms, by enlarging the configuration space, we enable the system to fluctuate more, and consequently we decrease the correlations at large distance.

Our task is now to check by explicit construction whether Eq. (10.26) is satisfied. In one dimension the analysis is rather simple. We consider a one-dimensional Ising model, with a finite-range Hamiltonian (i.e., only spins at maximal distance R are directly coupled). In order to prove Eq. (10.26) (when the internal and external boxes have sides L and $L + L_0$, respectively) we use the following trial configuration: all spins are positive in the internal box, all spins negative in the external box. As soon as L_0 is larger than the range of the forces (i.e., $L > R$), the difference in energy is L_0 independent: it is also finite for any value of L_0.

The argument that we have already presented implies that by consider-

10.2. No-Go Theorems

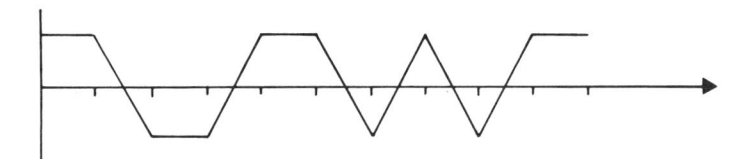

Figure 10.3. A configuration of the field $\varphi(X)$ for $n = 10$ which corresponds to the sequence of σ_n: 1, −1, −1, 1, 1, −1, 1, −1, 1, 1, ...

ing only the configurations

$$\varphi_i = \sigma_k \quad \text{for } L_0 k \le i < L_0(k+1), \tag{10.30}$$

we can determine that no spontaneous symmetry breaking can be present in this one-dimensional case (the same conclusion can be reached by the transfer matrix approach; see p. 223).

For the Ising model (in general models with only a finite number of equilibrium states), $\Delta E(L)$ increases as L^{D-1} if we consider the same configurations in higher dimensions. By enlarging the configuration set, we can also reach the same conclusion: spontaneous breaking of a discrete symmetry is possible as soon as D is greater than 1.

The study of the spontaneous breaking of a continuous symmetry requires more care: the first thing that we have to do is to look for more sophisticated interpolating configurations. Let us consider a smooth function $\psi_a(z)$ such that $\psi_a(0) = \varphi_a^+$ and $\psi_a(1) = \varphi_a^-$. If we set $L_n = 2^n L_0$ and

$$\varphi(x) = \psi\left(\frac{x - L_k}{L_k}\right) \quad L_k < x < L_{k+1} = 2L_k \tag{10.31}$$

when $\sigma_k = +1$, $\sigma_{k+1} = -1$, the only contribution to the energy (ΔE) comes from the kinetic term $(\partial_\nu \varphi)^2$. The integrand is proportional to L_k^{-2}, so that

$$\Delta E(L_k) = \int_{L_k < |x| < L_{k+1}} d^D x (\partial_\nu \varphi)^2 \sim L_k^{D-2} \quad \text{if } L_k \to \infty. \tag{10.32}$$

We have thus proved the needed result to complete the theorem on the absence of spontaneous breaking of a continuous symmetry in $D = 2$. The key ingredient we have used[14] is the fact that the interface energy between two regions of different magnetization is proportional to the surface of the interface in the Ising case, while in the Heisenberg case it is proportional to the surface divided by the thickness of the interface (as soon as the thickness is smaller than the radius of the interface), so it can be much smaller than the interface energy of the Ising model.[15]

10.3. Goldstone bosons

We have just seen that the spontaneous breaking of a continuous symmetry is quite different from the discrete case. We now investigate how this difference shows up in the perturbative expansion.

We consider an $O(N)$-invariant theory. If the external magnetic field is a constant $[h_a(x) = h_a]$, the free energy may depend on h only through $|h| = |\sum_1^N h_a^2|^{1/2}$. If the symmetry is spontaneously broken,[16]

$$f(h) = -|h|m - \frac{1}{2}\chi_L|h|^2, \qquad (10.33)$$

where m is the modulus of the spontaneous magnetization and χ_L is the longitudinal susceptibility. Indeed, if the point $h = 0$ is reached in the following way,[17]

$$h_N = h$$

$$h_\alpha = 0 \qquad \alpha = 1, \ldots, N-1 \qquad (10.34)$$

$$h \to 0^+,$$

we find that[18]

$$\langle \varphi_\alpha \rangle = 0 \qquad \alpha = 1, \ldots, N-1$$

$$\langle \varphi_N \rangle = m$$

$$\chi_{ab} \equiv -\frac{\partial f}{\partial h_a \partial h_b} = -\frac{\partial}{\partial h_a}\left(\frac{h_b}{|h|}\frac{\partial f}{\partial |h|}\right)$$

$$= -\frac{h_a h_b}{|h|^2}\frac{\partial^2 f}{\partial h^2} - \left(\delta_{ab} - \frac{h_a h_b}{|h|^2}\right)\frac{1}{|h|}\frac{\partial f}{\partial |h|} \qquad (10.35)$$

$$= [\chi_L(h)\delta_{a,N} + \chi_T(h)(1-\delta_{a,N})]\delta_{ab}$$

$$\chi_L(h) = -\frac{\partial^2 f}{\partial h^2}, \qquad \chi_T(h) = -\frac{1}{|h|}\frac{\partial f}{\partial h} = -\frac{m}{|h|}.$$

In other words, the transverse susceptibility[19] $\chi_T = \chi_{aa}$ $(a < N)$ becomes infinite when $h \to 0$ if $m \neq 0$: For small h the addition of a new external field may change completely the direction in which the magnetization points. The divergence of χ_T when $|h| \to 0$ implies the presence of long-range correlations, which are generated by the mechanism discussed in the preceding section.[20]

10.3. Goldstone Bosons

The divergence of χ_T stems from the $O(N)$ invariance of the free energy.[21] It should hold in each order of the perturbative expansion. If we generalize Eqs. (10.1)–(10.4) to the Hamiltonian

$$H[\varphi] = \int \left[\frac{1}{2} \sum_a^N (\partial_\nu \varphi_a)^2 + \frac{1}{2} \mu \left(\sum_a^N \varphi_a^2 \right) + \frac{g}{4!} \left(\sum_a^N \varphi_a^2 \right)^2 - h\varphi_N \right], \tag{10.36}$$

we find for $\mu < 0$, $h = 0^+$,

$$\varphi_a = \varphi_a^0 + \tilde{\varphi}_a$$

$$\varphi_a^0 = 0 \quad a < N$$

$$\varphi_N^0 = \varphi^0 \equiv \sqrt{\frac{-6\mu}{g}} + O(h) \quad \left(\mu \varphi_0 + \frac{g}{3!} \varphi_0^3 = h \right) \tag{10.37}$$

$$H[\varphi] = H[\varphi_0] + \tilde{H}[\tilde{\varphi}]$$

$$\tilde{H}(\tilde{\varphi}) = \int d^D x \left[\frac{1}{2} \sum_a^N (\partial_\nu \tilde{\varphi}_a)^2 + \frac{1}{2} \mu_T \tilde{\varphi}_N^2 + \frac{\mu_L}{2} \sum_a^{N-1} \tilde{\varphi}_a^2 + \frac{g_3}{3!} \tilde{\varphi}_N^3 \right.$$

$$\left. + \frac{g_3}{3!} \tilde{\varphi}_N \sum_1^{N-1} \tilde{\varphi}_a^2 + \frac{g_4}{4!} \left(\sum_a^N \tilde{\varphi}_a^2 \right)^2 \right]$$

$$\mu_L = \mu + \frac{1}{2} g\varphi_0^2 = 3 \frac{h}{\varphi_0} - 2\mu \qquad g_3 = g\varphi_0$$

$$\mu_T = \mu + \frac{1}{3!} g\varphi_0^2 = \frac{h}{\varphi_0} \qquad g_4 = g.$$

In the limit $g \to 0$ the correlation functions are, in momentum space,

$$G_{ab}^{(p)} = \delta_{ab}[G_L(p)\delta_{a,N} + G_T(p)(1 - \delta_{a,N})] \tag{10.38}$$

$$G_L(p) = \frac{1}{p^2 - 2\mu + 3h/\varphi_0}, \qquad G_T(p) = \frac{1}{p^2 + h/\varphi_0}.$$

Very often the pole at $p^2 = 0$ for $h = 0$ in the transverse correlation function is called the Goldstone boson.[22]

The existence of long-range correlations below T_c is a new phenomenon, which poses new problems, the first one being: are the diagrams of Fig. 10.4 infrared divergent in this case? Fortunately this does not happen in the relevant region $D > 2$. The argument is based on the observation that the connected correlation function of four transverse

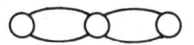

Figure 10.4. A possible dangerous infrared-divergent diagram, the blob being the four-field amputated correlation function at a given order in g.

fields at zero external momentum is given by the fourth derivative of f with respect to h:

$$G_{\alpha\beta\gamma\delta} = \frac{m}{|h|^3} [\delta_{\alpha\beta}\delta_{\gamma\delta} + \delta_{\alpha\gamma}\delta_{\beta\delta} + \delta_{\alpha\delta}\delta_{\beta\gamma}] \qquad \alpha, \beta, \gamma, \delta = 1, \ldots, N-1, \tag{10.39}$$

as can be seen by differentiating Eq. (10.35) twice. The contribution of the external legs to $G_{\alpha\beta\gamma\delta}$ is proportional to $(\chi_T)^4$, i.e., to $(m/h)^4$, consequently Eq. (10.39) implies that the amputated four-point correlation function at zero external momentum vanishes as h when $h \to 0$. Now if we consider the diagrams of Fig. 10.4 (the blob denotes the four-point amputated connected correlation function computed at a given order in g), the vanishing of the blob at zero external momentum strongly decreases the infrared divergences. At the first order in g we find that the blob is given by the two diagrams of Fig. 10.5, whose contribution is

$$\delta_{\alpha\beta}\delta_{\gamma\delta}A(p_1+p_2) + \delta_{\alpha\gamma}\delta_{\beta\delta}A(p_1+p_3) + \delta_{\alpha\delta}\delta_{\beta\gamma}A(p_1+p_4)$$

$$A(p_1) = -\frac{1}{3}g + \frac{\frac{1}{9}g^2 3}{p^2 - 2\mu + 3h/\varphi_0}$$

$$= -\frac{1}{3}g - \frac{6g\mu}{9} \frac{1}{p^2 - 2\mu + 3h/\varphi_0} \tag{10.40}$$

$$= -\frac{g}{3} \frac{p^2 + 3h/\varphi_0}{p^2 - 2\mu + 3h/\varphi_0} \qquad (\mu < 0).$$

At $h = 0$, $A(p)$ vanishes like p^2, decreasing the infrared divergences of diagrams of Fig. 10.4. Similar results persist at higher orders. The only

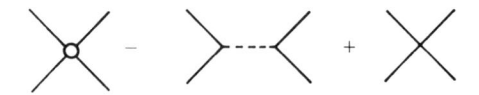

Figure 10.5. The blob of Figure 10.4 at zero loops, the dashed and solid lines being the longitudinal and transverse two-field correlation functions.

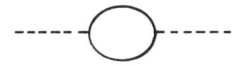

Figure 10.6. Infrared interesting one-loop correction to the longitudinal propagator.

new phenomenon to arise at the one-loop order is the divergence of the longitudinal susceptibility: The diagram of Fig. 10.6 gives

$$\chi_L \sim h^{(D-4)/2} \qquad h \to 0$$
$$G_L(p, h)|_{h=0} \sim p^{D-4} \qquad p \to 0. \qquad (10.41)$$

We immediately see that something disturbing would happen if Eq. (10.41) held in two dimensions; indeed, when $D = 2$ the vanishing of the four-point vertex at zero external momentum is not enough to tame the infrared divergences, as can be easily checked.

A careful analysis is needed to show that Eq. (10.41) is correct at all orders. The techniques introduced in the next section have been very useful in clearing up this point.

10.4. Near two dimensions

When the dimensions D of the space become equal to 2, two things happen simultaneously. The presence of Goldstone bosons induces infrared divergences in the perturbative expansion, and the symmetry-breaking transition disappears. Here we want to connect these two facts and to use the one to better understand the other.

We shall see that this approach, which was proposed about ten years ago,[23] works very well for Heisenberg models with N components [more generally for $O(N)$-symmetric models] in the case when N is larger than 2. When N is in this range the critical temperature and the critical exponents can be computed in powers of $\varepsilon = D - 2$. At exactly two dimensions, where no transition is supposed to be present, the susceptibility and the coherence length should remain finite at nonzero temperature. Their temperature dependence in the low-temperature region can be evaluated using the same techniques.

Let us work in $2 + \varepsilon$ dimensions. If the critical temperature $T_c(\varepsilon)$ goes to zero when $\varepsilon \to 0$ (e.g., $T_c(\varepsilon) \propto \varepsilon$), we should be able to use the low-temperature expansion to localize $T_c(\varepsilon)$. On the other hand, the low momentum correlation function of two and four transverse fields is g independent at fixed magnetization: the structure of the infrared singularities must be the same in the limit $g \to \infty$, where the Landau-Ginsburg model reduces to the Heisenberg model. These two facts together suggest that we should find the solution to our problems in the low-T expansion for the Heisenberg model.[24]

It is convenient to write the partition function of the Heisenberg model for small T in the limit $(h \to 0)$ [see Eq. (3.3)] as

$$\int \prod_a^{N-1} (d\pi_a(i)) \left(1 - T \sum_a^{N-1} \pi_a^2(i)\right)^{-1/2}$$

$$\times \exp\left(-\frac{1}{2} \sum_{i,k} J_{ik} \left\{ \sum_a^{N-1} (\pi_a(i) - \pi_a(k))^2 \right.\right.$$

$$\left.\left. + \frac{\left[\left(1 - T \sum_a^{N-1} \pi_a^2(i)\right)^{1/2} - \left(1 - T \sum_a^{N-1} \pi_a^2(k)\right)^{1/2}\right]^2}{T} \right\}\right) \quad (10.42)$$

$$\sigma_a(i) = T^{1/2} \pi_a(i) \qquad a = 1, \ldots, N-1$$

$$\sigma_N(i) = \left(1 - T \sum_a^N \pi_a^2(i)\right)^{1/2}.$$

In writing Eq. (10.42) we have implicitly assumed that there is a spontaneous magnetization in the Nth direction $\langle \sigma_N \rangle \neq 0$, and the spontaneous magnetization in the other directions is zero, the choice of direction being obviously arbitrary. This parametrization of the σ's is particularly convenient at low temperatures. Exactly at $T = 0$ we have $\langle \sigma_N \rangle = 1$, and all other connected correlation functions are zero: The system is frozen in the configuration where all the spins are parallel and the $O(N)$ symmetry is spontaneously broken. The exponent of Eq. (10.42) is just the old Hamiltonian (10.34), written as a function of the new variables, while the factor $(1 - T \sum_a^{N-1} \pi_a^2(i))^{-1/2}$ comes from the Jacobian of the transformation from the σ to the π variables. In the limit $T \to 0$ we find that the π distribution is Gaussian. The expectation value of σ_N and the correlation functions of the π's in momentum space are given, respectively, by

$$\langle \sigma_N(k) \rangle = 1 - \frac{T}{2} \left\langle \sum_a^N \pi_a^2(k) \right\rangle + O(T^2)$$

$$= 1 - \frac{T}{2} \frac{N-1}{(2\pi)^2} \int_B d^2 p \, \frac{1}{(2 - \cos p_x - \cos p_y)}, \quad (10.43)$$

$$C_{ab}(p) = \delta_{ab} \frac{1}{(2 - \cos p_x - \cos p_y)}, \qquad a, b = 1, \ldots, N-1.$$

10.4. Near Two Dimensions

When $D \to 2$ ($\varepsilon \to 0$), both $\langle \sigma_N \rangle$ and $\langle \sigma_a(x)\sigma_a(0)\rangle$ are infrared divergent. Fortunately no divergence is present at $D=2$ in the $O(N)$-symmetric correlation,[25]

$$G_I(l) \equiv \sum_a^N \langle \sigma_a(l)\sigma_a(0)\rangle = \sum_a^{N-1} T\langle \pi_a(l)\pi_a(0)\rangle + \langle \sigma_N(l)\sigma_N(0)\rangle$$

$$= \sum_a^{N-1} \delta_{aa} T/(2\pi)^2 \int_B dp \, \exp(ilp) \frac{1}{(2-\cos p_x - \cos p_y)} + 1$$

$$- \frac{1}{2} T \sum_a^{N-1} (\pi_a^2(l) + \pi_a^2(0)) + O(T^2)$$

$$= 1 - T(N-1)/(2\pi)^2 \int_B dp[1-\exp(ilp)] \frac{1}{2-\cos p_x - \cos p_y} .$$

$$(10.44)$$

The divergence of $\langle \sigma_N \rangle$ is a signal that we are doing an expansion around the wrong point $\langle \sigma_N \rangle \neq 0$; conversely, if we consider the $O(N)$-invariant correlation functions, we lose track of which non-$O(N)$-invariant configuration we have used as the starting point of the perturbative expansion.[26]

The evaluation of the critical exponents in $2 + \varepsilon$ dimensions is based on the observation that the theory, which has the following partition function in the continuum,

$$\int \prod_a^N d[\sigma] \exp\left[-\frac{1}{g}\int d^Dx \sum_a^N \frac{(\partial_\mu \sigma_a)^2}{2}\right] \delta\left(\sum_a^N \sigma_a^2 - 1\right)$$

$$\propto \int d[\pi] \exp\left(-\int d^Dx \frac{1}{2}\left\{\sum_a^{N-1}(\partial_\mu \pi_a)^2\right.\right.$$

$$\left.\left. + \frac{1}{g}\left[\partial_\mu\left(1 - g\sum_a^{N-1}\pi_a^2(x)\right)^{1/2}\right]^2\right\}\right) \times \left(1 - g\sum_a^{N-1}\pi_a^2(x)\right)^{-1/2}$$

$$(10.45)$$

$$\sigma_a(x) = g^{1/2}\pi_a(x), \quad \sigma_N(x) = (1 - g\pi_a^2(x))^{1/2},$$

must be renormalizable in two dimensions, because the coupling constant is dimensionless ($[\sigma] = [\pi] = (D-2)/2$, $[g] = 2 - D$). The theory turns out to be asymptotically free, and the methods of the preceding chapter can be used to derive the $2 + \varepsilon$ expansion. In order to implement this program we can study either the $O(N)$-noninvariant or the $O(N)$-invariant correlation functions.

We sketch the computation of the exponents and of T_c in the second approach. For simplicity we work in configuration space. Equation

(10.44) gives us, for large l,

$$G_I(l) \sim 1 - AT \ln l \qquad D = 2$$

$$G_I(l) \sim 1 + TA \frac{l^{-\varepsilon} - 1}{\varepsilon} \qquad D = 2 + \varepsilon \qquad (10.46)$$

$$A = (N-1)/(2\pi) + O(\varepsilon).$$

We now define a renormalized coupling constant (t_R) at the momentum scale $M = 1/l$:

$$t_R = -l \frac{\partial}{\partial l} \ln G_I(l)|_{l=1/M} = ATM^\varepsilon + O(T^2). \qquad (10.47)$$

We also define the b function as

$$M \frac{\partial}{\partial M} t_R = b(t_R), \qquad (10.48)$$

in the same way as in Eq. (9.26).

In principle, the function $b(t_R)$ depends on t_R and l/a. However, the theory is renormalizable in perturbation theory, and the renormalized correlation functions should go to a finite limit when a goes to zero at fixed renormalized coupling constant. This observation implies that the function $b(t_R, l/a)$ should go to a finite limit when a goes to zero, or better when l goes to infinity. In order to lighten the notation we write directly $b(t_R)$ instead of $\lim_{l/a \to \infty} b(t_R, l/a)$. A one-loop computation gives, for large M,

$$t_R(M) = ATM^\varepsilon - \frac{A(N-2)T^2 M^{2\varepsilon}}{2\pi} \left(\frac{1}{\varepsilon} + O(1)\right) + O(T^3)$$

$$b(t_R) = \left(\varepsilon t_R - \frac{N-2}{N-1} t_R^2\right). \qquad (10.49)$$

The same arguments as in the preceding chapter tell us that $b(t_R)$ remains finite in perturbation theory in the limit $D \to 2$, so that the ε expansion in $2 + \varepsilon$ dimensions can be constructed if $N > 2$. One finds that the critical exponent η is given by the condition[27]

$$D - 2 + \eta \equiv t_R^c \qquad b(t_R^c) = 0, \qquad (10.50)$$

which implies

$$\eta = \frac{\varepsilon}{N-2} + O(\varepsilon^2), \qquad t_R^c = \frac{N-1}{N-2} \varepsilon + O(\varepsilon^2). \qquad (10.51)$$

The ε expansion is poorly convergent[28] for $\varepsilon = 1$ [$\eta|_{\varepsilon=1}$ for $N = 3$ is about 0.03, not $O(1)$].[29]

10.4. Near Two Dimensions

In $2 + \varepsilon$ dimensions the $b(t)$ function has the shape shown in Fig. 9.2(c). If T is small enough we stay in the domain of attraction of the origin: the long-distance behavior can be computed using perturbation theory.[30] If T is greater than T_c we do not stay in the origin's domain of attraction: $t(M)$ will increase when M goes to zero. If the correlation functions are exponentially damped at large distance (finite correlation length), as we expect, we must have

$$G_l(x) \sim \exp\left(-\frac{x}{\xi}\right), \qquad t_R(M) \sim \frac{1}{M\xi} \quad (M \to 0)$$

$$b(t_R) \sim t_R \qquad \text{for } t \gg t_R^c. \tag{10.52}$$

However, this region cannot easily be reached using perturbation theory. The value of T (T_c), which separates the two regimes, is characterized by a different long-distance behavior, as given by Eq. (10.46).[31] The exponential decay of the correlation functions [Eq. (10.52)] follows if we assume that at $T > T_c$ no phase transitions are present. The usual high-temperature phase should extend up to T_c. The correlation length in this region can be computed by using the high-temperature expansion.

In the same spirit we could conjecture that in two dimensions no phase transitions are present for $N > 2$: the temperature at which the spontaneous magnetization appears goes to zero with $D = 2$. Moreover, we know that a transition to a phase with nonzero magnetization cannot be present in $D = 2$ (this is the main theorem of the preceding sections). This conjecture (which is very likely to be true) is not so innocent as it looks. We shall see soon that a phase transition is present for $D = 2$, $N = 2$, in spite of the absence of spontaneous magnetization.

We now compute the temperature dependence of the coherence length ξ near $T = 0$. Equations (10.47)–(10.50) imply that in the region of l where the correlation functions go to zero exponentially, the renormalized coupling t_R is given by l/ξ, and it is therefore of order 1 (at least). At a distance of a few lattice spacings t_R is very small, of order T ($AT + O(T^2)$). We can now use Eq. (10.48) to investigate the values of $M = 1/l$ at which t_R may become of order 1 (let us call it $\tilde{M}$). It seems reasonable to assume that the value of M^{-1} at which $t_R(M)$ becomes equal to 1 is proportional to the coherence length (this hypothesis will be verified later). We can now integrate Eq. (10.48).

We already know that $b_2 = (N-2)/(N-1)$. A two-loop computation gives $b_3 = (N-2)/(N-1)^2$. If for small values of t_R we have at $D = 2$

$$b(t_R) = -b_2 t_R^2 - b_3 t_R^3 + \tilde{b}(t_R) \qquad \tilde{b}(t_R) = O(t^4), \tag{10.53}$$

in the region of small T we get

$$\ln \frac{\tilde{M}}{a} = \int_t^1 \frac{dt_R}{b(t_R)} \simeq -\frac{1}{b_2 t} - \frac{b_3}{b_2^2} \ln t + \text{const.} + O(t) \qquad (10.54)$$

$$t \equiv AT.$$

We finally find

$$\xi \propto \tilde{M}^{-1} = \text{const. } \exp\left(\frac{1}{tb_2}\right) t^{b_3/b_2^2}[1 + O(t)]$$

$$= \text{const. } \exp\left(\frac{2\pi}{T(N-2)}\right) \cdot T^{1/(N-2)}[1 + O(T)]. \qquad (10.55)$$

The coherence length goes to infinity when $T \to 0$. The prefactor cannot be computed in perturbation theory because it requires a knowledge of $b(t)$ over the whole range of t. Alternatively we could say that if the lattice spacing goes to zero at fixed coherence length the temperature must approach zero logarithmically. The situation is very clear from the point of view of the theory defined on the continuum. The correlation functions of this theory satisfy the renormalization group equations of the preceding chapter, which imply that the continuum theory is essentially unique, apart from possible redefinitions of scale. Dimensionless quantities are therefore fixed in the continuum limit. For this reason we can safely assume that the value of M at which $t_R = 1$ (i.e., $\tilde{M}$) goes to a finite limit when a goes to zero, if everything is measured in units of the coherence length.

Equation (10.55) can be rederived (for simplicity we pay attention only to the exponential term) by noticing that the running coupling constant can be written in the large-momentum region as (see p. 152)

$$t_R(q^2) \equiv t_R(M)|_{M^2 = q^2} \simeq \frac{2}{b_2 \ln q^2/\tilde{M}^2}. \qquad (10.56)$$

It is possible to verify that the running coupling constant at values of q^2 at the end of the Brillouin zone [i.e., $q^2 = (\pi/a)^2$] and the temperature T coincide at low temperature; more precisely, we have

$$t_R\left(\frac{\pi}{a}\right) = AT + O(T^2), \qquad (10.57)$$

where higher-order corrections are computable in perturbation theory. Equation (10.57) is equivalent to Eq. (10.54).

A similar prediction can be obtained for the magnetic susceptibility:

$$\chi = \text{const. } T^{3/(N-2)} \exp\left(\frac{2\pi}{T(N-2)}\right)[1 + O(T)]. \qquad (10.58)$$

10.5. On Metastability

These predictions are very interesting. They have been checked against the results of the high-temperature expansion and the numerical simulations done with the Monte Carlo method (see Chapter 20).[32]

The situation for $N = 2$ in two dimensions is more complex: the function $b(t)$ vanishes not only at the first order in T but at all orders in the perturbative expansion, suggesting a power-law decay. Different techniques must be used to study this problem. After a certain amount of work the model can be reduced to a study of the behavior of a gas of charged particles in two dimensions, and this problem has been investigated by using the renormalization group equations.[33] At the end of a beautiful analysis, which the reader is strongly urged to look at in the original literature, one finds that

$$G_I(l) \xrightarrow[l \to \infty]{} l^{-\eta(T)} \quad T < T_c ; \quad G_I(l) \xrightarrow[l \to \infty]{} \exp\left[-\frac{l}{\xi(T)}\right] \quad T > T_c$$
(10.59)
$$\eta(T_c) = \frac{1}{4}, \quad \xi(T) \sim \exp[(T_c - T)^{-1/2}] \quad T \sim T_c .$$

Quite surprisingly it is possible to measure $\eta(T_c)$ in a reliable way:[34] the experimental results are in wonderful agreement with the theoretical value $\frac{1}{4}$.

Unfortunately it is not so clear what happens in $2 + \varepsilon$ dimensions for $N = 2$, although interesting suggestions have been put forward.[35]

10.5. On metastability

In this section we study the properties of the free energy in the metastable region. The case we study is a magnetic system with positive magnetization and negative small magnetic field in Ising-type models[36] ($N = 1$).

In this case at all orders in the perturbative expansion or in the low-temperature expansion one finds that $f(h)$ is equal to two different analytic functions $f_+(h)$ and $f_-(h)$, for h positive and negative, respectively $[f_+(h) = f_-(-h)]$, each $f_\pm$ also being an analytic function at $h = 0$. Here we argue that actually the function $f_+(h)$ has an essential singularity[37] at $h = 0$ (although it remains C^∞), i.e., a cut on the negative h axis, the imaginary part (the discontinuity on the cut) being exponentially small when $h \to 0$ (see Fig. 10.7).[38]

A fully rigorous proof of this result is still lacking; an heuristic argument may be constructed using the same technique as in the proof of the no-go theorems of Sec. 10.2. Here we consider only a portion of the configuration space, selected for our purposes: inside a sphere of very large radius R the magnetization is negative, while outside the magnetiza-

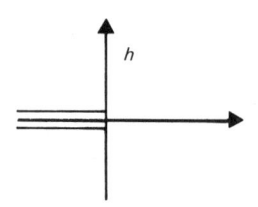

Figure 10.7. A cut on the negative h axis on the complex h plane.

tion is positive.[39] The contribution to the partition function of this class of configuration is[40] roughly given by

$$Z = Z_0 + \int_0^\infty \exp[-(R^{D-1} S_D \Sigma + 2mhR^D V_D)] \, dR$$

$$= Z_0 + \tilde{Z}(h), \quad (10.60)$$

where Σ is the increase in free energy due to the presence of the interface[41] and m is the value of the spontaneous magnetization. If $h < 0$, the integral is no longer convergent, signaling that the negative magnetization phase is the stable one. The analytic continuation of the integral representation from positive to negative h can be written

$$h = |h| \exp[i\theta]$$
$$R = |R| \exp\left[-\frac{i\theta}{D}\right]. \quad (10.61)$$

In this way we get

$$\tilde{Z}(|h|, \theta) = \int_0^\infty d|R| \exp\left[-\frac{i\theta}{D}\right] \exp\left[-2|R|^D V_D m|h|\right.$$
$$\left. - \Sigma S_D |R|^{D-1} \exp\left(-i\theta \frac{D-1}{D}\right)\right]. \quad (10.62)$$

The integral is convergent for all θ's but assumes different values for $\theta = \pm \pi$. The imaginary part may be estimated by using the saddle-point method, which corresponds to deforming the integration path in the complex R plane, as shown in Fig. 10.8. One finally gets

$$\operatorname{Im} \tilde{Z}(h, \pi) \simeq \exp\left[-\frac{c}{|h|^{D-1}}\right]$$

$$c = \frac{V_D (D-1)^D \Sigma^D}{(2m)^{D-1}} \quad (10.63)$$

$$R_c = \frac{(D-1)\Sigma}{m|h|}.$$

10.5. On Metastability

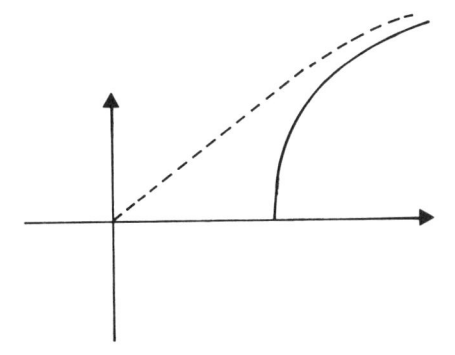

Figure 10.8. The integration path required by the saddle-point method.

We have thus found that the imaginary part of the free energy is proportional to the minimum as a function of R of the integrand in Eq. (10.50) at negative h ($R_c \sim 1/|h|$).

If we look at the real-time evolution of a system with a positive magnetization but negative magnetic field at the initial time,[42] we should see that spontaneous fluctuations form droplets of negative magnetization. If the radius of such a droplet is smaller than the critical radius R_c, after its formation the droplet would naturally tend to shrink, whereas if the radius is larger than R_c, it would expand (with very high probability) until it became as large as the whole system. The probability of creating by fluctuations a critical droplet is proportional to $\exp[-H(R_c)]$; the corresponding time we must wait to see such a fluctuation (i.e., the mean life of the metastable state) is thus proportional to $\exp H(R_c)$. The interface energy is proportional to $\exp(2\beta)$ for large β in the Ising model and to $1/g$ for small g in the Landau-Ginsburg model; in both cases for finite h Eq. (10.63) is exponentially small in the parameters characterizing the perturbative expansions (in $t = \exp(-4\beta)$ or g): it is not surprising that Im $Z(h, \pi)$ vanishes perturbatively at all orders.

Equation (10.63) is essentially correct, as far as the exponential term is concerned, but in order to obtain predictions for the prefactor as well, much more care[43] is needed. A careful analysis tells us that[44]

$$\text{Im } f_+(h, \pi) \sim B|h|^b \exp\left(-\frac{A}{h^{D-1}}\right)$$

$$b = \frac{(3-D)D}{2}, \quad 1 < D < 5, \quad D \neq 3 \quad (10.64)$$

$$b = -\frac{7}{3}, \quad D = 3.$$

The computation of the constant A is not simple, so it is more convenient to test the power of h of the prefactor. Arguments similar to those of Chapter 5 imply that

$$f_+(h) \simeq \sum_{n}^{\infty} f_n h^n, \qquad f_n \propto \left(-\frac{1}{A}\right)^{n/(D-1)} \Gamma\left(\frac{n}{D-1} - \frac{b-1}{D-1}\right). \quad (10.65)$$

It is possible to use the lengthy series of the low-temperature expansion for estimating the f_n at a given value of β not too near to T_c at $D = 2$. Now Eq. (10.65) implies (see p. 60)

$$b = \lim_{n \to \infty} b_n \quad (10.66)$$

$$b_n = 1 - n^2 \left[\frac{f_n^2}{f_{n-1} f_{n+1}} - 1\right].$$

We expect, therefore,

$$c_n = 1 - b_n = 0 + O\left(\frac{1}{n}\right). \quad (10.67)$$

The plot of the c_n versus $1/n$ is shown in Fig. 10.9. The agreement is impressive.

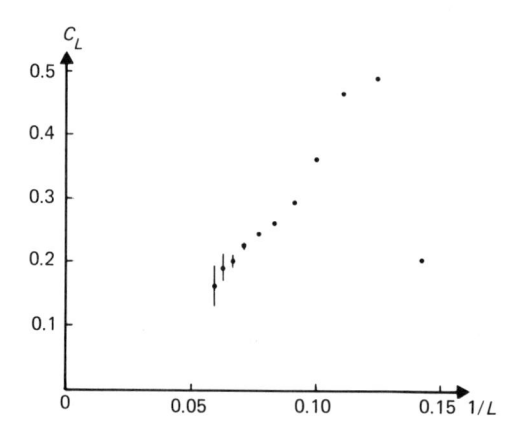

Figure 10.9: The coefficients C_n versus $1/n$.

Notes for Chapter 10

1. The procedure described here is the evaluation of an integral by developing it around the maximum point: the absolute maximum of the integrand $[\exp(-H)]$ corresponds to the absolute minimum of the Hamiltonian.
2. The structure of ultraviolet divergences is quite similar to that for $h = 0$, and indeed a theorem by Symanzik states that, under general hypothesis, the presence of an external field or spontaneous symmetry breaking cannot introduce new ultraviolet divergences.
3. C. De Dominicis, *J. Math. Phys.* 4, 255 (1962); C. De Dominicis and P. C. Martin, *J. Math. Phys.* 5, 14 (1963); G. Jona Lasinio, *Nuovo Cimento* 34, 1790 (1964).
4. See, for example, L. D. Landau and E. M. Lifshitz, *Statistical Physics*, Pergamon, Oxford (1968).
5. E. Brèzin, D. J. Wallace, and K. G. Wilson, *Phys. Rev. Lett.* 29, 591 (1972); *Phys. Rev. B* 7, 232 (1972).
6. One also finds $\langle(\theta(x) - \theta(y))^2\rangle \sim A \ln(x-y)^2$ for large distances at $D = 2$, suggesting that no spontaneous symmetry is present: indeed, in this approximation the correlation function $\langle \varphi_a(x)\varphi_b(y)\rangle$ is given by $\delta_{ab}\varphi_0^2 \exp[-\langle(\theta(x) - \theta(y))^2\rangle/2] \sim \delta_{ab}\varphi_0^2 |x-y|^{-A}$ and it goes to zero at infinity, indicating the absence of spontaneous symmetry breaking.
7. If long-range forces are present, the theorem no longer holds. Spontaneous magnetization is present in the one-dimensional Ising model if $J_{ik} \sim |i - k|^{-2}$ at large $|i - k|$. See P. W. Anderson and Y. Yuval, *J. Phys. C* 4, 607 (1971) and references therein.
8. See, for example, D. Ruelle, *Statistical Mechanics*, Benjamin, New York (1969) and references therein.
9. The full proof can be found in R. L. Dobrushin and S. B. Shlesman, *Commun. Math. Phys.* 53, 299 (1977).
10. Extreme sensitivity of the expectation values to boundary conditions, which is typical of the coexistence of two phases, must be absent.
11. In the one-dimensional model of note 7, $\Delta E(L) \sim \ln L$, so that the symmetry can be spontaneously broken. Indeed, we can easily see that if $J_{ik} \sim (i - k)^{-\alpha}$, spontaneous magnetization is possible only if $\alpha \leq 2$ (the condition that the total free energy does not increase more than the volume implies $\alpha > 1$).
12. We can take L_0 large enough that $\Delta E(L_0) \sim \Delta E(\infty)$.
13. The same train of argument would imply that if $H \sim (\varphi_i - \varphi_k)^4 J_{ik}$ on the lattice or $(\Sigma_\mu (\partial_\mu \varphi)^2)^2$ on the continuum, no spontaneous symmetry breaking is possible as soon as $D \leq 4$. (G. Jona Lasinio, S. Pierini, and A. Vulpiani, *Lett. Nuovo Cimento* 23, 353 (1978).)

14. We recall that the theorem is based on probabilistic ideas, and it may fail in the presence of second-quantized fermions (see Chapter 15).
15. Similar considerations arise in the study of block walls; see C. Kittel, *Introduction to Solid State Theory*, John Wiley and Sons, New York (1966).
16. Equation (10.33) is not exact because χ_L is infinite: the correct expression is given by Eq. (10.41). However, the difference is not important here.
17. There is an infinite number of equivalent directions; to simplify the notation we have chosen one.
18. The identities between different correlation functions (like $\chi_T = m/|h|$), which come from the invariance of the free energy with respect to a group of transformations, are called Ward identities.
19. The transverse and longitudinal susceptibilities (χ_L and χ_T) are proportional to the variation of the magnetization when a new external field is added in a direction transverse or longitudinal with respect to the magnetization.
20. The Goldstone theorem has been proven in statistical mechanics by V. M. Hugenholtz and D. Pines, *Phys. Rev. 116*, 489 (1959). The equivalence of quantum field theory with statistical mechanics implies that the theorem holds also in this case; such equivalence was not very clear at that time. The first two quantum field-theoretical models of spontaneous breaking of a symmetry (in which the existence of long-range correlations, i.e., massless particles was stressed) were Y. Nambu and G. Jona Lasinio, *Phys. Rev. 122*, 345 (1961); *Phys. Rev. 124*, 246 (1961); and J. Goldstone, *Nuovo Cimento 19*, 154 (1961). A full proof in this context appears later: J. Goldstone, A. Salam, and S. Weinberg, *Phys. Rev. 127*, 965 (1962).
21. This has been stressed by G. Jona Lasinio, *Acta Phys. Hung. 19*, 139 (1964); see also G. Parisi and M. Testa, *Nuovo Cimento 67A*, 13 (1970).
22. A telegraphic proof of the no-go theorem of the previous section runs as follows: the spontaneous breaking of a continuous symmetry implies the presence of Goldstone bosons. Goldstone bosons cannot exist in two dimensions ($1/p^2$ is not a Fourier transform of a function), ergo continuous symmetries are not spontaneously broken in two dimensions.
23. The $2 + \varepsilon$ expansion for the nonlinear σ model was begun by A. A. Migdal, *Zh. Eksp. Teor. Fis. 69*, 1457 (1976), and A. M. Poliakov, *Phys. Lett. B 59*, 79 (1975); a systematic treatment has been carried out in E. Brézin, J. C. Le Guillou, and J. Zinn-Justin, *Phys. Rev. Lett. 36*, 691 (1976) and *Phys. Rev. B 14*, 3110 (1976). We follow here a slightly modified approach suggested by D. J. Amit and G. Kotliar, *Nucl. Phys. B 170* [FS1] 187 (1980).
24. In the Ising model the critical temperature is proportional to ε in

1 + ε dimensions. The critical exponents can be computed by a careful analysis of the interface (which has ε dimensions) in such a strange situation.

25. This result was suggested by A. Jevicki, *Phys. Lett. B 71*, 327 (1977); a first (incomplete) proof was offered by S. Elitzur, *Phys. Rev. D 12*, 3978 (1978); a full proof can be found in F. David, *Nucl. Phys. B 190* [FS3] 205 (1981).
26. The cancellation of infrared divergences in Eq. (10.44) is trivial at this order. Each of the two divergent contributions is l independent; however, $(1/N)G_l(l)|_{l=0} = 1$. Thus no l-independent divergence can be present in $G(x)$. The full proof of the absence of infrared divergences is not simple. In the continuous version of this model [Eq. (10.45)] the absence of infrared divergences at $D = 2$ can be proven using the following general theorem: in a theory where the coupling constant is dimensionless, infrared divergences can be present only if local operators of dimension zero are available. In the present case the only nontrivial zero-dimension operators are not invariant under the group $O(N)$ (i.e., $\sigma_a(x)$ or $(\sigma_a(x)^k)$), while $O(N)$-invariant zero-dimension operators are trivial: $\sum_a^N \sigma_a^2 = 1$.
27. The critical exponent ν is given by

$$\nu(\varepsilon) = \frac{1}{\varepsilon} - 1 + \frac{1}{2}\varepsilon + O(\varepsilon^2).$$

28. The same bad convergence happens for the $1 + \varepsilon$ expansion of the Ising model:

$$T_c \propto \varepsilon$$

$$\nu(\varepsilon) = \frac{1}{\varepsilon} - \frac{1}{2}$$

$$(D - 2 + \eta(\varepsilon)) = \frac{8}{\pi} \varepsilon^{-(2+\varepsilon)/2} \exp\left\{-1 - 2\gamma - \frac{2}{\varepsilon}\right\}[1 + O(\varepsilon)],$$

where $\gamma \simeq 0.577$ is Euler's constant. $D - 2 + \eta(\varepsilon)$ is exponentially small when $\varepsilon \to 0$. For a review see D. Wallace in *Proceedings of the 1982 Les Houches Summer School*, ed. by J. B. Zuber and R. Stora, North-Holland, Amsterdam (1984).
29. For $N = 3$ we have $\eta = \varepsilon - 2\varepsilon^2 + 3\varepsilon^3 + O(\varepsilon^4)$.
30. Therefore the leading infrared singularities produced by the Goldstone bosons are the ones we have just computed in Sec. 10.3.
31. We also obtain $T_c = \varepsilon/(N - 2) + O(\varepsilon^2)$, as expected.
32. For example, see G. Martinelli, G. Parisi, and R. Petronzio, *Phys. Lett. B 100*, 485 (1981); G. Fox, R. Gupta, O. Martin, and S. Otto, *Nucl. Phys. B 205* [FS5], 188 (1982).

33. J. M. Kosterlitz and D. J. Thouless, *J. Phys. C* 6, 1181 (1973); J. M. Kosterlitz, *J. Phys. C* 7, 1046 (1974); J. Jose, L. P. Kadanoff, S. Kirkpatrick, and D. R. Nelson, *Phys. Rev. B* 16, 1217 (1977).
34. Thin films of superfluid helium are supposed to stay in the same universality class of the $N=2$ Hamiltonian. It is possible to prove that the superfluid density $\rho_s(T)$ must jump at the transition from a finite value ρ_c to zero. The theoretical prediction for ρ_c depends on $\eta(T_c)$ ($\rho_c = m^2 k T_c / (4\pi \hbar^2 \eta(T_c))$), m being the mass of the helium atom: $\rho_c / T_c = 3.491 \times 10^{-9}$ g/cm^2 K. The experimental result is $\rho_c / T_c = 3.5 \pm 0.5 \times 10^{-9}$ g/cm^2 K. For a review see P. R. Nelson, *Phys. Rep.* 49C, 255 (1979).
35. J. Cardy and H. Hamber, *Phys. Rev. Lett.* 45, 499 (1980).
36. If $N > 1$, Eqs. (10.33)–(10.41) tell us that $f(h)$ has a strong singularity at $h = 0$, which is not present for $N = 1$.
37. In the mean field approximation $f_+(h)$ has a singularity at $h = -|h_M|$, which signals the "end" of the metastable region (for supercooled water this seems to happen at a temperature of about $-40°C$).
38. Im $f_+(h)$ is of same order of magnitude as the inverse mean life of the system in the metastable phase. This purely classical result with a strong quantum-mechanical flavor is derived in J. Langer, *Ann. Phys. (N.Y.)* 41, 108 (1967).
39. Generally where we have coexistence of two phases (I and II) at a first-order transition, we consider a sphere of radius R of phase I, while outside we have phase II.
40. S_D and V_D are the constants giving the surface and the volume of the D-dimensional sphere

$$S_D = \frac{V_D}{D} \qquad S_D = \frac{(2\pi)^{D/2}}{\Gamma(D/2)}.$$

41. Σ is the increase in the free energy associated with a domain wall (divided by the surface of the domain wall); alternatively Σ is the surface tension of a large bubble of negative magnetization in a positively magnetized system.
42. Such a system may be realized by reversing the sign of the external magnetic field at $t = 0$. The time evolution of such a system is of particular interest here: we have already remarked that the mean life of the metastable state, which is produced by this procedure, is inversely proportional to the imaginary part of the free energy.
43. D. Wallace, in *Proceedings of the 1982 Les Houches Summer School*, ed. by J. B. Zuber and R. Stora, North Holland, Amsterdam (1984), p. 173.
44. One considers as droplets spheres centered in different points and regions which have the shape of a slightly deformed sphere.

CHAPTER 11

Other Models

11.1. The Ising model again

In the preceding chapters we constructed the diagrammatic perturbative expansion for the Ginsburg-Landau model. Here we study the Ising case. We have already seen the relations between the high-temperature expansions of the Ising-Gaussian and random walk models. To avoid painful combinatoric proofs, we proceed in a compact way.

We start from the Ising model Hamiltonian, which we write, as usual (see p. 23),

$$H = -\frac{1}{2} \sum_{i,k} J_{ik} \sigma_i \sigma_k - \sum_i h_i \sigma_i. \tag{11.1}$$

We now have (forgetting factors π):

$$\exp[-\beta H] = \det{}^{1/2}\left(\frac{J}{\beta}\right) \int d\varphi_i \exp\left[-\frac{\beta}{2} J_{ik}^{-1} \varphi_i \varphi_k - (\beta\varphi_i + \beta h_i)\sigma_i\right] \tag{11.2}$$

$$\sum_k J_{ik}^{-1} J_{kl} = \delta_{il}.$$

The sum over the configurations of the σ's can be trivially done. We get

$$Z = \det{}^{1/2}\left[\frac{J}{\beta}\right] \int \prod_i d\varphi_i \exp\left[-\frac{\beta}{2} \sum_{i,k} J_{ik}^{-1} \varphi_i \varphi_k\right] \sum_{\{\sigma\}} \exp\left[-\beta \sum_i \sigma_i(h_i + \varphi_i)\right]$$

$$= \det{}^{1/2}\left[\frac{J}{\beta}\right] \int \prod_i d\varphi_i \exp\left\{-\frac{\beta}{2} \sum_{i,k} J_{ik}^{-1} \varphi_i \varphi_k + \sum_i \ln[\cosh \beta(\varphi_i + h_i)]\right\}. \tag{11.3}$$

By differentiating Eq. (11.3) with respect to h and using the linear response theory we obtain

$$\langle \sigma_i \rangle = \langle \tanh[\beta(\varphi_i + h_i)]\rangle$$

$$\langle \sigma_i \sigma_j \rangle_c = \langle \tanh[\beta(\varphi_i + h_i)] \tanh[\beta(\varphi_j + h_i)]\rangle_c \qquad (11.4)$$

$$+ \delta_{ij} \left\langle \frac{1}{\cosh^2[\beta(\varphi_i + h_i)]} \right\rangle,$$

where the expectation values on the r.h.s. of Eq. (11.4) are evaluated with the Hamiltonian

$$\beta H[\varphi] = \frac{\beta}{2} \sum_{i,k} J^{-1}_{i,k} \varphi_i \varphi_k - \ln \cosh[\beta(\varphi_i + h_i)] . \qquad (11.5)$$

If we look for the minimum of $H[\varphi]$ in the presence of a constant field ($h_i = h$), we find

$$\sum_k J^{-1}_{ik} \varphi_k = \tanh[\beta(\varphi_i + h)] , \qquad (11.6)$$

which can also be written as

$$\beta\varphi \equiv \beta\varphi_k = \sum_k J_{ik} \tanh[\beta(\varphi_k + h)] = 2D \tanh(\beta\varphi + h) . \qquad (11.7)$$

If we substitute for the expectation values in Eq. (11.4) their values evaluated at the minimum of $H[\varphi]$, the final result for $\langle \sigma_i \rangle$ can readily be seen to be equivalent to the old mean-field equations.

A better understanding may be obtained by developing $H[\varphi]$ in powers of φ (this may be justified for high temperatures). We obtain at $h = 0$

$$\beta H[\varphi] = \sum_{i,k} \frac{\beta}{2} (J^{-1}_{ik} - \beta\delta_{ik})\varphi_i \varphi_k + \sum_i \left(\frac{\beta^4}{12} \varphi_i^4 - \frac{\beta^6}{45} \varphi_i^6 + O(\varphi_i^8 \beta^8) \right)$$

$$= \sum_{i,k} \frac{1}{2} (J^{-1}_{ik} - \beta\delta_{ik})\tilde{\varphi}_i \tilde{\varphi}_k + \sum_i \left[\frac{\beta^2 \tilde{\varphi}_i^4}{12} - \frac{\beta^3 \tilde{\varphi}_i^6}{45} + O(\beta^4 \tilde{\varphi}_i^8) \right] \qquad (11.8)$$

$$\tilde{\varphi}_i = \beta^{1/2} \varphi_i .$$

By comparing Eq. (11.8) with the results of Chapter 4 we see that considering only the quadratic term in $H[\varphi]$ corresponds to the random-walk approximation, while higher orders in φ give interactions of the type

11.1. The Ising Model Again

we have already studied. If we neglect all powers greater than 4, we get a lattice Landau-Ginsburg-type model, with $g = \frac{1}{2}\beta^2$; the propagator at $g = 0$ given by

$$G(p) = \frac{\sum_{\nu}^{D} 2\cos(p_\nu)}{1 - 2\beta \sum_{\nu}^{D} \cos(p_\nu)} = \frac{1}{J^{-1}(p) - \beta}$$

$$J(p) = \sum_{\nu}^{D} 2\cos p_\nu \qquad (11.9)$$

$$\langle \tilde{\varphi}_0 \tilde{\varphi}_k \rangle = \frac{1}{(2\pi)^D} \int_B d^D p \, G(p) \exp(ipk) \,.$$

In this way we can reconstruct the high-temperature expansion using the rules of the standard perturbative expansion and developing $G(p)$ in powers of β.

The main difference between the Ising and Landau-Ginsburg models is the presence of higher powers of $\tilde{\varphi}$ in the first case. We shall argue later in this chapter that these higher powers do not change the critical exponents and the fixed-point Hamiltonian.

As an exercise we can sum all the diagrams of Fig. 11.1 for the self-energy in the high-temperature phase at $h = 0$. If we write $\ln \cosh \beta^{1/2} \tilde{\varphi} = \sum_n c_n \beta^{n/2} \tilde{\varphi}^n$, it is evident that the self-energy is given by

$$\Sigma = -\sum_{n}^{\infty} n(n-1)c_n \beta^{n/2} \langle \tilde{\varphi}_i^{n-2} \rangle = \beta \langle \tanh^2(\beta^{1/2}\tilde{\varphi}) \rangle$$
$$G(p) = \frac{1}{J(p)^{-1} - \beta - \Sigma}, \qquad (11.10)$$

where the expectation values are taken with the Hamiltonian (11.8) at $\beta = 0$. The simplest way to evaluate $\langle g(\tilde{\varphi}_i) \rangle_0$, g being an arbitrary function, is to notice that, if the Hamiltonian is quadratic, the probability

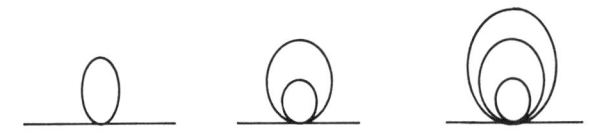

Figure 11.1. Examples of diagrams whose sum is given by Eq. (11.10).

distribution of $\tilde{\varphi}_i$ is a Gaussian with variance $\langle \tilde{\varphi}^2 \rangle_0$. We thus obtain

$$\langle g(\tilde{\varphi}_i) \rangle_0 = \frac{1}{[2\pi \langle \tilde{\varphi}^2 \rangle_0]^{1/2}} \int_{-\infty}^{\infty} dz \, \exp\left[-\frac{z^2}{2\langle \varphi^2 \rangle_0}\right] g(z) \quad (11.11)$$

$$\langle \tilde{\varphi}^2 \rangle_0 = \frac{1}{(2\pi)^D} \int_B d^D p \, \frac{1}{J(p)^{-1} - \beta}.$$

The "gap" equation that corresponds to the sum of all diagrams of the shape shown in Fig. 11.2 is

$$\Sigma = \langle \tanh^2(\beta^{1/2} \tilde{\varphi}^a) \rangle$$

$$= \frac{1}{(2\pi \langle \tilde{\varphi}^2 \rangle)^{1/2}} \int_{-\infty}^{\infty} dz \, \exp\left(-\frac{z^2}{2\langle \tilde{\varphi}^2 \rangle}\right) \tanh^2(\beta^{1/2} z) \quad (11.12)$$

$$\langle \tilde{\varphi}^2 \rangle = \frac{1}{(2\pi)^D} \int_B d^D p \, \frac{1}{J^{-1}(p) - \beta - \Sigma}.$$

This equation is a generalization of Eq. (6.6).

The careful reader has certainly already noticed that Eqs. (11.2)–(11.3) do not make sense; the functional integral representation (11.2) is not convergent because the matrix J is not a positive operator [$J(p)$ is not positive]; none of the functional representations we have written are convergent. The remedy to this difficulty is rather simple in momentum space: We must integrate over the φ fields from $-\infty$ to $+\infty$ when $-\pi/2 \leq p \leq \pi/2$ as before; however, the integration path should be deformed in the complex plane and go from $-i\infty$ to $+i\infty$ when $-\pi \leq p \leq -\pi/2$ or $\pi/2 \leq p \leq \pi$.

One can ignore the integration paths if one perturbs around the

Figure 11.2. An example of a diagram that contributes to Eq. (11.12).

11.2. The Real Gas

Gaussian integrals and uses the rule

$$\frac{\int dx \, \exp(-x^2/2\alpha) \, x^2}{\int dx \, \exp(-x^2/2\alpha)} = \alpha \tag{11.13}$$

even for negative α, as we have implicitly done in the previous equations.

A more subtle problem is the following: the interaction Hamiltonian does not change (apart from a trivial additive factor) if we add to it a term $-\lambda \sum_i \sigma_i^2$, because $\sigma_i^2 \equiv 1$. However, if such a term is included in the Gaussian transformation, the form of the matrix $J(p)$ changes; it becomes

$$\sum_\nu 2 \cos p_\nu + \lambda \, . \tag{11.14}$$

Obviously if we cut the perturbative expansion at any given order, the results depend on λ; in particular, the critical temperature is shifted.[1]

The motivations for sticking to the case $\lambda = 0$ arise from the need to minimize the diagrams with loops. Indeed, at $\beta \simeq 0$, $\lambda = 0$, $\langle \varphi^2 \rangle = 0$, and the diagrams of Fig. 11.1 give a zero contribution. Moreover, if $\lambda = 0$ (or if λ does not increase with D) when $D \to \infty$, all the diagrams of Fig. 11.2 go to zero and we recover the mean-field result. This can be seen by noticing that if $J(p)$ is given by Eq. (11.9) we have that

$$\frac{1}{(2\pi)^D} \int_B d^D p \, J(p) = \frac{1}{(2\pi)^D} \int_B d^D p \, J^3(p) = 0$$

$$\frac{1}{(2\pi)^D} \int_B d^D p \, J^2(p) = 2D \qquad \frac{1}{(2\pi)^D} \int_B d^D p \, J^4(p) = 2D(2D+1) \, . \tag{11.15}$$

The use of similar identities shows that all the loop diagrams vanish when $D \to \infty$, so that mean-field theory becomes exact in this limit. Of course the infinite-D limit must be taken at fixed $\tilde{\beta} = 2D\beta$ (see p. 36).

The detailed proof of this last statement can be left to the willing reader; the reader could also show how to use the loop expansion to generate the $1/D$ expansion. These considerations are not crucial here. Our aim was to show how the Ising model can be cast in a form reminiscent of the Landau-Ginsburg model.[2]

11.2. The real gas

If we have N real classical pointlike particles in a box of volume L^D the Hamiltonian can be written as

$$H = \sum_{1}^{N}{}_i \frac{1}{2m} p_i^2 + U_N(x)$$
$$U_N(x) = \frac{1}{2} \sum_{i,j} V(x_i - x_j),$$
(11.16)

where we suppose that only two-body interactions are present. The partition function is given by

$$Z_N = \int \prod_i d^D p_i \, d^D x_i \, \exp[-\beta H(p, x)] = \left(\frac{2m\pi}{\beta}\right)^{DN/2} \tilde{Z}_N$$

$$\tilde{Z}_N = \int \prod_i d^D x_i \, \exp[-\beta U_N(x)].$$
(11.17)

The thermodynamic limit is obtained when N goes to infinity at fixed density $\rho = N/L^D$.[3] For practical reasons it is more convenient to define the so-called grand partition function

$$Z_G(\beta, z) = \sum_N \frac{z^N}{N!} \tilde{z}_N(\beta),$$
(11.18)

directly in the very large box, where $z = \exp(\beta\mu)$ is the fugacity, and μ is the chemical potential. The use of the grand partition function, the related physical problems, and the expansion of Z_G in powers of z (the virial expansion) are described in many textbooks of statistical mechanics.[4] Here we want only to rewrite Eq. (11.15) as a functional integral. This can be done by using the following identity:

$$\exp\left[-\frac{1}{2} \sum_{k,j}^{N} \beta V(x_k - x_j)\right] = (\det V)^{-1/2} \int d[\varphi] \exp\left[-\int \tfrac{1}{2} V^{-1}(x - y)\right.$$
$$\left. \times \varphi(x)\varphi(y) \, d^D x \, d^D y + i \sum_{k=1}^{N} \beta^{1/2} \varphi(x_k)\right]$$
(11.19)

$$\int dy \, V(x - y) V^{-1}(y - z) = \delta^D(x - z).$$

Equation (11.19) is the standard Gaussian integral formula. Apart from overall constants we can now write

$$Z_G(\beta, z) \propto \int d[\varphi]$$

$$\times \left\{ \sum_N \frac{z^N}{N!} \int \prod_{k=1,N} [dx_k \exp(i\beta^{1/2}\varphi_k)] \exp\left[-\frac{1}{2}\int d^Dx\, d^Dy \right.\right.$$

$$\left.\left.\times \varphi(x)\varphi(y)V^{-1}(x-y)\right]$$

$$= \int d[\varphi] \exp\left[z \int d^Dx \exp(i\beta^{1/2}\varphi(x))\right.$$

$$\left.+ \frac{1}{2}\int d^Dx\, d^Dy\, V^{-1}(x-y)\varphi(x)\varphi(y)\right]. \tag{11.20}$$

We again find a functional integral representation where the $\beta = 0$ correlation function is

$$\langle \varphi(x)\varphi(y)\rangle = V(x-y).$$

We also obtain

$$\rho = \frac{\langle N \rangle}{V} = \frac{1}{V}\frac{\partial}{\partial z}\ln Z_G(\beta, \mu) = \langle \exp(i\beta^{1/2}\varphi(x))\rangle. \tag{11.21}$$

The main differences between this and the Ising case are that all powers (of even and odd) are present and the Hamiltonian depends on two parameters. We expect, therefore, a critical point β_c and μ_c, which is the endpoint of a line of first-order transitions. The model is very similar to the Landau-Ginsburg model in the presence of a magnetic field (if powers of φ greater than 4 are neglected, the correspondence is exact after a shift in φ ($\varphi = \varphi_0 + \tilde{\varphi}^2$, which kills the cubic term).

The first-order transition may be identified with the gas-liquid transition, and the critical point with the endpoint of such a transition. It is clear that after the correct identifications are made, the critical exponents must be the same as for the Ising model, in good agreement with experimental data. It is possible to discuss the details of the phenomenology of the gas-liquid transition and to study the liquid-solid transition in this language. This is beyond the aim of this book.[5] Here we wish only to show that the concept that we have developed to study magnetic systems has a much wider domain of applications.

11.3. On universality

We show here that the presence of extra terms like φ^6 does not change the value of the critical exponents if they are small enough. In other words, we must check that our fixed point (restricted to the space of even interactions) has only one unstable eigenvalue. To this end we consider a theory with the following Hamiltonian:

$$\beta H[\varphi] = \int d^D x \left[\frac{1}{2} (\partial_\nu \varphi)^2 + \frac{\mu}{2} \varphi^2 + \frac{g_4}{4!} \varphi^4 + \frac{g_6 \varphi^6}{6!} \right]. \quad (11.22)$$

We want to prove that for small g_6 (positive) nothing changes qualitatively (with respect to the case $g_6 = 0$) in the small m^2 region where the correlation length becomes infinite. We define the renormalized field φ and the dimensionless renormalized couplings $\lambda_4 \equiv \lambda$ as before, and λ_6 as

$$\lambda_6 = -m^{2(D-3)} \Gamma_R^6(p)|_{p=0}, \quad (11.23)$$

where $\Gamma_R^6(p)|_{p=0}$ is the amputated connected one-particle irreducible correlation function of the renormalized field evaluated at zero external momenta. In order to compare the theory with $g_6 = 0$ and $g_6 \neq 0$, it is convenient to define

$$\tilde{\lambda}_6(g_4, g_6) = \lambda_6(g_4, g_6) - \lambda_6(g_4, 0), \quad (11.24)$$

where the cases $g_6 \neq 0$ and $g_6 = 0$ correspond to different values of μ but to the same values of m^2.

We saw in Chapter 8 that the limit $m^2 \to 0$ can be studied by considering equations of the form

$$m^2 \frac{\partial}{\partial m^2} \lambda_4 \Big|_{g_4, g_6 \text{ fixed}} = b_4(\lambda_4, \tilde{\lambda}_6), \quad m^2 \frac{\partial}{\partial m^2} \tilde{\lambda}_6 \Big|_{g_4, g_6 \text{ fixed}} = b_6(\lambda_4, \tilde{\lambda}_6). \quad (11.25)$$

As a consequence of dimensional analysis, no explicit dependence on m can be present on the r.h.s. of Eqs. (11.25) after the introduction of the dimensionless coupling constants λ_4 and $\tilde{\lambda}_6$. If we assume that in the limit $m^2 \to 0$ both λ_4 and $\tilde{\lambda}_6$ have a finite limit (as a consequence of the scaling laws of Chapter 8), which we call λ_4^c and $\tilde{\lambda}_6^c$, we have

$$b_4(\lambda_4^c, \tilde{\lambda}_6^c) = b_6(\lambda_4^c, \tilde{\lambda}_6^c) = 0. \quad (11.26)$$

The solutions of Eq. (11.26) will be called fixed points. A fixed point of coordinates $\lambda_4^f, \tilde{\lambda}_6^f$ is stable if

$$\lim_{t \to \infty} \lambda_4(t) = \lambda_4^f, \quad \lim \tilde{\lambda}_6(t) = \lambda_6^f, \quad (11.27)$$

where $\lambda_4(t), \tilde{\lambda}_6(t)$ are the solutions of the differential equations

11.3. On Universality

$$\frac{d}{dt}\lambda_4(t) = -b_4(\lambda_4, \tilde{\lambda}_6); \quad \frac{d}{dt}\tilde{\lambda}_6(t) = -b_6(\lambda_4, \tilde{\lambda}_6), \quad (11.28)$$

with boundary conditions $\lambda_4(0)$ and $\tilde{\lambda}_6(0)$ in a neighbor of λ_4^f and $\tilde{\lambda}_6^f$. (In the notation of Chapter 7, a stable fixed point is an attractive fixed point.) It is evident (see p. 120) that the stability of a fixed point is equivalent to the condition that the matrix

$$\begin{vmatrix} \dfrac{\partial b_4}{\partial \lambda_4} & \dfrac{\partial b_4}{\partial \tilde{\lambda}_6} \\ \dfrac{\partial b_6}{\partial \lambda_4} & \dfrac{\partial b_6}{\partial \tilde{\lambda}_6} \end{vmatrix} \quad (11.29)$$

have eigenvalues with positive real parts.

Where are the fixed points of Eq. (11.29)? The origin is an obvious fixed point, which is unstable in fewer than four dimensions; indeed, for small couplings we have

$$\begin{aligned} b_4(\lambda_4, \tilde{\lambda}_6) &= (D-4)/2\lambda_4, \\ b_6(\lambda_4, \tilde{\lambda}_6) &= (D-3)\tilde{\lambda}_6. \end{aligned} \quad (11.30)$$

The definition (11.24) implies also that $\tilde{\lambda}_6(g_4, 0)$ is identically zero; therefore

$$b_6(\lambda_4, 0) = 0. \quad (11.31)$$

In other words, a trajectory starting on the line $\tilde{\lambda}_6 = 0$ will remain on this line. We also have

$$b_4(\lambda_4, 0) = b(\lambda_4), \quad (11.32)$$

where the function $b(\lambda_4)$ has been studied in Chapter 8. Equations (11.31) and (11.32) imply that $(\lambda_4^c, 0)$ is a fixed point if $b_4(\lambda_4^c) = 0$. At the fixed point the stability matrix has the form

$$\begin{vmatrix} \dfrac{\partial b}{\partial \lambda_4} & \dfrac{\partial b_4}{\partial \tilde{\lambda}_6} \\ 0 & \dfrac{\partial b_6}{\partial \tilde{\lambda}_6} \end{vmatrix}, \quad (11.33)$$

with eigenvalues $\partial b/\partial \lambda_4$ and $\partial b_6/\partial \tilde{\lambda}_6$. The stability condition of the old fixed point with respect to a φ^6 perturbation implies that

$$b_6^c \equiv \left.\frac{\partial b_6}{\partial \tilde{\lambda}_6}\right|_{\lambda_4=\lambda_4^c,\ \tilde{\lambda}_6=0} > 0. \tag{11.34}$$

Indeed, if Eq. (11.34) is satisfied we have

$$\tilde{\lambda}_6 \sim (m^2)^{b_6^c} \quad \text{when} \quad m^2 \to 0.$$

In other words, if $b_6^c > 0$ the effect of having a small φ^6 term in the Hamiltonian disappears as $m^{2b_6^c}$; on the other hand, if b_6^c is negative, the presence of a quite small term has a very strong effect: the fixed point would be unstable and the critical behavior would be controlled by other fixed points.

We proceed now to a one-loop computation to see the sign of b_6^c. At one-loop order the only diagrams contributing to $\tilde{\lambda}_6$ are shown in Fig. 11.3. After some combinatoric analysis we obtain [cf. Eq. (8.15)]

$$\Gamma_6(0) = g_6 - \frac{15}{2} \frac{g_4 g_6}{(2\pi)^D} \int \frac{d^D k}{(k^2+m^2)^2}$$

$$= g_6\left[1 - \frac{15}{2}\frac{1}{(4\pi)^{D/2}} \lambda_4 \Gamma\!\left(2-\frac{D}{2}\right)\right]$$

$$\tilde{\lambda}_6 = g_6 \cdot m^{2(D-3)}\left[1 - \frac{15}{2}\frac{\lambda_4}{(4\pi)^{D/2}}\Gamma\!\left(2-\frac{D}{2}\right)\right]$$

$$b_6(\lambda_4, \tilde{\lambda}_6) = -(D-3)\tilde{\lambda}_6 + \tilde{\lambda}_6 \frac{15}{2}\frac{\lambda_4}{(4\pi)^{D/2}}\Gamma\!\left(3-\frac{D}{2}\right) \tag{11.35}$$

$$= \tilde{\lambda}_6\left[-(D-3) + \frac{15}{2}\frac{\lambda_4}{A}\right]$$

$$A = \frac{(4\pi)^{D/2}}{\Gamma(3-D/2)}$$

$$\lambda_4^c = \frac{4-D}{3A}.$$

By differentiating with respect to $\tilde{\lambda}_6$ we finally arrive at

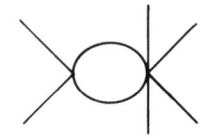

Figure 11.3. The diagram contributing to Eq. (11.35).

$$b_6'(\lambda_4) = \frac{\partial}{\partial \lambda_6} b_6(\lambda_4, \tilde{\lambda}_6)|$$
$$= -(D-3) + \frac{5\lambda_4}{2\lambda_4^c}(D-4). \tag{11.36}$$

We see that when $D < 3$ we start to be in danger: $b_6'(\lambda_4)$ is negative in the small λ_4 region. Fortunately the next term $[O(\lambda_4^2)]$ is large and positive and compensates for the negative term for not too small λ_4. Indeed, if we substitute the value of λ_4^c computed at one loop we find that

$$b_6^c \equiv b_6'(\lambda_4)|_{\lambda_4=\lambda_4^c} = -(D-3) + \frac{5}{2}(D-4)$$
$$= 1 + \frac{3}{2}(4-D) \qquad D < 4. \tag{11.37}$$

Similar results could also be obtained in the framework of the ε expansion at the order ε;[6] in an N-component theory we get

$$b_6^c = 1 + \frac{N+26}{2(N+8)}(4-D). \tag{11.38}$$

This computation can be extended to higher orders in ε by treating together all operators of dimensions near 6 for $D \simeq 4$ [e.g., $(\Delta\varphi)^2$ or $(\partial_\mu \varphi)^2 \varphi^2$]. The estimates (11.37) and (11.38) are probably not very accurate, having been made at only one loop; however, the large positive value we find for b_6^c strongly suggests that it will be positive. The evaluations of the critical exponents presented in the previous chapters are thus consistent. The critical exponents must be the same for a large class of Hamiltonians, and they are stable with respect to perturbations.[7]

Notes for Chapter 11

1. If λ is larger than $2D$, the quadratic term in Eq. (11.3) becomes positive and the deformation of the integration path in the complex plane is not needed [see T. H. Berlin and M. Kac, *Phys. Rev.* **86**, 821 (1952)].
2. The relevance of the results of this and the next section has been stressed by A. M. Poliakov, *Sov. Phys. JEPT* **28**, 533 (1969).

3. If the potential V is attractive, the thermodynamic limit does not exist; conversely, if the potential is repulsive ($V \geq 0$) everything goes smoothly. If the potential changes sign as a function of the distance, a more careful analysis of the potential is needed. See D. Ruelle, *Statistical Mechanics*, Benjamin, New York (1969) and references therein.
4. See, for example, K. Huang, *Statistical Mechanics*, John Wiley and Sons, New York (1963).
5. For example, we could recover the virial expansion by developing $Z(\beta, z)$ in powers of z and by resuming the diagrams, applying the same techniques that were used to derive Eq. (11.11).
6. See, for example, D. Amit, *Field Theory, the Renormalization Group, and Critical Phenomena*, McGraw-Hill, New York, 1978, Chapter X.
7. For other interactions it is possible that the fixed point obtained in the ε expansion becomes unstable when one decreases the dimensions: see, for example, F. Fucito and G. Parisi, *J. Phys. A* 14, L507 (1981).

CHAPTER 12

The Transfer Matrix

12.1. One-dimensional models

The transfer matrix is a crucial tool, which can be used not only to obtain the exact solution of one- or two-dimensional models, but also to connect statistical and quantum mechanics.[1] As an introduction, let us consider a one-dimensional example,

$$\beta H = \frac{\beta}{4} \sum_{(i,k)} (\sigma_i - \sigma_k)^2 + \sum_i V(\sigma_i), \qquad (12.1)$$

where the σ's are continuous variables and the potential $V(\sigma)$ goes to infinity in such a way that the partition function is well defined ($V(\sigma) \propto |\sigma|^\alpha$, $\alpha > 0$ for large σ is certainly enough). We define the integral operator[2] $\hat{T}$:

$$\hat{T}\psi(\sigma) = \int d\mu \, T(\sigma, \mu)\psi(\mu)$$
$$T(\sigma, \mu) = T(\mu, \sigma) = \exp\left[-\frac{\beta}{4}(\sigma - \mu)^2 - \frac{1}{2}V(\sigma) - \frac{1}{2}V(\mu)\right]. \qquad (12.2)$$

The operator $\hat{T}$ (the transfer matrix) is a Hilbert-Schmidt operator (i.e., $\operatorname{tr} \hat{T}^2 < \infty$) acting on functions belonging to $L^2(-\infty, \infty)$. Indeed,[3]

$$\operatorname{Tr} \hat{T}^2 = \int d\mu \, d\sigma \, T^2(\mu, \sigma) < \infty.$$

We can define the eigenvalues and eigenvectors of $\hat{T}$:

$$\hat{T}\psi_n(\sigma) = t_n \psi_n(\sigma) \qquad (t_n \geq t_{n+1})$$
$$\operatorname{Tr} \hat{T}^2 = \sum_n t_n^2 \qquad (12.3)$$
$$T(\sigma, \mu) = \sum_{n=0}^{\infty} t_n \psi_n(\sigma)\psi_n(\mu).$$

221

The spectrum is discrete, and the t_n have 0 as the only accumulation point. We want now to compute the correlation functions directly from the t_n and the ψ_n. Before doing so, we notice that in the Ising case where the σ's takes only two values, the equivalent of Eq. (12.2) is

$$T = \begin{vmatrix} \exp(-V(1)) & \exp\left[-\beta - \frac{1}{2}(V(1) + V(-1))\right] \\ \exp\left[-\beta - \frac{1}{2}(V(1) + V(-1))\right] & \exp(-V(-1)) \end{vmatrix}. \tag{12.4}$$

In other words, we can consider the two-dimensional space of functions of the variable $\sigma = \pm 1$; a basis on this space is given by the vectors $|+\rangle$ and $|-\rangle$, which corresponds, respectively, to the functions $f_\pm(\sigma)$ defined by

$$f_+(1) = 1 \quad f_+(-1) = 0$$
$$f_-(1) = 0 \quad f_-(-1) = 1. \tag{12.5}$$

The matrix T [Eq. (12.4)] is essentially the integral operator (12.2) evaluated in this basis. Generally speaking, if the σ's can take only discrete values, T is a finite-dimensional matrix, while for continuous distributions, $\hat{T}$ acts on an infinite-dimensional space. The difference is negligible for most purposes; Hilbert-Schmidt operators have very similar properties to finite-dimensional operators.

If t_0 is the largest eigenvalue of t (in modulus), it is easy to see that the free energy is given by

$$f = -\frac{1}{\beta} \lim_{N \to \infty} \frac{1}{N} \ln Z_N = -\frac{1}{\beta} \ln(t_0) + O(t_1/t_0)^N \tag{12.6}$$

(t_0 is positive), where Z_N is the partition function of a chain of N spins with periodic boundary conditions. Indeed,

$$Z_N = \int \prod_1^N d\sigma_i \exp\left\{\sum_1^N \left[-\frac{\beta}{4}(\sigma_i - \sigma_{i+1})^2 - V(\sigma_i)\right]\right\}, \tag{12.7}$$

where periodic boundary conditions are enforced by the convention: $(i+1)|_{i=N} = 1$. Just by inspection we see that

$$Z_N = \text{Tr}(\tilde{T} \cdot \hat{T} \cdots \hat{T})_{(N \text{ times})} = \text{Tr}\, \hat{T}^N$$
$$= \sum_0^\infty t_n^N, \tag{12.8}$$

12.1. One-Dimensional Models

if we write $\hat{T}^N$ using N times the multiplication rule:

$$A \cdot B(\sigma, \mu) = \int d\sigma' \, A(\sigma, \sigma') B(\sigma', \mu) . \tag{12.9}$$

In the limit $N \to \infty$, Eq. (12.8) yields Eq. (12.6).

Using this technique we can prove that no phase transition (i.e., a non-analyticity of the free energy) as a function of β may be present. That is, $f = (-1/\beta) \ln t_0$ is an analytic function of β if V depends analytically on β (e.g., linearly) and does not have singularities. Indeed, a well-known theorem states that if the eigenvalues are not degenerate, they must be analytic functions; the only possible singularities may be connected with level crossings.

The proof may be sketched as follows: if $t_n(\beta)$ is an eigenvalue of $\hat{T}$ nondegenerate, perturbation theory tells us that it satisfies the following equation:

$$\frac{d}{d\beta} t_n(\beta) = \left\langle \psi_n \left| \frac{d\hat{T}}{d\beta} \right| \psi_n \right\rangle , \tag{12.10}$$

where we have assumed that $T(\sigma, \mu)$ is a differentiable function of β. Equation (12.10) holds also if we assign an imaginary part to β; if the imaginary part of β is not too large, no singularity may be produced. The function $t_n(\beta)$ is differentiable also continued to complex β, and therefore it must be analytic.[4]

The crucial step for applying this theorem is to prove that the largest eigenvalue of $\hat{T}$, t_0, is not degenerate. This can be done by noticing that

$$t_0 = \sup_{\psi(\sigma)} \frac{\int d\sigma \, d\mu \, T(\sigma, \mu) \psi(\sigma) \psi(\mu)}{\int d\sigma \, \psi^2(\sigma)} = \frac{\langle \psi | T | \psi \rangle}{\langle \psi | \psi \rangle} . \tag{12.11}$$

It is evident from the definition that $T(\sigma, \mu) > 0$. This positivity condition implies that the maximum in (12.11) is reached for a positive ψ. Indeed, if we decompose ψ into its positive and negative components,

$$\psi(\sigma) = \psi_+(\sigma) - \psi_-(\sigma)$$

$$\psi_+(\sigma) > 0 \quad \psi_-(\sigma) > 0 \tag{12.12}$$

$$\psi_+(\sigma) \psi_-(\sigma) = 0 ,$$

we see immediately that the function

$$\tilde{\psi}(\sigma) = \psi_+(\sigma) + \psi_-(\sigma) \tag{12.13}$$

is such that

$$\langle \tilde{\psi}|\tilde{\psi}\rangle = \langle \psi|\psi\rangle$$

$$\langle \tilde{\psi}|\hat{T}|\tilde{\psi}\rangle \geq \langle \psi|\hat{T}|\psi\rangle \qquad (12.14)$$

$$\frac{\langle \tilde{\psi}|\hat{T}|\tilde{\psi}\rangle}{\langle \tilde{\psi}|\tilde{\psi}\rangle} \geq \frac{\langle \psi|\hat{T}|\psi\rangle}{\langle \psi|\psi\rangle}.$$

The equality in (12.14) holds only if $\psi_- = 0$, in the general case, provided that $T(\sigma, \mu)$ is never zero. The maximum eigenvector $\psi_0(\sigma)$ must be positive [more precisely, $\psi_0(\sigma)$ must not change sign, but by convention we say that $\psi_0(\sigma)$ is positive]:

$$\psi_0(\sigma) > 0. \qquad (12.15)$$

The corresponding eigenvalue cannot be degenerate, otherwise there would be two orthonormal positive functions, and this is not possible.

By introducing the transfer matrix we have reduced the evaluation of the partition function of a linear chain to the computation of the largest eigenvalue of a linear operator, this result allows us to use many powerful theorems of analysis, as we have just seen. The transfer matrix is a bridge between two apparently unrelated fields, probability theory and linear operators on Hilbert spaces.

We proceed now to the computation of the correlation functions:

$$\langle g_1(\sigma_i)g_2(\sigma_{i+l})\rangle \equiv C_{g_1,g_2}(l), \qquad (12.16)$$

g_1 and g_2 being two generic functions. Equation (12.8) can be generalized to

$$C_{g_1,g_2}(l) = \lim_{N\to\infty} \frac{\mathrm{Tr}(\hat{g}_1 \hat{T}^l \hat{g}_2 \hat{T}^{N-l})}{\mathrm{Tr}(\hat{T}^N)}, \qquad l \geq 0, \qquad (12.17)$$

where $\hat{g}_1$ is the operator of multiplication by $g_1(\sigma)$, i.e.,

$$\hat{g}_1 \cdot \psi(\sigma) = g_1(\sigma)\psi(\sigma). \qquad (12.18)$$

The integral kernel of the operator $\hat{g}_1$ is thus $\delta(\sigma - \mu)g_1(\sigma)$. In the limit $N \to \infty$ we obtain

$$C_{g_1,g_2}(l) = \sum_n \langle \psi_0|\hat{g}_1|\psi_n\rangle \langle \psi_n|\hat{g}_2|\psi_0\rangle \left(\frac{t_n}{t_0}\right)^l$$

$$= \langle \psi_0|\hat{g}_1 \hat{R}^l \hat{g}_2|\psi_0\rangle, \qquad \hat{R} = \frac{\hat{T}}{t_0}. \qquad (12.19)$$

12.2. A Few Examples

In other words, when N goes to infinity $\hat{R}^N$ becomes the projector on ψ_0,[5]

$$\lim_{N\to\infty} \hat{R}^N = |\psi_0\rangle\langle\psi_0|. \tag{12.20}$$

A similar representation can be derived for higher-order correlation functions, e.g.,

$$\langle g_1(\sigma_i)g_2(\sigma_{i+l_1})g_3(\sigma_{i+l_1+l_2})g_4(\sigma_{i+l_1+l_2+l_3})\rangle$$

$$\xrightarrow[N\to\infty]{} \langle\psi_0|\hat{g}_1\hat{R}^{l_1}\hat{g}_2\hat{R}^{l_2}\hat{g}_3\hat{R}^{l_3}\hat{g}_4|\psi_0\rangle, \quad l_i \geq 0. \tag{12.21}$$

The coherence length ξ (see Chapter 3) is given by

$$\xi = \left[\ln\left(\frac{t_0}{t_1}\right)\right]^{-1}, \tag{12.22}$$

and it is always finite. A byproduct of this analysis is that

$$\langle g(\sigma_i)g(\sigma_{i+l})\rangle > 0 \tag{12.23}$$

for even l always, and for odd l if the eigenvalues t_n of T are all positive.[6]

12.2. A few examples

The simplest case is the one-dimensional Ising model at zero magnetic field, the Hamiltonian being

$$\beta H = \frac{\beta}{2}\sum_i[(\sigma_i - \sigma_{i+1})^2 - 2] = -\beta\sum_i \sigma_i\sigma_{i+1}. \tag{12.24}$$

The transfer matrix (12.4) becomes now

$$\begin{vmatrix} \exp\beta & \exp-\beta \\ \exp-\beta & \exp\beta \end{vmatrix}. \tag{12.25}$$

The two eigenvectors and the corresponding eigenvalues are

$$\psi_0 = \frac{1}{\sqrt{2}}(|+\rangle + |-\rangle) \quad t_0 = 2\cosh\beta$$
$$\psi_1 = \frac{1}{\sqrt{2}}(|+\rangle - |-\rangle) \quad t_1 = 2\sinh\beta. \tag{12.26}$$

We thus find the well-known results [cf. Eq. (4.48)]

$$\lim_{N\to\infty} \frac{1}{N} \ln Z_N = \ln t_0 = \ln(2\cosh\beta)$$

$$\langle \sigma_i \sigma_{i+l} \rangle \underset{N\to\infty}{=} K^l$$

$$K = \frac{t_1}{t_0} = \tanh\beta,$$

$$\xi = -(\ln\tanh\beta)^{-1},$$

(12.27)

where we have used the relations

$$\hat{\sigma}|\pm\rangle = \pm|\pm\rangle$$

$$\hat{\sigma}|\psi_0\rangle = |\psi_1\rangle, \qquad \hat{\sigma}|\psi_1\rangle = |\psi_0\rangle.$$

(12.28)

In a similar way we obtain

$$\langle \sigma_i \sigma_{i+l_1} \sigma_{i+l_2+l_1} \sigma_{i+l_1+l_2+l_3} \rangle = K^{l_1+l_3}, \qquad l_i \geq 0$$

$$\langle \sigma_i \sigma_{i+l_1} \sigma_{i+l_1+l_2} \sigma_{i+l_1+l_2+l_3} \rangle_c = -2K^{l_1+2l_2+l_3}.$$

(12.29)

In the one-dimensional Ising case ξ goes to infinity only for $K \to 1$ (i.e., $\beta \to \infty$). The two-spin correlation function is given in momentum space by

$$G(p) = \frac{1}{1 - K\exp(ip)} + \frac{1}{1 - K\exp(-ip)} - 1$$

$$= \frac{1 - K^2}{1 + K^2 - 2K\cos p},$$

(12.30a)

which (for small momenta and K close to 1) can be written as

$$G(p) \simeq \frac{2(1-K)}{(1-K)^2 + p^2} \qquad \xi \simeq (1-K)^{-1}$$

$$z_1 \simeq 2(1-K).$$

(12.30b)

In the same manner we find that when $K \to 1$, $G_4(0)$ (the Fourier transform of the four-point connected correlation at zero external momenta) is given by

$$G_4(0) \sim \frac{24}{(1-K)^3}.$$

(12.31)

The dimensionless renormalized coupling constant [cf. Eq. (8.4)]

12.2. A Few Examples

$$\lambda = \frac{G_4 z_1^2}{(G_2(0))^4 m^{4-D}} \simeq 6$$

$$m = \xi^{-1} \quad D = 1 \tag{12.32}$$

goes to a finite limit ($\lambda_c = 6$) when K goes to one, in agreement with the scaling laws.[7]

Another simple example is the Gaussian model:

$$\beta H = \sum_i \left[\frac{\beta}{4} (\sigma_i - \sigma_{i+1})^2 + \frac{\mu}{2} \sigma_i^2 \right]. \tag{12.33}$$

The transfer matrix is now

$$T(\sigma, \nu) = \exp\left[-\frac{\beta}{4} (\sigma - \nu)^2 - \frac{\mu}{4} (\sigma^2 + \nu^2) \right]. \tag{12.34}$$

If we look for an eigenvalue of $\hat{T}$ of the form

$$\psi_0(\sigma) \propto \exp\left(-\frac{c}{2} \sigma^2 \right), \tag{12.35}$$

we find that

$$\int d\nu \, T(\sigma, \nu) \exp\left[-\frac{c}{2} \nu^2 \right] = \left[\frac{4\pi}{2c + \mu + \beta} \right]^{1/2} \exp\left[-\frac{c'}{2} \sigma^2 \right]$$

$$c' = \frac{1}{2} \frac{2c\beta + 2c\mu + \mu^2 + 2\beta\mu}{2c + \beta + \mu}. \tag{12.36}$$

The function in (12.35) is an eigenvector of $\hat{T}$ only if $c' = c$, which implies

$$c = \frac{(\mu^2 + 2\beta\mu)^{1/2}}{2}$$

$$t_0 = \frac{4\pi}{[2(\mu^2 + 2\mu\beta)^{1/2} + \mu + \beta]^{1/2}}. \tag{12.37}$$

We can now try to find other eigenvalues of the form

$$\psi_n(\sigma) \propto P_n(\sigma) \exp\left[-\frac{c\sigma^2}{2} \right]. \tag{12.38}$$

The existence of the identity[8]

$$\int_{-\infty}^{+\infty} dy \, \exp[-(x-y)^2] H_n[zy] = \pi^{1/2} (1-z^2)^{n/2} H_n\left(\frac{z}{(1-z^2)^{1/2}} x \right) \tag{12.39}$$

suggests we choose $P_n(\sigma) = H_n(b\sigma)$. Indeed, we find after some algebra that the eigenvalues and eigenvectors of $\hat{T}$ are given by

$$\psi_n(\sigma) \propto H_n(b\sigma) \exp\left(-\frac{c\sigma^2}{2}\right)$$

$$t_n = t_0 \exp(-mn)$$

$$b = \frac{1}{2}\left(\frac{\Delta^2 - \beta^2}{\Delta}\right)^{1/2} \quad (12.40)$$

$$m = \ln\left(1 + \frac{2c + \mu}{\beta}\right) = \ln\left(\frac{\Delta}{\beta}\right), \quad \Delta = 2c + \mu + \beta.$$

The fact that $\ln t_n$ is linear in n should not be a surprise. The two-field correlation function computed in the previous chapters is (in the one-dimensional case) in momentum and configuration space, respectively,

$$\tilde{G}(p) = \frac{1}{\mu + \beta(1 - \cos p)}$$

$$G(l) = \frac{1}{(\mu^2 + 2\mu\beta)^{1/2}} \exp(-ml), \quad (12.41)$$

$$m = \ln\left(1 + \frac{\mu}{\beta} + \left(\frac{\mu^2}{\beta^2} + \frac{2\mu}{\beta}\right)^{1/2}\right).$$

The correlation function

$$G_{q_1,q_2}(l) = \langle \sigma^{q_1}(l)\sigma^{q_2}(0)\rangle, \quad (q_1 \geq q_2) \quad (12.42)$$

is a polynomial in $G(0)$ and $G(x)$ and is of the form

$$G_{q_1,q_2}(l) = \sum_{k=0}^{q_2} [G(l)]^k [G(0)]^{[(q_1+q_2)/2-k]} c_k(q_1, q_2), \quad (12.43)$$

where the $c_k(q_1, q_2)$ are combinatorial factors not relevant here. The presence of only terms proportional to $\exp(-kml)$ with integer k, is crucial. Indeed, if we compare Eqs. (12.19) and (12.43) we get

$$\frac{t_n}{t_0} = \exp(-nm). \quad (12.44)$$

Equation (12.43) makes it possible for us to compute the eigenvalues of $\hat{T}$ in a fast way. The same technique may be used to compute the ψ_n. We should first reconstruct $\langle \psi_n | \sigma^q | \psi_m \rangle = \sigma_{n,m}^q$ by comparing Eqs. (12.19) and (12.43) and extract the eigenvectors from the $\sigma^q_{n,m}$. For the ground state $\psi_0(\sigma)$ the computation is very simple. We know that

$$\langle \sigma^n(0) \rangle = (n-1)!!\, G(0)^{n/2}, \qquad n \text{ even}$$

$$\langle \sigma^n(0) \rangle = 0, \qquad n \text{ odd} \qquad (12.45)$$

$$G(0) = \langle \sigma^2(0) \rangle = \frac{1}{(\mu^2 + 2\mu\beta)^{1/2}} \equiv A.$$

On the other hand, we have

$$\langle \sigma^n(0) \rangle = \int d\sigma \, \psi_0^2(\sigma) \sigma^n. \qquad (12.46)$$

This moment problem has only one solution:[9]

$$\psi_0 = (2\pi A)^{-1/4} \exp\left(-\frac{\sigma^2}{4A}\right). \qquad (12.47)$$

We have thus recovered the results for c and m in an indirect way. We leave to the interested reader the long exercise of checking that if we use the explicit form of the ψ_n [Eq. (12.40)] in Eq. (12.19), we obtain Eq. (12.42) for the $G_{q_1,q_2}(l)$. [Hint: the $\psi_n(\sigma)$ can be considered as the eigenvalues of an harmonic oscillator, and σ can be written as a combination of the creation and destruction operators (a and a^+) of this oscillator....]

12.3. The zero lattice spacing limit

We have already seen the power (and the weakness) of the approach based on the transfer matrix. Now we study what happens when the lattice spacing goes to zero as in Chapter 5.[10] In this case Eqs. (12.1) and (12.2) become

$$H = \sum_i \left[\frac{(\sigma_i - \sigma_{i+1})^2}{2a} + a V(\sigma_i) \right]$$

$$T_a(\sigma, \nu) = \exp\left[\frac{-1}{2a}(\sigma - \nu)^2 - \frac{a}{2}(V(\sigma) + V(\nu)) \right]. \qquad (12.48)$$

We are interested in computing

$$\hat{T}_t = \lim_{a \to 0} (c(a) \hat{T}_a)^{t/a} \qquad t > 0, \qquad (12.49)$$

where $c(a)$ is an opportune normalization factor we introduce in order to have a finite limit when $a \to 0$.[11] After the $a \to 0$ limit, the correlation function in a box of side L can be written as

$$\langle \sigma(t)\sigma(0) \rangle = \frac{\text{Tr } \hat{T}_t \hat{\sigma} \hat{T}_{L-t} \hat{\sigma}}{\text{Tr } T_L}. \tag{12.50}$$

Now $\hat{T}_t$ can also be written as $\exp(-t\hat{\mathcal{H}})$ ($\hat{\mathcal{H}} = \lim_{a \to 0}(-1/a)\ln \hat{T}_a$). If E_0 is the smallest eigenvalue of $\mathcal{H}$ ($\mathcal{H}|\psi_0\rangle = E_0|\psi_0\rangle$), we find that

$$\lim_{L \to \infty} \langle \sigma(t)\sigma(0) \rangle = \langle \psi_0 | \hat{\sigma} \exp[-t(\hat{\mathcal{H}} - E_0)]\hat{\sigma}|\psi_0\rangle$$

$$= \langle \psi_0 | \hat{\sigma} \hat{R}_t \hat{\sigma} | \psi_0 \rangle \tag{12.51}$$

$$\hat{R}_t = \exp[-t(\hat{\mathcal{H}} - E_0)] = \hat{T}_t \exp(tE_0).$$

If we change $c(a)$, we only add a constant to $\hat{\mathcal{H}}$, modifying E_0 but not $E_n - E_0$ or $\hat{R}_t$. How can we find $\hat{\mathcal{H}}$? After the limit $a \to 0$, $\hat{T}_t$ satisfies the equation

$$\frac{d}{dt}\hat{T}_t = -\hat{\mathcal{H}}\hat{T}_t. \tag{12.52}$$

For finite a we have

$$\hat{T}_{t+a} = c(a)\hat{T}_a \hat{T}_t \Rightarrow \frac{\hat{T}_{t+a} - \hat{T}_t}{a} = c(a)\frac{\hat{T}_a - 1}{a}\hat{T}_t. \tag{12.53}$$

By comparing Eqs. (12.52) and (12.53) we see that $\hat{\mathcal{H}} = \lim_{a \to 0}(\hat{T}_a - 1)/a$. We can use the fact that in operator notation

$$\hat{T}_a = (2\pi a)^{1/2} \exp\left[-\frac{1}{2}a\hat{V}\right] \exp\left[\frac{a}{2}\frac{\hat{d}^2}{d\sigma^2}\right] \exp\left[-\frac{1}{2}a\hat{V}\right]. \tag{12.54}$$

The operator $\exp(t/2)(\hat{d}^2/d\sigma^2)$ is an integral operator with kernel

$$\left(\frac{1}{2\pi t}\right)^{1/2} \exp\left[-\frac{(\sigma - \nu)^2}{2t}\right] \equiv G_t(\sigma - \nu). \tag{12.55}$$

This can be seen either by using the Fourier transform with respect to σ

12.3. The Zero Lattice Spacing Limit

or by noticing that

$$\lim_{t \to 0} G_t(\sigma - \nu) = \delta(\sigma - \nu)$$

$$\frac{d}{dt} G_t(\sigma - \nu) = \left(\frac{d}{d\sigma}\right)^2 G_t(\sigma - \nu).$$

(12.56)

Now it is evident that

$$\exp\left[-\frac{a}{2}\hat{A}\right]\exp[-a\hat{B}]\exp\left[-\frac{a}{2}\hat{A}\right] = 1 - a[\hat{A} + \hat{B}] + 0(a^2). \quad (12.57)$$

Substituting in Eq. (12.53) and sending $a \to 0$, we get (if $c(a) = (2\pi a)^{-1/2}$)

$$\frac{d}{dt}\hat{T}_t \equiv -\hat{\mathcal{H}}\hat{T}_t = \left[-\frac{1}{2}\left(\frac{\hat{d}}{d\sigma}\right)^2 + \hat{V}(\sigma)\right]\hat{T}_t \Rightarrow \hat{\mathcal{H}} = -\frac{1}{2}\left(\frac{\hat{d}}{d\sigma}\right)^2 + \hat{V}(\sigma).$$

(12.58)

In other words, $\hat{T}_a$ is given by $\exp(-a\hat{\mathcal{H}} + 0(a^2))$; therefore $(\hat{T}_a)^{t/a} = \exp(-t\hat{\mathcal{H}} + 0(a))$.

Equation (12.58) is the standard Schrödinger equation of quantum mechanics (apart from a crucial factor i in the time derivative). If the potential V goes fast enough to infinity when σ goes to infinity, the spectrum of $\mathcal{H}$ is discrete, with an accumulation point at infinity:

$$\hat{\mathcal{H}}|\psi_n\rangle = E_n|\psi_n\rangle. \quad (12.59)$$

We finally get

$$\int d\mu[\sigma] g_1(\sigma(t)) g_2(\sigma(0)) \equiv \langle g_1(\sigma(t)) g_2(\sigma(0)) \rangle$$

$$= \sum_n \langle \psi_0|\hat{g}_1|\psi_n\rangle\langle \psi_n|\hat{g}_2|\psi_a\rangle \exp[-(E_n - E_0)t]$$

(12.60)

$$\int d\mu[\sigma] = 1$$

$$d\mu[\sigma] \propto d[\sigma]\exp\left[-\int\left(\frac{1}{2}\left(\frac{d\sigma}{dt}\right)^2 + V(\sigma)\right)dt\right].$$

Equation (12.60) gives a precise relation between the eigenvalues and eigenvector of the time-independent Schrödinger equation and a functional integral over a field; a knowledge of the correlation functions of statistical mechanics is sufficient to extract the spectrum of $\mathcal{H}$.

In the next chapter we shall try to sharpen this apparently surprising connection between quantum and statistical physics.

12.4. Higher dimensions

The use of the transfer matrix is not limited to one-dimensional systems. Let us consider a D-dimensional system that is infinite in one direction and finite in the other direction, the points of the lattice being labeled as $(i, i_\perp)$; the first index refers to one direction, the other to the $D-1$ directions in which the system is finite. We can consider the Hamiltonian

$$H = \sum_i \sum_{i_\perp} \left\{ \sum_{k_\perp} J_{i_\perp,k_\perp} (\sigma_{i,i_\perp} - \sigma_{i,k_\perp})^2 \frac{a}{4} \right.$$
$$\left. + \frac{1}{2a} (\sigma_{i,i_\perp} - \sigma_{i+1,i_\perp})^2 + aV(\sigma_{i,i_\perp}) \right\}, \quad (12.61)$$

where $J_{i_\perp,k_\perp}$ is the usual nearest-neighbor interaction in the transverse direction (if $a=1$ we recover the usual isotropic model). The transfer matrix is now

$$T(\sigma, \nu) = \exp\left\{ -\frac{a}{4} \sum_{i_\perp k_\perp} J_{i_\perp,k_\perp}[(\sigma_{i_\perp} - \sigma_{k_\perp})^2 + (\nu_{i_\perp} - \nu_{k_\perp})^2] \right.$$
$$\left. + \sum_{i_\perp} \left[\frac{a}{2} (V(\sigma_{i_\perp}) + V(\nu_{i_\perp})) + \frac{1}{2a} (\sigma_{i_\perp} - \nu_{i_\perp})^2 \right] \right\}. \quad (12.62)$$

Here $\hat{T}$ acts on the functions of L^{D-1} variables. Everything is similar to the previous section: when $a \to 0$, $\hat{T}_t = [c(a)T_a]^{t/a}$ satisfies the differential equation (12.58) with

$$\mathcal{H} = \sum_{i_\perp} \left\{ -\frac{1}{2} \left(\frac{\partial}{\partial \sigma_{i_\perp}} \right)^2 + V(\sigma_{i_\perp}) + \sum_{k_\perp} J_{i_\perp,k_\perp} \frac{1}{2} (\sigma_i - \sigma_k)^2 \right\}. \quad (12.63)$$

We thus have a system of L^{D-1} coupled unharmonic oscillators.

Nothing special need be noticed; we only wish to point out here that in two dimensions the transfer matrix technique is very powerful for discrete spin systems and that many models have been solved in this way.[12]

Notes for Chapter 12

1. Application of transfer matrix techniques to the solution of the one- and two-dimensional Ising models can be found in K. Huang, *Statistical Mechanics*, John Wiley and Sons, New York (1963).

Notes for Chapter 12   233

2. In order to avoid ambiguities a hat will often be used to identify operators.
3. Hilbert-Schmidt operators have only a discrete spectrum. The trace of an integral operator is given by $\text{Tr } \hat{A} = \int d\sigma\, A(\sigma, \sigma)$; see, for example, A. F. Taylor and D. C. Coy, *Introduction to Functional Analysis*, John Wiley and Sons, New York, (1980).
4. Differentiability in the complex plane implies analyticity.
5. The largest eigenvalue of R is 1 by construction.
6. An operator is positive if all its eigenvalues are positive or equivalently if $\langle \psi | A | \psi \rangle > 0$ for any nonzero vector $|\psi\rangle$. The transfer matrix (12.2) is indeed a positive operator: $\int d\sigma\, d\nu\, \psi(\sigma)\psi(\nu) \exp[-\tfrac{1}{2}V(\sigma) - \tfrac{1}{2}V(\nu)]\, \exp(-(\beta/4)(\sigma - \mu)^2] = \int d\sigma\, d\nu\, \psi_V(\sigma)\psi_V(\nu) \exp[-(\beta/4)(\sigma - \mu)^2] = (4\pi/\beta)^{1/2} \int dp\, |\tilde{\psi}_V(p)|^2\, \exp(-p^2/\beta) > 0$ where $\tilde{\psi}_V(p)$ is the Fourier transform of $\psi_V(\sigma) \equiv \exp[-\tfrac{1}{2}V(\sigma)]\psi(\sigma)$.
7. The scaling laws [Eq. (7.43)] are satisfied and the critical exponents are $\eta = 1$ and $\nu = \infty$; indeed, in the region of large ξ we have $\xi \propto \exp(2\beta)$ correspondingly: $\xi\, d\beta/d\xi \xrightarrow[\xi \to \infty]{} \text{const.} = \xi^0$ [cf. Eq. (7.32)].
8. I. S. Gradshteyn and I. M. Ryzhik, *Table of Integrals, Series, and Products*, Academic Press, New York (1965), p. 837.
9. See, for example, on the uniqueness of the solution of the moment problem, G. A. Baker, *Essentials of Padé Approximants*, Academic Press, New York (1975).
10. A similar presentation can be found in J. Kogut, *Rev. Mod. Phys.* 51, 659 (1979) or S. Shenker, in *Proceedings of the 1982 Les Houches Summer School*, ed. by J. B. Zuber and R. Stora, North-Holland, Amsterdam (1984).
11. If $c(a)$ is not specified, there is an ambiguity in the definition of $\hat{T}_x$ which corresponds to a free multiplicative factor $\exp[\beta|x|]$, as can be seen by setting $c'(a) = c(a)[1 + Ba]$. This ambiguity is irrelevant for the evaluation of the correlation functions and enters only as an additive factor in the free energy.
12. A recent review of the exact solutions of two-dimensional models is R. J. Baxter, *Exactly Solved Models in Statistical Mechanics*, Academic Press, New York (1982).

CHAPTER 13

Path Integrals for Quantum Mechanics

13.1. Quantum mechanics

We recall here some of the results of quantum mechanics. We consider for simplicity a one-dimensional system with Hamiltonian

$$\mathcal{H} = \frac{1}{2} p^2 + V(q) = -\frac{1}{2} \hbar^2 \left(\frac{d}{dq}\right)^2 + V(q), \qquad (13.1)$$

and we assume that $V(q)$ goes to plus infinity at infinity, so that the spectrum of H is discrete. We also assume that $V(q)$ is a sufficiently smooth function.

$\mathcal{H}$ is a self-adjoint operator. Its eigenvalues and eigenvectors are E_n and ψ_n:

$$\mathcal{H}|\psi_n\rangle = E_n|\psi_n\rangle$$

$$E_n < E_{n+1}.$$

If we work in the Heisenberg representation the states do not depend on time,[1] while the operators evolve according to the law

$$\dot{O}(p, q) = \frac{1}{i\hbar} [O, \mathcal{H}] \equiv \frac{1}{i\hbar} [O\mathcal{H} - \mathcal{H}O]. \qquad (13.2)$$

Equation (13.2) can be easily integrated:

$$O(q, p|t) = \exp\left(\frac{it\mathcal{H}}{\hbar}\right) O(q, p|0) \exp\left(-\frac{it\mathcal{H}}{\hbar}\right). \qquad (13.3)$$

For practical purposes it is convenient to introduce the states $|x, t\rangle$, which are "eigenstates"[2] of $q(t)$,

13.1. Quantum Mechanics

$$q(t)|x, t\rangle = x|x, t\rangle$$

$$\langle x, t|x', t\rangle = \delta(x - x') \qquad (13.4)$$

$$|x, t\rangle = \exp\left(\frac{it\mathcal{H}}{\hbar}\right)|x, 0\rangle .$$

Green's function $G(y, x|t) \equiv \langle x, t|y, 0\rangle = \langle x, 0|\exp[-i\mathcal{H}t/\hbar]|y, 0\rangle$ is the probability amplitude for having a particle at point x at time t, if it was localized at y at time zero. Green's function satisfies the time-dependent Schrödinger equation:

$$i\hbar \frac{\partial}{\partial t} G(y, x|t) = \mathcal{H}_x G(y, x|t)$$

$$= \left[-\frac{\hbar^2}{2} \frac{\partial^2}{\partial x^2} + V(x)\right] G(y, x|t) , \qquad (13.5)$$

with boundary condition $G(y, x|0) = \delta(x - y)$. The solution is

$$G(y, x|t) = \sum_n \exp\left(-\frac{iE_n t}{\hbar}\right) \psi_n(x) \psi_n(y) . \qquad (13.6)$$

If we write

$$G(y, x|t) = \exp\left[\frac{iA(y, x|t)}{\hbar}\right], \qquad (13.7)$$

it is easy to see that A satisfies the differential equation

$$-\partial_t A = \frac{1}{2}(\partial_x A)^2 + V(x) - i\hbar \partial_x^2 A . \qquad (13.8)$$

In the limit $\hbar \to 0$, Eq. (13.8) becomes the Hamilton-Jacobi equation,[3] whose solution is given by the total action computed as a function of the initial and final points:

$$A(y, x|t) \xrightarrow[\hbar \to 0]{} S_c(y, x|t) \equiv \int_0^t L(\omega_c(t')) \, dt'$$

$$L(\omega_c(t')) = \frac{1}{2} \frac{d^2}{dt'^2} \omega_c(t') - V(\omega_c(t')) , \qquad (13.9)$$

where $\omega_c(t)$ is a trajectory starting from y at $t = 0$ and arriving at x at t, satisfying the classical equations of motion, L is the Lagrangian, and S_c is the action corresponding to such a trajectory.[4]

The function A is thus the quantum analog of the classical action, as a function of the initial and final positions and of the time; in the next subsection we shall carry this analogy a bit further.[5]

13.2. Path integrals

The superposition principle is one of the basic axioms of quantum mechanics. An operative version is the following:

$$G(y, x|t) = \int dz\, G(y, z|t_1) G(z, x|t - t_1) \qquad 0 \le t_1 \le t. \quad (13.10)$$

Stated simply, the amplitude for going from x to y is the integral over z of all amplitudes for going from x to y with the condition that the particle be at point z at the time t_1. Equation (13.10) is automatically assumed in the formalism of quantum mechanics: the states $|z, t_1\rangle$ form a complete set, i.e., $\int dz |z, t_1\rangle\langle z, t_1| = 1$; we thus have

$$\langle x, t | y, 0 \rangle = \int dz \langle x, t | z, t_1 \rangle \langle z, t_1 | y, 0 \rangle, \quad (13.11)$$

which coincides with Eq. (13.10).[6] If we divide the integral $0 - t$ into N subintervals by defining the times $t_n = t_n/N$, $n = 0, \ldots, N$ ($t_n - t_{n-1} = t/N \equiv \Delta t$), Eq. (13.10) generalizes to

$$G(y, x|t) = \int dz_1\, dz_2 \cdots dz_{n-1} \quad (13.12)$$

$$G(y, z_1|\Delta T) G(z_1, z_2|\Delta t) \cdots G(z_{n-1}, x|\Delta t) \equiv \int d^N[\omega]_{x,y}^t \exp\left[\frac{iA}{\hbar}[\omega]\right],$$

where $\int d^N[\omega]_{x,y}^t$ is the integral over all trajectories with the constraints that

$$\omega(t_n) = z_n, \qquad 1 \le n < N$$

$$\omega(0) = x \equiv z_N, \quad (13.13)$$

$$\omega(t) = y \equiv z_0,$$

with measure $\prod_{1}^{N-1} dz_n$ and

$$A[\omega] = \sum_{0}^{N-1} A(z_n, z_{n+1}, \Delta t). \quad (13.14)$$

Although the notation is suggestive, $\int d^N[\omega]_{x,y}^t$ is obviously a finite-dimensional integral and not a functional integral; however, when $N \to \infty$ and Δt goes to zero we should obtain a real functional integral, the dimensions of the space on which we integrate being $N - 1$. In order to study the limit $N \to \infty$, we must attempt to control $G(y, x|\Delta t)$ in the limit

13.2. Path Integrals

$\Delta t \to 0$. This can be done by observing that $G(y, x|\Delta t)$ is the kernel of the integral operator $\exp -i(\Delta t/\hbar)\mathcal{H}$. For small Δt we have

$$\exp\left(-i\mathcal{H}\frac{\Delta t}{\hbar}\right) = \exp\left[-i\frac{\Delta t}{\hbar}\frac{V(q)}{2}\right]\exp\left[+i\frac{\Delta t}{\hbar}\left(\frac{d}{dq}\right)^2\right]$$

$$\times \exp\left(-\frac{i\Delta}{\hbar}\frac{V(q)}{2}\right) + O((\Delta t)^2)$$

$$\langle x, 0|\exp\left(-\frac{i\mathcal{H}\Delta t}{\hbar}\right)|y, 0\rangle = \exp\left[-\frac{i\Delta t}{\hbar}\left(\frac{V(x)}{2} + \frac{V(y)}{2}\right.\right.$$
$$\left.\left. - \frac{(x-y)^2}{2(\Delta t)^2}\right)\right] \cdot \left(\frac{i}{2\pi\hbar\Delta t}\right)^{1/2},$$

(13.15)

where we have used the fact that

$$\langle x, 0|\exp -i\mathcal{H}\frac{\Delta t}{\hbar}|y, 0\rangle = \left(\frac{i}{2\pi\hbar\Delta t}\right)^{1/2}\exp\left[\frac{i}{\hbar}\frac{(x-y)^2}{2\Delta t}\right] \quad \text{if } V=0.$$

(13.16)

We finally obtain

$$A[\omega] = \Delta t\sum_{1}^{N}\left\{\frac{(z_n - z_{n-1})^2}{2(\Delta t)^2} - \frac{V(z_n) + V(z_{n-1})}{2}\right\} - i\hbar\frac{N}{2}\ln\left(\frac{i}{2\pi\hbar\Delta t}\right)$$
$$+ O(\Delta t)$$

$$= \int_0^t dt'\left\{\frac{1}{2}(\dot\omega(t))^2 - V(\omega)\right\} - i\hbar\frac{N}{2}\ln\left(\frac{i}{2\pi\hbar\Delta t}\right) + O(\Delta t). \quad (13.17)$$

The functional integral representation is thus

$$G(y, x|t) = \int d[\omega]_{y,x}^t \exp\left\{\frac{iS[\omega]}{\hbar}\right\}$$

$$S[\omega] = \int_0^t dt'\left\{\frac{1}{2}(\dot\omega(t))^2 - V(\omega)\right\},$$

(13.18)

where the measure $d[\omega]_{y,x}^t$ is defined as

$$\lim_{N\to\infty}\left[\left(\frac{i}{2\pi\hbar\Delta t}\right)^N\prod_1^{N-1} dz_n\right], \quad z_n = \omega\left[\frac{nt}{N}\right]. \quad (13.19)$$

In more physical terms: When $\Delta t \to 0$, quantum effects become negligible in the evaluation of A, and we can replace A with S_c.[7]

Equation (13.18) is thus the superposition principle plus the correspondence principle at small time: It states that the amplitude for going from y to x in time t is the sum of the amplitudes of all the possible trajectories of the particle, each of them receiving a weight $\exp\{iS[\omega]/\hbar\}$. Equation (13.19) looks very similar to Eq. (12.55) of the previous chapter, the main difference being an extra i in the exponent, with the identification of $\hbar$ with $1/kT$.

It is thus natural to ask what happens to the following quantities, which are the analogues of the statistical correlation functions:

$$G^T_{g_1,g_2}(t) \equiv \frac{\int d\nu[\omega]^T g_1(\omega(0))g_2(\omega(t))}{\int d\nu[\omega]^T}$$

$$d\nu[\omega]^T = \int dx\, d[\omega]^t_{x,x} \exp\left\{\frac{iS[\omega]}{\hbar}\right\}, \tag{13.20}$$

if we sum over all the periodic trajectories of period T.[8] We obtain[9]

$$\int d\nu[\omega]^T = \int dx\, G(x,x|T) = \operatorname{Tr} \exp\left(-i\frac{\mathcal{H}}{\hbar}T\right)$$

$$= \sum_n \exp\left(-\frac{iE_n T}{\hbar}\right)$$

$$\int d\nu[\omega]^T g_1(\omega(0))g_2(\omega(t)) = \int dx\, dy\, g_1(x)g_2(y)G(x,y|t)G(y,x|T-t) \tag{13.21}$$

$$= \operatorname{Tr}\left[g_1(q)\exp\left(-\frac{i\mathcal{H}t}{\hbar}\right)g_2(q)\right.$$

$$\left.\times \exp\left(-\frac{i\mathcal{H}(T-t)}{\hbar}\right)\right]$$

$$= \sum_{n,m} \langle \psi_n|\hat{g}_1|\psi_m\rangle\langle\psi_m|\hat{g}_2|\psi_n\rangle$$

$$\times \exp\left\{-i\frac{[TE_n + t(E_m - E_n)]}{\hbar}\right\}.$$

When T goes to infinity, we obtain oscillating terms from the denominator and the numerators, so that Eq. (13.20) is an oscillating function without limit. In order to define the limit, it is convenient to set $T = |T|\exp(i\theta)$, at an intermediate stage, to perform first the limit $|T|\to\infty$ and then $\theta\to 0^+$. In this way the oscillations are damped and we

13.2. Path Integrals

find that the r.h.s. of Eq. (13.21) becomes, when $T \to \infty$,

$$\sum_m \langle \psi_0 | g_1(q) | \psi_m \rangle \langle \psi_m | g_2(q) | \psi_0 \rangle \exp[-i(E_n - E_0)t]$$
$$= \langle \psi_0 | g_1(q(0)) g_2(q(t)) | \psi_0 \rangle . \tag{13.22}$$

The previous equations are valid for $t > 0$; if $t < 0$ we get

$$\lim_{T \to \infty} G^T_{g_1, g_2}(t) = \langle \psi_0 | g_2(q(t)) g_1(q(0)) | \psi_0 \rangle . \tag{13.23}$$

Equations (12.22) and (12.23) can be written in a unified way by using the time-ordered product τ defined as

$$\tau[A(t)B(0)] \equiv B(0)A(t), \quad t > 0$$
$$\tau[A(t)B(0)] \equiv A(t)B(0), \quad t < 0. \tag{13.24}$$

We can thus write for both signs of t

$$\lim_{T \to \infty} G^T_{g_1, g_2}(t) = \langle \psi_0 | \tau[g_1(q(0)) g_2(q(t))] | \psi_0 \rangle . \tag{13.25}$$

The correlation functions of the functional integral approach become (in the limit $T \to \infty$) the expectation values of the time-ordered products of operators on the ground state. The same results hold for more than two operators, the time-ordered products being defined as

$$\tau[A(t_1)B(t_2)C(t_3)] = A(t_1)B(t_2)C(t_3), \quad t_1 < t_2 < t_3$$
$$\tau[A(t_1)B(t_2)C(t_3)] = C(t_3)A(t_1)B(t_2), \quad t_3 < t_1 < t_2, \tag{13.26}$$

and so on. The θ-dependent damping factor can be introduced at the level of the functional integral representation by setting $t = |t| \exp(-i\theta)$. We have

$$dt \to d|t| \exp(-i\theta)$$
$$\frac{d}{dt} \to \frac{d}{d|t|} \exp(i\theta) . \tag{13.27}$$

Putting everything together, Eq. (13.20) becomes

$$G_{g_1 g_2}^{|T|\exp(-i\theta)}(|t|\exp(-i\theta)) = \frac{\int d\nu_\theta[\omega]^{|T|} g_1(\omega(0)) g_2(\omega(t))}{\int d\nu_\theta[\omega]^{|T|}}$$

$$d\nu_\theta[\omega]^{|T|} = \int dx \, d\omega_{x,x}^{|T|} \exp\left\{\frac{i}{\hbar} S_\theta[\omega]\right\}$$

$$S_\theta[\omega] = \exp(i\theta) \int_0^{|T|} dt' \left[\frac{1}{2}\left(\frac{d\omega}{dt'}\right)^2 - \exp(-2i\theta) V(\omega(t'))\right]. \quad (13.28)$$

We can thus define for $0 < \theta < \pi$

$$G_{g_1,g_2}^\theta(t) = \lim_{|T|\to\infty} G_{g_1 g_2}^{T\exp(-i\theta)}(t\exp(-i\theta)) \qquad \text{if } t > 0$$

$$= \langle \psi_0 | g_1(q(0)) g_2(q(t\exp(-i\theta))) | \psi_0 \rangle$$

$$= \sum_n \langle \psi_0 | g_1(q) | \psi_n \rangle \langle \psi_n | g_2(q) \psi_0 \rangle \exp\left[-\frac{it\exp(-i\theta)}{\hbar}(E_n - E_0)\right].$$

$$(13.29)$$

In the limit $\theta \to 0^+$ we recover the previous equations; for $\theta = \pi/2$, we obtain the correlation functions of statistical mechanics.

In this way we see that the correlation functions of statistical mechanics are the expectation values of the time-ordered products in the ground state analytically continued to imaginary time [cf. Eq. (12.51)],

$$C_{g_1,g_2}(t) \equiv G_{g_1,g_2}^{\pi/2}(t) = G_{g_1,g_2}^{0^+}(it). \quad (13.30)$$

In the same way the functional integral of statistical mechanics is the analytic continuation to imaginary times of the path integral representation of quantum mechanics and vice versa. The real-time and the imaginary-time theories are thus equivalent in the sense that we can switch from one to the other theory by doing an analytic continuation.[10] However, from the point of view of the functional integral, the theory for $\theta = 0^+$ is much more complicated because it deals with oscillating integrals that are not absolutely convergent (this creates many mathematical difficulties, some of which we have already seen in defining the limit $T \to \infty$).[11]

The opposite case $\theta = \pi/2$ is the best: $d\nu_{\pi/2}(\omega)$ is a probability measure, and all the techniques we have developed in statistical mechanics can be used. Although the path integral representation can be rigorously defined even for real time,[12] it is much simpler to use it only at real times and to continue analytically the results to imaginary times.

The relation between the correlation function at real and imaginary times is also simple for their Fourier transforms. If we define $\tilde{G}^{0^+}_{g_1,g_2}(E)$ as

$$\tilde{G}^{0^+}_{g_1,g_2}(t) = \frac{1}{2\pi} \int_{-\infty}^{+\infty} dE \exp\left[-\frac{iEt}{\hbar}\right] \tilde{G}^{0^+}_{g_1,g_2}(E), \quad (13.31)$$

we get

$$\tilde{G}^{0^+}_{g_1,g_2}(E) = \lim_{\varepsilon \to 0} \sum_n \frac{c_n}{-E^2 + (E_n - E_0)^2 + i\varepsilon}$$

$$c_n = \frac{1}{2}(E_n - E_0)\langle \psi_0 | g_1(q) | \psi_n \rangle \langle \psi_n | g_2(q) | \psi_0 \rangle, \quad (13.32)$$

as can be checked by using the identities

$$\lim_{\varepsilon \to 0^+} \frac{1}{2\pi} \int_{-\infty}^{+\infty} \frac{\exp(-iEt/\hbar)}{E - E_0 \pm i\varepsilon} = \exp(-iE_0 t/\hbar)\theta(\pm t), \quad (13.33)$$

where $\theta(t) = 1$ if $t > 0$ and $\theta(t) = 0$ if $t < 0$. In the same way we obtain

$$G^{\pi/2}_{g_1,g_2}(l) = \frac{1}{2\pi} \int_{-\infty}^{\infty} d\Omega \exp\left[-\frac{i\Omega l}{\hbar}\right] \tilde{G}^{\pi/2}_{g_1,g_2}(\Omega)$$

$$\tilde{G}^{\pi/2}_{g_1,g_2}(\Omega) = \sum_n \frac{c_n}{\Omega^2 + (E_n - E_0)^2}, \quad (13.34)$$

with the same expression for the c_n as in Eq. (13.32).

Thus we see that $\tilde{G}^{0^+}(E)$ is the analytic continuation of $\tilde{G}^{\pi/2}(\Omega)$ evaluated at $\Omega = \exp(-i\theta)E$. The $i\varepsilon$ prescription tells us that the poles are not crossed by deforming the integration path in Eq. (13.31) from real to imaginary E.

13.3. Equation of motion

In the Heisenberg representation the operator q satisfies the classical equation of motion

$$\ddot{q} = -\frac{\partial V}{\partial q} \quad (13.35)$$

and the equal-time commutation relations[13]

$$[q, p] = [q, \dot{q}] = i\hbar. \quad (13.36)$$

Equations (13.35) and (13.36) have simple consequences for the time-ordered products. At first sight we might think that

$$\frac{\partial^2}{\partial t^2} \tau[q(t)q(0)] = -\tau\left[\frac{\partial V}{\partial q}(t)q(0)\right]. \tag{13.37}$$

Equation (13.37) is not exact; indeed the time derivative does not commute with the symbol of time ordering, i.e., in the general case we have

$$\frac{\partial}{\partial t} \tau[g_1(t)g_2(0)] = \tau\left[\frac{\partial g_1(t)}{\partial t} g_2(0)\right] + \delta(t)[g_1(t), g_2(0)], \tag{13.38}$$

as can be seen from the definition (13.37). If we apply Eq. (13.38) to Eq. (13.37) we obtain

$$\frac{\partial^2}{\partial t^2} \tau[q(t)q(0)] = \frac{\partial}{\partial t} \tau[\dot{q}(t)q(0)]$$

$$= -\tau\left[\frac{\partial V}{\partial q}(t)q(0)\right] + i\hbar\delta(t). \tag{13.39}$$

Equation (13.39) contains information from both the equation of motion and the canonical commutation relations. In the same way we get

$$\frac{\partial^2}{\partial t_1^2} \tau[q(t_1)q(t_2)q(t_3)q(t_4)] = -\tau\left[\frac{\partial V}{\partial q}(t_1)q(t_2)q(t_3)q(t_4)\right]$$

$$+ i\hbar\{\delta(t_1 - t_2)\tau[q(t_3)q(t_4)]$$

$$+ \delta(t_1 - t_3)\tau[q(t_2)q(t_4)]$$

$$+ \delta(t_1 - t_4)\tau[q(t_2)q(t_3)]\}. \tag{13.40}$$

If we sandwich Eqs. (13.39) and (13.40) between two ground-state wave functions, we obtain differential equations for the expectation values of the time-ordered products on the ground state e.g.,

$$\frac{\partial^2}{\partial t^2} \langle\psi_0|\tau[q(t)q(0)]|\psi_0\rangle + \langle\psi_0|\tau\left[\frac{\partial V}{\partial q}(t)q(0)\right]|\psi_0\rangle = i\hbar\delta(t). \tag{13.41}$$

If we consider the analytic continuation at imaginary times we see that the statistical mechanics correlation functions satisfy similar equations:

$$-\frac{\partial^2}{\partial t^2} \langle\sigma(t)\sigma(0)\rangle + \langle V(\sigma(t))\sigma(0)\rangle = \hbar\delta(t) \tag{13.42}$$

and so on.

We have already remarked that these relations for the correlation function can be obtained very simply from the functional integral formalism if we integrate parts of Chapter 9, e.g. Eqs. (13.41) and (13.42) can be derived from the identity[14]

$$\int d[\omega] \frac{\delta}{\delta\omega(t)} \left\{ \omega(0) \exp\left[\frac{-1}{\hbar} S^\theta[\omega] \right] \right\} = 0. \tag{13.43}$$

This example is interesting because it shows how subtly the commutation relations are transcribed in the functional formalism.

13.4. An example: the harmonic oscillator

No book on quantum mechanics would be complete without a detailed study of the harmonic oscillator. In this case the Hamiltonian is

$$\mathcal{H} = \frac{p^2}{2} + r^2 \frac{q^2}{2}$$

$$S_{\pi/2}[\omega] = \int dt \left\{ \frac{1}{2} \left(\frac{d\omega}{dt} \right)^2 + \frac{r^2 \omega^2}{2} \right\}. \tag{13.44}$$

In this way we recover an old friend, the Gaussian model. Let us now compute the ground state wave function[15] $\psi_0(x)$ and the Green's function $G(x, y|t)$ for imaginary times, using the functional integral representation (13.28). The simple fact that the harmonic oscillator has as its functional representation the Gaussian model tells us before doing any explicit computation that both $\psi_0(x)$ and $G(x, y)$ have the following form:[16]

$$\psi_0(x) = \left(\frac{1}{2\pi c} \right)^{1/4} \exp\left(-\frac{x^2}{4c} \right)$$

$$G(x, y|l) = b(l) \exp[-(x^2 + y^2)d(l) + g(l)xy], \tag{13.45}$$

where we have used the symmetry $x \leftrightarrow -x$ of the Hamiltonian to avoid unwanted terms (e.g., a linear term in x in the exponential). We have only to compute the parameters $b(l)$, $d(l)$, and $g(l)$. This can easily be done. We know already that

$$\langle \omega^2(0) \rangle = G(0)$$

$$\langle \omega(0)\omega(l) \rangle = G(l) \tag{13.46}$$

$$G(l) = \frac{1}{2r} \exp(-r|l|).$$

We can compare Eqs. (13.46) with the following relations:

$$\langle \omega^2(0) \rangle = \int dx\, \psi_0^2(x) x^2$$

$$\langle \omega(0)\omega(l) \rangle = \int dx\, dy\, G_S(x, y|l) \psi_0(x) \psi_0(y) xy \qquad (13.47)$$

$$G_S(x, y|l) \equiv G(x, y|l) \exp\left(|l| \frac{E_0}{\hbar}\right).$$

If we use the identity

$$\int dy\, G_S(x, y|l) \psi_0(y) = \psi_0(x), \qquad (13.48)$$

we find after some algebra that

$$C = \frac{1}{2r}$$

$$b(l) = \left[\frac{r}{2\pi \sinh(rl)}\right]^{1/2}$$

$$d(l) = \frac{r}{[2\tanh(rl)]} \qquad (13.49)$$

$$g(l) = \frac{r}{\sinh(rl)}.$$

By comparing Eq. (13.44) with Eqs. (13.6) and (13.41), we see after some algebra that $E_n - E_0 = nr$. We can also check that we have

$$\int dx\, G(x, x|l) = \frac{1}{2\sinh(rl/2)} = \exp\left(-\frac{1}{2} rl\right) \cdot \left[\sum_0^\infty \exp(-nrl)\right], \qquad (13.50)$$

as it should be. It is clear that this comparison method cannot give us the value of E_0. However, using that fact that E_0 is the "free energy" of the system, i.e.,

$$E_0 = \lim_{T \to \infty} \frac{-1}{T} \ln\left[\int dv_{\pi/2}^T(\omega)\right], \qquad (13.51)$$

we obtain

$$\frac{\partial E_0}{\partial r^2} = \left\langle \frac{1}{2} \omega^2(0) \right\rangle = \frac{1}{4r}. \qquad (13.52)$$

13.4. An Example: The Harmonic Oscillator

If $E_0(r)$ is normalized in such a way that $E_0(0) = 0$, we get the well-known result

$$E_0(r) = \frac{1}{2} r. \tag{13.53}$$

A direct evaluation of $G(x, y|l)$ is also possible. We must use the formula

$$G(x, y|l) = \int d[\omega]_{x,y}^{l} \exp\{-S_{\pi/2}[\omega]\}, \tag{13.54}$$

where $\omega(0) = y$, $\omega(l) = x$, and compute explicitly the functional integral. The first step consists in defining

$$\omega(t) = \omega_c(t) + \tilde{\omega}(t), \tag{13.55}$$

where

$$\tilde{\omega}(0) = \tilde{\omega}(l) = 0$$
$$S_{\pi/2}[\omega] = S_{\pi/2}[\omega_c] + O(\tilde{\omega}^2). \tag{13.56}$$

In other words, $\omega_c(t)$ is the function that minimizes $S_{\pi/2}[\omega]$, given the constraints; it satisfies the differential equation

$$\frac{\delta}{\delta \omega(t)} S_{\pi/2}[\omega] = -\frac{d^2}{dt^2} \omega(t) + r^2 \omega(t) = 0, \tag{13.57}$$

whose solution is

$$\omega_c(t) = a \exp(rl) + b \exp(-rl), \tag{13.58}$$

where a and b are adjusted to satisfy the constraints.

After some algebra we find that $S_{\pi/2}[\omega_c]$ is just the exponent of Eq. (13.42). We have to integrate over the $\tilde{\omega}$ to get the normalization constant $b(l)$, which is formally given by

$$b(l) = \int d[\omega]_{0,0}^{l} \exp\left\{-\int dt \left[\frac{\dot{\omega}^2}{2} + \frac{r^2 \omega^2(t)}{2}\right]\right\}$$

$$= \left\{\prod_{1}^{\infty}{}_n \left[\frac{1}{2} \frac{\pi n^2}{l^2} + \frac{r^2}{2\pi}\right]\right\}^{-1/2},$$

$$\propto \det^{-1/2}\left[-\left(\frac{d}{dt}\right)^2 + r^2\right], \tag{13.59}$$

where we have used the explicit form of the eigenvalues of $-(d/dt)^2$ on the internal $0-1$, with the zero boundary conditions

$$-\left(\frac{d}{dt}\right)^2 \psi_n(t) = \left(\frac{\pi n}{l}\right)^2 \psi_n(t)$$

$$\psi_n(t) \propto \sin\left(\frac{\pi n}{l} t\right).$$
(13.60)

We have already remarked that the absolute normalization functional integral is divergent unless a regularization prescription integral is used: only the ratio between the denominator and the numerator of Eq. (13.28) is well defined; e.g., if we use the lattice regularization, this divergence is compensated for by the factor $[c(a)]^{t/a}$ of Eq. (12.49). Regardless of the procedure used to define the functional integral, the r-dependent part of the determinant in Eq. (13.59) is finite and equal to

$$\prod_{1}^{\infty}{}_{n}\left(1 + \frac{r^2 l^2}{n^2 \pi^2}\right) \prod_{1}^{\infty}{}_{n} \frac{n^2 \pi^2}{2l^2} = \frac{\sinh(rl)}{rl} \exp[I(l)]$$

$$I(l) = \sum_{1}^{\infty}{}_{n} \ln \frac{n^2 \pi^2}{2l^2}.$$
(13.61)

As it stands, $I(l)$ is terribly divergent. However, as we said, this divergence disappears if we introduce a lattice regularization and take care of the appropriate $c(a)$ factors: the computation would be straightforward, but somewhat long (some tricks must be used to extract the finite term as the ratio of two quantities which diverge when a goes to zero). Here we try to be more general and to obtain a finite result without specifying the regularization procedure we use. As an intermediate step we define a regularized determinant as follows:

$$b_\Lambda^{-2}(l, r) = \prod_{1}^{\infty}{}_{\infty} f_\Lambda\left[\frac{\pi n^2}{2l^2} + \frac{r^2}{2\pi}\right],$$
(13.62)

where $f_\Lambda(z)$ is an arbitrary C^∞ function such that

$$f_\Lambda(z) = z + \exp\left(-\frac{\Lambda}{z}\right) \qquad z \sim 0$$

$$f_\Lambda(z) = 1 + O\left(\exp\left(-\frac{z}{\Lambda}\right)\right) \qquad z \to \infty$$
(13.63)

$$\lim_{\Lambda \to \infty} f_\Lambda(z) = z.$$

By construction, the determinant (13.62) is finite; however, it becomes

13.4. An Example: The Harmonic Oscillator

divergent again when the regularization is removed by sending Λ to infinity. We would be quite satisfied if we could prove that

$$b_\Lambda^{-2}(l, r) \underset{\Lambda \to \infty}{\approx} \exp[D_f(\Lambda)l + R(l, r)], \qquad (13.64)$$

where $D_f(\Lambda)$ (which depends on f) diverges when Λ goes to infinity and $R(l, r)$ is f independent. Indeed, adding a term $\exp(Al)$ in $G(x, y|l)$ corresponds to changing the absolute value of the ground-state energy. If proper definitions are used (as we have done with the lattice regularization) for the normalization factors in front of the function integral, the term $D_f(\Lambda)l$ should be cancelled by these normalization factors.

Let us begin the evaluation of (13.62) for large Λ. Clearly we have

$$-2 \ln b_\Lambda(l, z) = \sum_n \ln f_\Lambda\left(\frac{n^2\pi}{2l^2} + \frac{r^2}{2\pi}\right), \qquad (13.65)$$

where the sum is rapidly convergent for $n \to \infty$ due to the condition (13.63). We can now use the Euler-MacLaurin formula,[17]

$$\delta \sum_{n=1}^{N} g(\delta n) = \int_\delta^{N\delta} g(x)\, dx + \frac{\delta}{2}[g(N\delta) + g(\delta)]$$

$$+ \sum_{k=1}^{\infty} \frac{\delta^{2k} B_{2k}}{(2k)!}[g^{(2k-1)}(N\delta) - g^{(2k-1)}(\delta)] \qquad (13.66)$$

$$g^{(i)}(x) \equiv \left(\frac{d}{dx}\right)^i g(x),$$

where the B_k are the Bernoulli numbers.[18] After a careful computation (which the interested reader can find in the appendix), we get for large Λ

$$b_\Lambda^{-2}(l, r) \simeq \frac{2\pi \sinh(rl)}{r} \exp\left[l \int_0^\infty dy \ln\left[f_\Lambda\left(\frac{\pi y^2}{2}\right)\right]\right], \qquad (13.67)$$

as was anticipated in Eq. (13.64), in agreement with Eqs. (13.45)–(13.49). The reader (who has not skipped the boring appendix) can appreciate how much labor can be saved by a careful use of the Gaussian integrals and by avoiding direct calculation of determinants as far as possible, as was done in Sec. 13.2.

A simple way to obtain the same result starting from Eq. (13.61) is based on the relation $\int dx\, G(x, x|l) = \sum_n \exp[-(E_n l)]$. As long as the difficult term to be evaluated depends only on l, not on r, it is fixed by this relation. The only ambiguity may be a multiplicative factor $\exp(-Al)$, which corresponds to an overall shift of the energy levels; the value of A may also be fixed by the condition that in the absence of a potential, i.e., in the limit r going to zero, the ground-state energy goes to zero.

13.5. Path integrals in phase space

While classical mechanics in the Hamilton formulation treats the q and p variables in a symmetric way, this symmetry is lost in Eq. (13.18) because the trajectories are functions only of q; this symmetry may be restored by writing

$$G(x, y|t) \propto \int d[q]_{x,y}^t \, d[p]^t \exp[iS(p, q)/\hbar]$$

$$S[p, q] = \int_0^t dt'[p\dot{q} - \mathcal{H}(q, p)]$$

$$= \int_0^t dt'\left[-\frac{p^2}{2} - V(q) + p\dot{q}\right]. \quad (13.68)$$

No boundary condition has been imposed on the function $p(t')$.

The integral over the p is Gaussian and can be trivially done, reproducing Eq. (13.18). However, Eq. (13.68) differs from Eq. (13.18) if $\mathcal{H}$ is not a quadratic function of p^2. [In this case Eq. (13.68) is the correct formulation.][19] We stress that Eq. (13.68) is not (contrary to its appearance) invariant under canonical transformations, as can be seen by a careful definition of the functional integration by discretizing the time.[20] It is well known that different results are obtained if we quantize a classical Hamiltonian after or before a canonical transformation: in quantum mechanics p and q do not commute, and a classical expression like $g(q)p^2$ has many corresponding quantum versions.[21]

What happens if we try to define $G(x, y; p_1|t)$ by fixing the value of $p(t')$ at $t' = 0$ ($p(0) = p_1$)? In this case we obtain

$$G(x, y; p_1|t) = 0; \quad (13.69)$$

because of the uncertainty principle we cannot fix p and q together. This can be seen by computing at an intermediate stage

$$G_\delta(x, y; p_1|t) = \int d[q]_{x,y}^t \, d[p]^t \exp[iS[p, q]/\hbar], \quad (13.70)$$

with the constraint

$$\frac{1}{\delta}\int_0^\delta p(t') \, dt' = p_1. \quad (13.71)$$

An explicit computation (which we omit) gives for small δ

$$G_\delta(y, x; p_1|t) = \delta G(y, y + \delta p_1|\delta)G(y + \delta p_1, x|t - \delta)$$

$$\simeq \delta \cdot G(y, y + \delta p_1|\delta)G(y, x|t). \quad (13.72)$$

Equation (13.72) states that $\int_0^\delta p(t')\, dt'$ may be approximated by $\int_0^\delta dq/dt'\, dt' = q(\delta) - q(0)$ (the two expressions are equivalent in the classical limit); the factor δ in (13.72) is necessary in order to satisfy the normalization condition,

$$\int_{-\infty}^{+\infty} G_\delta(x, y; p_1|t)\, dp_1 = G(x, y|t). \tag{13.73}$$

Using the explicit formula for the Green's function at small times, we get

$$G_\delta(x, y; p_1|t) \sim (\pi\delta)^{1/2} \exp\left(\frac{i\delta p_1^2}{2}\right) G(x, y|t). \tag{13.74}$$

In other words, values of p_1 up to $\delta^{-1/2}$ have nearly the same amplitude.

This phenomenon reflects the fact that the trajectories which dominate the functional integral are not differentiable (cf. footnote 7 and Chapter 19),

$$\left\langle \frac{(\omega(0) - \omega(\delta))^2}{\delta} \right\rangle_{\delta \to 0} \sim O(1). \tag{13.75}$$

The expectation value of $\dot\omega^2$ is infinite and the Heisenberg principle is thus saved.

This formulation has also another disadvantage: the exponent of the imaginary-time version of (13.68) is not real, so that at imaginary times we do not obtain a probability distribution over the trajectories in phase space. This problem is probably connected to the lack of any positivity requirement for the wave function of the ground state in momentum space: i.e., $\tilde\psi_0(p) = \int dx \exp(ipx)\psi_0(x)$ is not positive definite in the general case (an elementary example is provided by the ground-state wave function for a square-well potential).

Appendix to Chapter 13

In this appendix we present for completeness the computations leading to Eq. (13.67).

We begin our analysis by applying Eq. (13.66) to the evaluation of Eq. (13.65). If we set $\delta = 1$, we get

$$b_\Lambda^{-2}(l,r) = \int_1^\infty \ln\left[f_\Lambda\left(\frac{\pi x^2}{2l^2} + \frac{r^2}{2\pi}\right)\right] dx + \frac{1}{2}\ln\left[f_\Lambda\left(\frac{\pi}{2l^2} + \frac{r^2}{2\pi}\right)\right]$$

$$-\sum_{k}^\infty \frac{B_{2k}}{(2k)!}\left(\frac{d}{dx}\right)^{2k-1} \ln\left[f_\Lambda\left(\frac{\pi x^2}{2} + \frac{r^2}{2\pi}\right)\right]\bigg|_{x=1}$$

(A13.1)

where we have used Eq. (13.66) in the limit $N \to \infty$. The correction terms come from the upper integration and are negligible due to the condition (13.63), e.g., $g^{(2k-1)}(N)$.

All the terms but the integral have a finite limit when Λ goes to infinity, and we can substitute there z for $f_\Lambda(z)$. It is convenient to manipulate the integral as follows:

$$\int_1^\infty dx \ln f_\Lambda\left(\frac{\pi x^2}{2l^2} + \frac{r^2}{2\pi}\right) = l\left\{\int_0^\infty dy \ln\left[f_\Lambda\left(\frac{\pi y^2}{2}\right)\right]\right\}$$

$$+ l\int_0^\infty dy\left\{\ln\left[f_\Lambda\left(\frac{\pi y^2}{2} + \frac{r^2}{2\pi}\right)\right]\right.$$

$$\left. - \ln\left[f_\Lambda\left(\frac{\pi y^2}{2}\right)\right]\right\} - \int_0^1 dx \ln f_\Lambda\left(\frac{\pi x^2}{2l^2} + \frac{r^2}{2\pi}\right).$$

(A13.2)

Everything but the first integral is finite when $\Lambda \to \infty$. Assembling all components, we obtain

$$-2\ln[b_\Lambda(l,r)] \simeq l\int_0^\infty dy \ln\left[f_\Lambda\left(\frac{\pi y^2}{2}\right)\right] + l\int_0^\infty dy \ln\left(1 + \frac{r^2}{\pi^2 y^2}\right)$$

$$- \int_0^1 dx \ln\left(\frac{\pi x^2}{2l^2} + \frac{r^2}{2\pi}\right) + \frac{1}{2}\ln\left[\frac{\pi}{2l^2} + \frac{r^2}{2\pi}\right]$$

$$- \sum_k \frac{B_{2k}}{(2k)!}\left(\frac{d}{dx}\right)^{2k-1} \ln\left(\frac{\pi x^2}{2l^2} + \frac{r^2}{2\pi}\right).$$

(A13.3)

Equation (A13.3) has just the needed form [Eq. (13.64)] with the identification

$$D_f(\Lambda) = \int_0^\infty dy \ln\left[f_\Lambda\left(\frac{\pi}{2} y^2\right)\right].$$

(A13.4)

The evaluation of $R(l, r)$ from Eq. (13.3) is not easy; it can be simplified if we use Eq. (13.61), which tells us how (A13.3) depends on r:

$$R(l,r) = \ln\left[\frac{\sinh(rl)}{rl}\right] + \tilde{R}(l).$$

(A13.5)

Appendix to Chapter 13

Equation (A13.5) can be used to fix $\tilde{R}(l)$ by computing $R(l, r)$ at only one given value of r. The convenient choice is to consider the case $r \to \infty$. In this situation the corrections coming from the Bernoulli terms are negligible. The integrations are elementary. We finally find

$$R(l, r) \sim rl - \frac{1}{2} \ln\left(\frac{\pi^2}{l^2} + r^2\right) + 2 - \frac{rl}{\pi} \operatorname{arctanh}\left(\frac{2\pi}{lr}\right). \quad (A13.6)$$

Comparing Eq. (A13.6) with (A13.5), we obtain

$$\tilde{R}(l) = \ln 2\pi l, \quad (A13.7)$$

which implies:

$$R(l, r) = \ln[2\pi \sinh(lr)/r]. \quad (A13.8)$$

We have now reached our goal, i.e., to prove Eq. (13.67). It is interesting that the whole computation could also have been done directly at $r = 0$ using a slightly different strategy.

If k is a large number ($k/l\Lambda$ still being very small), we have

$$-2\ln[b_\Lambda(l, 0)] \simeq \sum_1^k \ln\left[f_\Lambda\left(\frac{n^2\pi}{2l^2}\right)\right] + \int_{k+1/2}^\infty dp \ln\left[f_\Lambda\left(\frac{p^2\pi}{2l^2}\right)\right]$$

$$\simeq \sum_1^k \ln\left[\frac{n^2\pi}{2l^2}\right] - \int_0^{k+1/2} dp \ln\left[\frac{p^2\pi}{2l^2}\right]$$

$$+ \int_0^\infty dp \ln\left[f_\Lambda\left(\frac{p^2\pi}{2l^2}\right)\right], \quad (A13.9)$$

where the correction terms vanish when k goes to infinity.

We can now safely take the limit $\Lambda \to \infty$: apart from the usual divergent factor $l \int_0^\infty dq \ln[f_\Lambda((q^2/2)\pi)]$ we get

$$-2\ln[b_\infty(l, 0)] = C - \frac{1}{2} \ln\left(\frac{\pi}{2l^2}\right)$$

$$C = \lim_{k \to \infty} \left[\sum_1^k \ln n^2 - \int_0^{k+1/2} dp \ln p^2\right] \quad (A13.10)$$

in perfect agreement with the previous result.

The evaluation of the constant C is rather simple:

$$C = 2 \lim_{k \to \infty} \left[\ln(k!) - \int_0^{k+1/2} dp \ln p \right] = \ln(2\pi), \quad (A13.11)$$

as follows from the Stirling approximation for the factorial.

We finally get

$$b_\infty^2(l, 0) = \frac{1}{2^{3/2} \pi^{1/2} l}. \quad (A13.12)$$

The careful reader has certainly noticed that Eq. (A13.12) agrees with Eq. (A13.8), but both disagree (for the normalization) with the correct result [i.e., Eq. (13.49)]. Indeed, the technique we have used to separate the divergent part from the finite part is rather *ad hoc*, and we cannot hope to get as well the correct normalization: a disagreement by a multiplicative factor [i.e., $(\pi/2)^{1/4}$] should not be a surprise. A multiplicative factor that is l and r independent can easily be found by imposing the basic condition

$$\int dy\, G(x, y|t_1) G(y, z|t_2) = G(x, y|t_1 + t_2). \quad (A13.13)$$

Let us go through the exercise of rederiving the correct result by using an explicit lattice regularization (see pp. 236–237). No problems arise in the evaluation of the r-dependent part, so we can safely work at $r = 0$. In this case we have

$$b(l, 0) = \int d\omega_1 \cdots d\omega_{N-1} (a2\pi)^{-N/2} \exp\left[-\sum_i^N \frac{1}{2a} (\omega_i - \omega_{i-1})^2 \right],$$

$$N = \frac{l}{a}, \quad (A13.14)$$

with the boundary conditions $\omega_0 = \omega_N = 0$.

The integration over the ω's can be done sequentially (first ω_1, later ω_2, and so on). As the reader can easily verify, we obtain

$$b(l, 0) = (2\pi Na)^{-1/2} = (2\pi l)^{-1/2}, \quad (A13.15)$$

which is the correct result, as it should be.

As a check of our ability to evaluate integrals, we can rederive this last result using the same approach as Eqs. (A13.9)–(A13.12).

The argument of the exponential in Eq. (A13.14) can be written as

$$\sum_{i,k}^{N-1} \omega_i A_{ik} \omega_k. \quad (A13.16)$$

It is easy to check that $\hat{A}$ is a self-adjoint operator, whose eigenvalues are given by

$$\lambda_n = \frac{1 - \cos\left(\frac{\pi n}{N}\right)}{2a}, \quad n = 1, \ldots, N-1, \quad \text{(A13.17)}$$

the corresponding eigenvectors being

$$\omega_i^{(n)} = \sin\left(\frac{\pi n i}{N}\right).$$

The Gaussian integral over the ω's is now trivial, and we get

$$b(l, 0) = \left(\frac{a}{2\pi}\right)^{1/2} a^{-N} \exp\left\{-\sum_{n=1}^{N-1} \ln\left[\frac{2 - 2\cos(\pi a n/l)}{a^2}\right]\right\}. \quad \text{(A13.18)}$$

Using the same manipulations as before, the argument of the exponential can be written approximately as (for large k)

$$-\sum_{n=1}^{k} \ln\left(\frac{\pi^2 n^2}{l^2}\right) + \int_0^{k+1/2} \ln\left(\frac{x^2 \pi^2}{l^2}\right) dx$$

$$-\int_0^N dx \ln\left[\frac{2 - 2\cos(\pi a x/l)}{a^2}\right] + \frac{1}{2} \ln\left(\frac{4}{a^2}\right). \quad \text{(A13.19)}$$

The last term arises from the contribution of the upper end of the sum, and it would be missing in the previous analysis.

We finally obtain for the argument of the exponential

$$\frac{1}{2} \ln\left(\frac{\pi^2}{l^2}\right) - C - \frac{l}{a} \int_0^1 dx \ln\left[\frac{2 - 2\cos(\pi x)}{a^2}\right] + \frac{1}{2} \ln\left(\frac{4}{a^2}\right) \quad \text{(A13.20)}$$

where C is $\ln 2\pi$ (as before).

If we combine all terms we get[22]

$$\left(\frac{1}{2\pi l}\right)^{1/2}, \quad \text{(A13.21)}$$

in perfect agreement with Eq. (A13.15).

The motivation for this appendix was to familiarize the reader with the various techniques available for the evaluation of functional integrals in the Gaussian case, where everything is under control. Similar, but slightly more complex manipulations, must be performed in the study of the Casimir effect.[23]

Notes for Chapter 13

1. In the opposite Schrödinger representation the operators do not depend on time, while the states evolve according to the law

$$(d/dt)|\psi(t)\rangle = -i\mathcal{H}/\hbar|\psi(t)\rangle \quad (|\psi(t)\rangle = \exp[-it\mathcal{H}/\hbar]|\psi(0)\rangle \, .$$

2. These "eigenstates" are not normalizable because the corresponding "eigenvalues" belong to the continuous spectrum of $q(t)$.
3. See any book on classical mechanics for the definition and properties of the Hamilton-Jacobi equation (e.g., H. Goldstein, *Classical Mechanics*, Addison-Wesley, Reading, MA (1950). The relation between the Hamilton-Jacobi equation and the function G in the limit $h \to 0$ is stressed by P. A. M. Dirac, *The Principles of Quantum Mechanics*, Clarendon Press, Oxford, 1958, section 32.
4. We recall that in classical mechanics one associates each trajectory $\omega(\tau)$ going from x to y in time t with the action $S[\omega] = \int_0^t d\tau \, L(\omega(\tau))$, the actual trajectory $\omega_c(\tau)$ being fixed by the condition $\delta S/\delta \omega = 0$, i.e., $S[\omega_c + \delta\omega] = S[\omega_c] + O((\delta\omega)^2)$ if $\delta\omega(0) = \delta\omega(t) = 0$.
5. This was done by Dirac in *Phys. Z. Soviet Union*, Band 3, Heft I (1933), and it is the basis of the later Feynman path-integral approach (see R. P. Feynman and A. R. Hibbs, *Quantum Mechanics and Path Integrals*, McGraw-Hill, New York, 1965).
6. Equation (13.10) can be used as the starting point for computing the interference in a typical diffraction experiment, e.g., an electron can go through one of two small holes (we recall that the probability is $|G^2|$); applications to optics of similar formulae can be found in W. Pauli, *Optics and the Theory of the Electrons*, MIT Press, Cambridge (1973).
7. Both $(2\pi\hbar \Delta t)^{-1/2} \exp[i/(2\Delta t \hbar)(x-y)^2]$ and $G(x, y|\Delta t)$ become distributions concentrated at $x - y = 0$ in the limit $\Delta t \to 0$, i.e., $\lim_{\Delta t \to 0} \int dy \, G(x, y|\Delta t) g(y) = g(x)$ for any smooth function $g(y)$. The contribution of the integration region with $x \neq y$ is damped by the fast oscillations of the integrand. For small Δt the relevant integration region is $(x - y)^2 = O(\Delta t)$ [cf. Eq. (13.75)]. In this region $V(x) = V(y) + O((\Delta t)^{1/2})$, and the action of the classical trajectory (which is of order Δt) is given by $\Delta t\{(x - y)^2/2(\Delta t)^2 - [V(x) + V(y)]/2\}$ with a relative accuracy of order $(\Delta t)^{1/2}$. It is evident that the trajectories dominating the functional integral have $[\omega(\Delta t) - \omega(0)]^2 \sim O(\Delta t)$. These trajectories are continuous but not differentiable: in mathematical terminology they are Holder functions of class $\frac{1}{2}$

(see Chapter 19). For a mathematical study of Holder functions, see, for example, H.P. McKean, *Stochastic Integrals*, Academic Press, New York (1969).
8. Periodicity has been imposed by setting $\omega(-T/2) = \omega(T/2)$.
9. We recall that $G(x, y|t)$ is the integral kernel of the operator $\exp[it\mathcal{H}/\hbar]$.
10. The relation between imaginary and real time has been stressed by K. Symanzik in *Local Field Theory*, ed. by R. Jost, Academic Press, New York (1969); see also E. S. Fradkin, *Dokl. Acad. Nauk. USSR* 125, 311 (1959).
11. We must be very careful when we deal with oscillatory integrals that are not absolutely convergent: formal manipulations may lead to wrong results; the very definition of the integral may be problematic.
12. For a review see S. Albeverio and R. Høegh-Krohn, *Mathematical Theory of Feynman Path Integrals*, Springer, Berlin (1976).
13. The canonical commutation relations can be considered as the boundary conditions for the second-order differential equations (13.35).
14. The derivation is sound as long as $\theta \neq 0$. It may be a very dangerous operation to use integration by parts for an integral that is not absolutely convergent.
15. Here x is the equivalent of σ (or ν) of the previous chapter; it is not the coordinate of the imaginary time (which is denoted by t)!
16. We recall that $\langle A(\omega(0))\rangle = \int dx\, \psi_0(x)^2 A(x)$ and that $\langle A(\omega(0))B(\omega(t))\rangle = \int dx\, dy\, A(x)B(y)\psi_0(x)\psi_0(y)G(x, y|t)$. The probability distribution of $\omega(0) \equiv x$ is $\psi_0^2(x)$ and the conjoint probability distribution of $\omega(0) \equiv x$ and $\omega(t) \equiv y$ is $\psi_0(x)\psi_0(y)G(x, y|t)$. For the harmonic oscillator, $S[\omega]$ is a quadratic functional, and each of the $\omega(t)$'s has a Gaussian probability distribution, hence Eq. (13.45).
17. The proof of the Euler-Maclaurin sum formula can be found in many places, for example, A. Ralston and P. Rabinowitz, *A First Course in Numerical Analysis*, McGraw-Hill, New York (1965).
18. Equation (13.66) is true as an asymptotic expansion in powers of δ. We shall use Eq. (13.66) at $\delta = 1$ only for large r (as we shall see), where the terms proportional to δ^{2k} become proportional also to r^{-2k}. Equation (13.66) at $\delta = 1$ can be interpreted as an asymptotic expansion in powers of $1/r$.
19. This point is discussed in E. S. Abers and B. W. Lee, *Phys. Rep. 9C*, 1 (1973).
20. A careful study of the transformation of the functional integral from rectangular (x, y, p_x, p_y) to polar $(r, \theta, p_r, p_\theta)$ coordinates can be found in I. K. Edwards, *Am. T. Phys. 47*, 153 (1979).
21. It is usual to quantize theory in rectangular coordinates; the need to choose a "quantization frame" is considered by some physicists as an annoying feature of the formalism of quantum mechanics.

22. I. S. Gradshteyn and I. M. Ryzhik, *Table of Integrals, Series, and Products*, Academic Press, New York (1965), Eq. (4.384, 18) can be the starting point for deriving the useful result

$$\int_0^1 dx \ln[2 - 2\cos(\pi x)] = 0.$$

23. For a similar discussion of the Casimir effect see C. Itzykson and J. B. Zuber, *An Introduction to Field Theory*, McGraw-Hill, New York (1980). More recent results on the Casimir effect for a non-Gaussian theory can be found in K. Symanzik, *Nucl. Phys. B 190* [FS3], 1 (1981).

CHAPTER 14

Semiclassical Methods

14.1. The method of the stationary phase

In this chapter we shall study some features of the limit $\hbar \to 0$, or, equivalently, the behavior of the system for energies much higher than the ground-state energy. Various applications of the techniques introduced in earlier chapters are presented; the hurried reader may prefer to skip this chapter.

If we take the limit $\hbar \to 0$ for real time, we face the problem of evaluating a rapid oscillatory integral. This can be done by using the method of the stationary phase which we describe here.[1] We consider the integral

$$\int_{-\infty}^{+\infty} dz \, \exp\left[\frac{if(z)}{\hbar}\right] \equiv I(\hbar), \qquad (14.1)$$

where $f(z)$ is a real analytic function of z. We want to estimate the integral in the limit $\hbar \to 0$. There are two possibilities: (a) there is one (or more) real value of z, (z_0) such that $f'(z_0) = 0$; (b) $f'(z_0) \neq 0$ for any real z.

In the first case we know that the integration regions far from z_0 give rapidly oscillating contributions that average to zero. The leading term comes from the region near z_0, where the phase of the integrand is stationary. We can thus approximate $f(z)$ with

$$f(z_0) + \frac{1}{2} f^{(2)}(z_0)(z - z_0)^2 + \frac{1}{6} f^{(3)}(z_0)(z - z_0)^3$$

$$+ \frac{1}{24} f^{(4)}(z_0) (z - z_0)^4 \cdots . \qquad (14.2)$$

If only the first two terms are included we obtain a Gaussian integral:

$$I(\hbar) \simeq I_0[\hbar](1 + O(\hbar))$$

$$I_0[\hbar] = \exp\left[\frac{if(z_0)}{\hbar}\right]\left[\frac{i2\hbar\pi}{f''(z_0)}\right]^{1/2}.$$

(14.3)

The higher-order terms can be treated in perturbation theory; we easily find

$$I(\hbar) = I_0[\hbar]\left[1 + i\hbar\left(\frac{1}{8}\frac{f^{(4)}(z_0)}{(f^{(2)}(z_0))^2} + \frac{5}{24}\frac{(f^{(3)}(z_0))^2}{(f^{(2)}(z_0))^3}\right) + O(\hbar^2)\right]. \quad (14.4)$$

Higher-order corrections in $\hbar$ can also be computed.

If there are more values of z for which $df/dz = 0$ we must sum over the contributions of all of them. In the most difficult case, where $df/dz = 0$ has no real solution, $I(\hbar)$ is exponentially small $[O(\exp(-A/\hbar))]$. A precise evaluation of $I(\hbar)$ may be obtained by deforming the integration path in the complex z plane, as was done for the exponentially small imaginary parts in Chapter 8. Fortunately in the following we shall not need such an evaluation. We note that only in this last case must f be analytic, while Eqs. (14.3) and (14.4) could be derived under the condition of a sufficiently smooth f. In the next subsections we shall boldly apply the same method to the evaluation of functional integrals; a rigorous justification of this procedure has been obtained only quite recently.[2]

14.2. The classical limit

Let us look again at the path integral representation of $G(x, y|t)$ [Eq. (13.18)] for real times: we should like to evaluate it in the limit $\hbar \to 0$.

In order to use the stationary phase method we must find the trajectory that makes the classical action stationary; such a trajectory is (because of the classical action principle) the classical trajectory satisfying Newton's equations. The trajectories dominating the path integral are those close to the classical ones. Thus the old result $G(x, y|t) \sim \exp[iS_c(x, y|t)/\hbar]$, which we rederive here, appears in a new light.

We can now approximate the functional integral by integrating only over those trajectories close to the classical one; more precisely, we write

$$\omega(\tau) = \omega_c(\tau) + \delta\omega(\tau), \quad \omega_c(0) = x, \quad \omega_c(t) = y, \quad \delta\omega(0) = \delta\omega(t) = 0,$$

(14.5)

where $\omega_c(\tau)$ is the classical trajectory [which satisfies the classical equation of motion $\ddot{\omega} = -V'(\omega)$] and $\delta\omega$ is the displacement with respect to

14.2. The Classical Limit

the classical trajectory (which is, as we shall see, of order $\hbar^{1/2}$). Expanding the action in powers of $\delta\omega$ we readily find

$$S[\omega] = S_c[\omega] + \int_0^t d\tau \left[\frac{1}{2} (\delta\dot\omega(\tau))^2 - \frac{1}{2} V''(\omega(\tau))(\delta\omega(\tau))^2 \right] + O(\delta\omega)^3, \tag{14.6}$$

$$S_c = \int_0^t d\tau \left[\frac{1}{2} \dot\omega^2 - V(\omega) \right],$$

where the terms linear in $\delta\omega$ are absent because the functional action is stationary with respect to small variations of the trajectory near the classical one (see Appendix, Chapter 2).

The functional representation for $G(x, y|t)$ has a form quite similar to Eq. (14.1). We finally find, with a relative error of $O(\hbar)$,

$$\begin{aligned}G(x, y|t) &= \int d[\omega]_{x,y}^t \exp\left[\frac{iS[\omega]}{\hbar}\right] \\ &= \int d[\delta\omega]_{0,0}^t \exp\left[\frac{i}{\hbar} S_c(x, y|t)\right. \\ &\quad + \frac{i}{\hbar} \int_0^t d\tau(\delta\omega(\tau)\mathcal{D} \cdot \delta\omega(\tau)) + O(\delta\omega)^3 \bigg] \\ &\propto \exp\left[\frac{iS_c}{\hbar}(x, y|t)\right](\det \mathcal{D})^{-1/2}[1 + O(\hbar)], \end{aligned} \tag{14.7}$$

where the operator $\mathcal{D}$ is defined as

$$\mathcal{D} \cdot g(\tau) = -\frac{d^2}{d\tau^2} g(\tau) - V''(\omega(\tau))g(\tau) \tag{14.8}$$

on the space of functions g satisfying $g(0) = g(t) = 0$.

As we saw in Chapter 13, the determinants of differential operators are not well defined: they are formally infinite. Indeed, we have forgotten the normalization factor of the functional integral. We can, however, use the computation of the preceding chapter for $\det(-d^2/dt^2)$, in which we have taken care of the normalization (see p. 245 et seq.). We finally find

$$\begin{aligned}G(x, y|t) &\simeq N \det\left(-\frac{d^2}{d\tau^2}\right)^{-1/2} \det\left[1 + \left(\frac{d^2}{d\tau^2}\right)^{-1} V''(\omega(\tau))\right]^{-1/2} \\ &\quad \times \exp\left[\frac{iS_c(x, y|t)}{\hbar}\right] \\ &= \left(\frac{i}{2\pi\hbar t}\right)^{1/2} \det\left[1 + \left(\frac{d^2}{d\tau^2}\right)^{-1} V''(\omega(\tau))\right]^{-1/2} \exp\left[\frac{iS_c(x, y|t)}{\hbar}\right],\end{aligned} \tag{14.9}$$

where N is the formally infinite normalization factor. In the general case it would be impossible to evaluate the determinant of Eqs. (14.7) and (14.9) in an analytic way; however, the fact that $\omega(t)$ is a solution of the classical equation of motion, and therefore $[-(\partial/\partial\tau)^2 + V''(\omega)]\dot{\omega}(\tau) \equiv \mathcal{D}\dot{\omega} = 0$, makes it possible for us to evaluate the determinant exactly.[3] A fast alternative way to obtain the correct result is to use Eq. (13.8). In both cases one gets, after some manipulations,

$$G(x, y|t) \simeq \left[\frac{i}{2\pi\hbar\dot{\omega}(t)\dot{\omega}(0)} \frac{\partial E_c}{\partial t}(x, y|t) \right]^{1/2} \exp\left[\frac{iS_c(x, y|t)}{\hbar} \right], \quad (14.10)$$

where $E_c(x, y|t)$ is the energy of the particle on the classical trajectories going from x to y in time t. From Eq. (14.10) the probability of a particle's going from x to y in time t is given by

$$\frac{1}{2\pi\hbar\dot{\omega}(t)\dot{\omega}(0)} \frac{\partial E_c}{\partial t}(x, y|t). \quad (14.11)$$

The factor $[\dot{\omega}(0)\dot{\omega}(t)]^{-1}$ is rather intuitive: The time spent by a particle in a region is inversely proportional to its classical velocity. Higher-order corrections in $\hbar$ can be obtained either by considering higher-than-quadratic terms in the functional integral or by using Eq. (13.9) in a recursive way.

14.3. The density of eigenvalues

We consider now a D-dimensional Hamiltonian,

$$\mathcal{H}(p, q) = \sum_{\nu}^{D} \frac{p_\nu^2}{2} + V(q) \qquad V(q) \geq 0. \quad (14.12)$$

If the potential $V(q)$ goes to infinity with q, the spectrum of $\mathcal{H}$ is discrete:

$$\mathcal{H}|\psi_n\rangle = E_n|\psi_n\rangle \qquad E_{n+1} \geq E_n, \; E_0 \geq 0. \quad (14.13)$$

If $D > 1$, the eigenvalues can be degenerate as happens when the potential $V(q)$ is symmetric (e.g., in the case of rotational symmetry).

We can define a function $N(E)$ that is equal to the number of eigenvalues of $\mathcal{H}$ less than E. In the same way we can define a density of eigenvalues

$$\rho(E) = \frac{d}{dE} N(E). \quad (14.14)$$

$N(E)$ and $\rho(E)$ are the sums of the step and δ functions, respectively.

14.3. The Density of Eigenvalues

However, when E goes to infinity we may think that we can approximate $N(E)$ and $\rho(E)$ by a continuous function of E, which we shall now compute analytically.

The strategy we follow is based on the introduction of the resolvent of $\mathcal{H}$:

$$R(E) = \frac{1}{\mathcal{H} - E} = \sum_n |\psi_n\rangle\langle\psi_n| \frac{1}{E_n - E}. \tag{14.15}$$

The resolvent has remarkable properties. If we denote by $R(x, y|E)$ its kernel and by $r(E)$ its trace, we get

$$R(x, y|E) = \sum_n \psi_n(x)\psi_n(y) \frac{1}{E_n - E}$$

$$= \int_0^\infty \frac{dt}{\hbar} G^{\pi/2}(x, y|t) \exp\left(\frac{tE}{\hbar}\right), \quad \text{Re}(E) < 0$$

$$= i \int_0^\infty \frac{dt}{\hbar} G^{0^+}(x, y|t) \exp\left(\frac{itE}{\hbar}\right), \quad \text{Im } E > 0$$

$$r(E) = \sum_n \frac{1}{E_n - E} = \int_0^\infty \frac{dE' \, \rho(E')}{E - E'} \tag{14.16}$$

$$= \int_0^\infty dt \, \text{Tr}(\exp(-t\mathcal{H} + tE)), \quad \text{Re } E < 0$$

$$\rho(E) = \frac{1}{2\pi} \text{Im}\left[r(E)\right],$$

as can be seen by using the explicit forms of R and G (we could also use the identity $A^{-1} = \int_0^\infty dt \exp(-tA)$, which is valid for a positive A operator). The trace of the resolvent $r(E)$ is an analytic function with poles on the positive real E axis, and it satisfies a dispersion relation[4] (the sum over n being convergent only if E_n increases fast enough with n).

Now the large-E behavior of $r(E)$ for $E \to -\infty$ is related to the behavior when $E \to +\infty$ by the dispersion relations; moreover, for large and negative E the small t region dominates the integral representation (14.16) and approximate methods may be used. In the simplest approach we can use the commutation relations $[q, p] = i\hbar$ to develop Eq. (14.16) in powers of $\hbar$. Indeed, if for small $\hbar$ commutators are neglected, we get, for $\text{Re}(E) < 0$,

$$r(E) = \int_0^\infty dt \exp(tE) \, \text{tr}\left[\exp\left(-\frac{t}{2}\sum_\nu^D p_\nu^2 - tV(q)\right)\right]$$

$$= \int_0^\infty dt \exp(tE) \, \text{tr}\left[\exp\left[-\frac{t}{2}\sum_\nu^D p_\nu^2\right]\exp[-tV(q)]\right](1 + O(\hbar))$$

$$= \int_0^\infty dt \exp(tE) \int d^D x \langle x| \exp\left(-\frac{t}{2}\sum_\nu^D p_\nu^2\right)$$

$$\times \exp[-tV(q)]|x\rangle(1 + O(\hbar))$$

$$= \int d^D x \int_0^\infty dt \exp[tE - tV(x)](2\pi t \hbar^2)^{-D/2}$$

$$= \int d^D q \, d^D p \, \frac{(2\pi\hbar)^{-D}}{(\mathcal{H}(q, p) - E)}$$

$$= (2\pi\hbar^2)^{-D/2}\Gamma\left(1 - \frac{D}{2}\right)\int d^D q(-E + V(q))^{D/2-1} \, . \qquad (14.17)$$

Correspondingly we find

$$\rho(E) = \frac{(2\pi)^{-D/2}\hbar^{-D}}{\Gamma(D/2)} \int_{V(q) \le E} d^D q(-V(q) + E)^{D/2-1}$$

$$= (2\pi\hbar)^{-D}\int d^D p \, d^D q \, \delta(\mathcal{H}(p, q) - E) \qquad (14.18)$$

$$N(E) = (2\pi\hbar)^{-D}\int d^D p \, d^D q \, \theta(E - \mathcal{H}(p, q)) \, .$$

It is important that in Eq. (14.18) the p's and q's are numbers, not operators.

Equation (14.18) [cf. Eq. (1.13)] is the leading result when $\hbar \to 0$. Higher-order corrections can be obtained from the Baker-Hausdorf relation,[5]

$$\exp(A + B) = \exp(A)\exp\left\{-\frac{1}{2}[A, B] + \frac{1}{6}[A, [A, B]]\right.$$

$$\left. + \frac{1}{6}[B, [B, A]] + \cdots\right\}\exp(B) \qquad (14.19)$$

for the exponential of two noncommuting operators, if we set $A = t\sum_\nu^D \frac{1}{2}p_\nu^2$, $B = tV(q)$, where the terms neglected contain three or more commutators. Each commutator gives an extra factor $\hbar$. At order $\hbar^2$ a

14.3. The Density of Eigenvalues

careful computation[6] gives[7]

$$r(E) = (2\pi\hbar)^{-D} \int d^D p \, d^D q \left\{ (\mathcal{H}(q,p) - E)^{-1} - \frac{\hbar^2}{4} \right.$$

$$\left. \cdot \left[\sum_{\nu\,1}^{D} \frac{\partial^2 V/\partial q_\nu^2}{(\mathcal{H}(q,p) - E)^3} + \frac{Dp^2 \sum_{\nu\,1}^{D} \partial^2 V/\partial q_\nu^2 + \sum_{\nu\,1}^{D} (\partial V/\partial q_\nu)^2}{(\mathcal{H}(p,q) - E)^4} \right] \right.$$

$$\left. + O(\hbar^4) \right\}$$

$$\rho(E) = (2\pi\hbar)^{-D} \left\{ \int d^D p \, d^D q \, \delta(\mathcal{H}(p,q) - E) \right.$$

$$+ \frac{\hbar^2}{8} \frac{d^2}{dE^2} \int d^D p \, d^D q \, \delta(\mathcal{H}(p,q) - E) \sum_{\nu\,1}^{D} \frac{\partial^2 V}{\partial q_\nu^2}$$

$$+ \frac{\hbar^2}{24} \frac{d^3}{dE^3} \int d^D p \, d^D q \, \delta(\mathcal{H}(p,q) - E) \left[Dp^2 \sum_{\nu\,1}^{D} \frac{\partial^2 V}{\partial q_\nu^2} \right.$$

$$\left. + \sum_{\nu\,1}^{D} \left(\frac{\partial V}{\partial q_\nu} \right)^2 \right] + O(\hbar^4) \right\}$$

$$= \frac{(2\pi)^{-D/2} \hbar^{-D}}{\Gamma(D/2)} \left\{ \int_{V(q) \leq E} d^D q (E - V(q))^{D/2 - 1} \right.$$

$$- \frac{\hbar^2}{8} \frac{d^2}{dE^2} \int_{V(q) \leq E} d^D q \left[(E - V(q))^{D/2 - 1} \sum_{\nu\,1}^{D} \frac{\partial^2 V}{\partial q_\nu^2} \right]$$

$$+ \frac{\hbar^2}{24} \frac{d^3}{dE^3} \int_{V(q) \leq E} d^D q \left[(E - V(q))^{D/2 - 1} \left((E - V(q)) \sum_{\nu\,1}^{D} \frac{\partial^2 V}{\partial q_\nu^2} \right. \right.$$

$$\left. \left. + \sum_{\nu\,1}^{D} \left(\frac{\partial V}{\partial q_\nu} \right)^2 \right) \right] + O(\hbar^4) \right\}. \tag{14.20}$$

The terms with higher powers in $\hbar$ become smaller and smaller when E goes to infinity; indeed, the large-E limit is controlled by the short-time behavior of Green's function $G(x, x|t)$. The corrections coming from the commutator of $tp^2/2$ with $tV(q)$ obviously vanish when t goes to zero. Equation (14.18) is thus correct for $E \to +\infty$.

Similar results can be found if one studies the following related problem: We consider the Hamiltonian

$$\mathcal{H} = \frac{1}{2} \sum_{\nu\,1}^{D} p_\nu^2 + gV(q), \quad V(q) < 0, \tag{14.21}$$

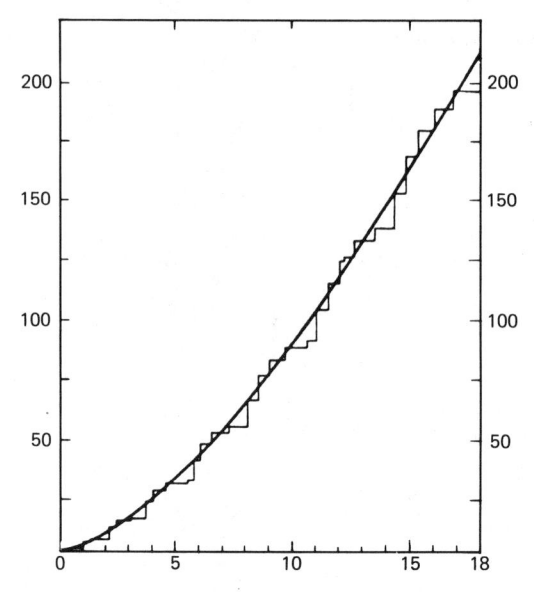

Figure 14.1. The exact function $N(g)$ and the approximate form Eq. (14.22) for the attractive potential described in the text (from E. Brézin and G. Parisi, *J. Stat. Phys.* **25**, 273 (1978)).

where $V(q)$ goes to zero at infinity. The number of bound states (discrete negative levels) of energy less than E ($E < 0$) is a function of g, and it is denoted as $N_E(g)$. Using the same approximations, we find for $N_E(g)$ the asymptotic expression in the large-g region:

$$N_E(g) = (2\pi\hbar)^{-D} \int d^D p \, d^D q \, \theta\left(-\sum_1^D \frac{p_\nu^2}{2} - E - gV(q)\right) = (2\pi\hbar)^{-D} S_D$$

$$\times \int_{E > gV(q)} d^D q (E - gV(q))^{D/2} \sim g^{D/2} (2\pi\hbar)^{-D} \int d^D q \, \theta(-V(q)). \tag{14.22}$$

A rigorous elementary proof of the correctness of Eq. (14.22), with precise error estimates, can be found in a paper of A. Martin.[8]

In Fig. 14.1 we plot Eq. (14.22) against the exact (numerically evaluated) $N(g)$ in a typical three-dimensional case, $V(x) = -3\varphi^2(x)$, where φ is the unique positive spherical symmetric solution (vanishing at infinity) of the equation $-\Delta\varphi + \varphi = \varphi^3$. The agreement is quite good.[9]

14.4. The WKB method and its generalization

While the true function $N(E)$ is integer valued, this cannot happen in any of the approximations of the last subsection. When we study the

14.4. The WKB Method and Its Generalization

behavior of $r(E)$ for $E \to -\infty$, possible oscillations in $\rho(E)$ are washed out. It is thus natural to ask if we can find extra oscillatory terms for $\rho(E)$. This can be done by using the integral representation (14.17) for nearly real positive E ($E = E + i\varepsilon$), where $\rho(E)$ is written as an integral over the real-time Green's function:

$$\rho(E) = \lim_{\varepsilon \to 0^+} \frac{1}{\pi} \operatorname{Im} i \int d^D x \int_0^\infty \frac{dt}{\hbar} \cdot \exp\left[-\frac{\varepsilon t}{\hbar} + \frac{iEt}{\hbar}\right] G^{0^+}(x, x|t). \quad (14.23)$$

Let us now try to obtain an approximate expression for Eq. (14.23). Using Eq. (14.6), neglecting prefactors, we can write

$$G(x, x|t) \simeq \sum_n^{M(t)} \exp[iS_c^n(t)], \quad (14.24)$$

where n labels the classical trajectories [whose number is $M(t)$] that go from x to x in time t and S_n^c is the corresponding classical action. If t is small, $M(t) = 1$, while for higher values of t, $M(t)$ may be larger, i.e.,

$$M(t) = k \quad \text{for } t_{k-1} < t < t_k \quad (t_0 = 0). \quad (14.25)$$

In the same way that in the previous section the leading contribution comes from $t = 0$, here the leading contributions to Eq. (14.23) come from the region $t = t_k$. We finally obtain

$$\rho(E) \simeq \operatorname{Re} \sum_n C_n(E) \exp\left(\frac{iEt_n(E)}{\hbar}\right), \quad (14.26)$$

where we have explicitly indicated the dependence of the classical periods on E. A more accurate computation is needed to find the prefactors $C_n(E)$.

The crucial point we want to stress is that the periods of classical closed trajectories correspond to oscillations in the density of eigenvalues;[10] a beautiful application of these ideas is shown in Fig. 14.2.

In the D-dimensional case the classification of periodic trajectories is not simple; fortunately, in the one-dimensional case the trajectories may be labeled by only their energy E_c.[11] If we call $\tau(E_c)$ and $S(E_c)$ the periods of the trajectory of energy and the corresponding action, we can approximate Eq. (14.23) by

$$\rho(E) \simeq \lim_{\varepsilon \to 0} \operatorname{Im} \int_0^\infty \sum_n \exp(-t\varepsilon + itE) \, dt \cdot \int dE_c \exp\left[\frac{inS(E_c)}{\hbar}\right] \delta(t - n\tau(E_c))$$

$$= \lim_{\varepsilon \to 0} \frac{1}{\pi} \operatorname{Im}\left\{\sum_n \int_0^\infty dE_c \exp\left[+n\tau(E_c)(iE - \varepsilon) - \frac{inS(E_c)}{\hbar}\right]\right\}, \quad (14.27)$$

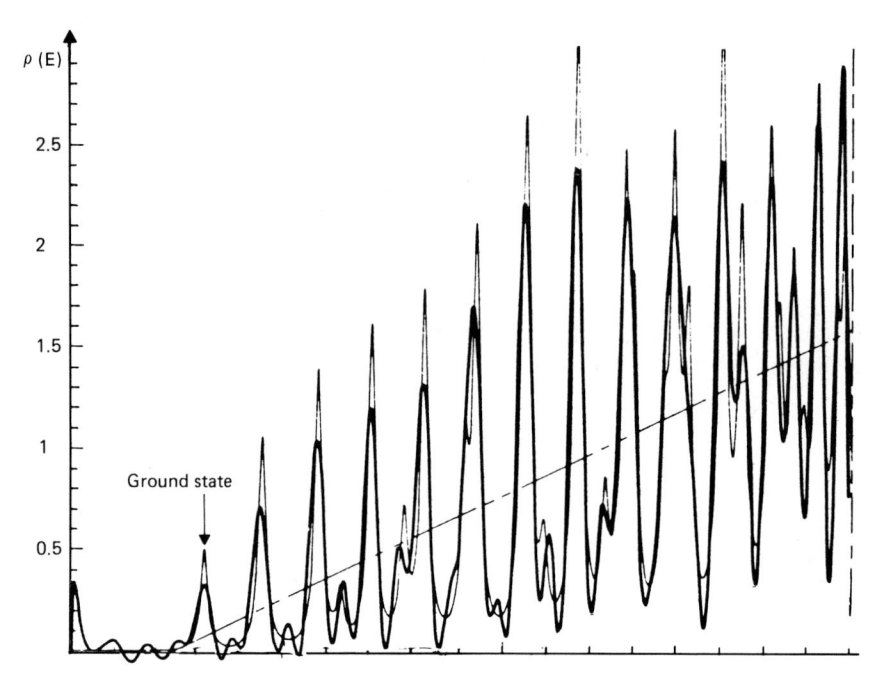

Figure 14.2. The smoothed (exact) function $\rho(E)$ (fine line) and the approximate form (heavy line) obtained by summing over many classical trajectories for a particle confined to move inside a sphere; for more details the reader should see R. Balian and C. Bloch, *Ann. Phys. (N.Y.)* **69**, 76 (1972).

where we have used the fact that we can construct a trajectory of period $n\tau(E_c)$ by repeating n times the same trajectory of period $\tau(E_c)$. For high values of the energy (or equivalently for small $\hbar$ at fixed E), the integrand is rapidly oscillating, and we can use the technique of the stationary phase to integrate over E_c. The final result is

$$\rho(E) = \lim_{\varepsilon \to 0}\left(\text{Im}\left\{\sum_{n}^{\infty} \exp\left[n\tau(E_c(E))(iE - \varepsilon) - \frac{inS(E_c(E))}{\hbar}\right]\right\}\right) \quad (14.28)$$

where $E_c(E)$ is fixed by the condition

$$\left.\frac{d\tau/dE_c}{dS/dE_c}\right|_{E_c = E_c(E)} = E . \quad (14.29)$$

A simple exercise in classical mechanics will show that the solution of Eq. (14.29) is given by

$$E_c(E) = E . \quad (14.30)$$

If the prefactors are carefully evaluated one finds[12]

14.4. The WKB Method and Its Generalization

$$\rho(E) = \text{Im}\left\{\frac{i\tau(E)}{\hbar\pi}\sum_0^\infty{}_n \exp\left[\frac{inW(E) + i\varepsilon E}{\hbar}\right]\right\}$$

$$= \text{Im}\left\{\frac{\tau(E)}{\hbar\pi}\frac{1}{1 + \exp[iW(E)]/\hbar}\right\}$$

$$= \sum_0^\infty{}_n \delta\left[W(E) - \hbar\left(n + \frac{1}{2}\right)\pi\right]\tau(E) = \sum_n \delta(E_n - E) \quad (14.31)$$

$$W(E) = S_c(E) - E\tau(E) = N_0(E)$$

$$N_0(E) = 2^{-1/2}\int_{V(q)<E} dq(E - V(q))^{1/2} = \frac{1}{2}\oint_{V(q)<E} p(q|E)\,dq$$

$$\frac{1}{2}p(q|E)^2 = E - V(q),$$

where the E_n are fixed by the condition

$$W(E_n) = \left(n + \frac{1}{2}\right)\pi\hbar \quad (14.32)$$

and we have used the relation $dW/dE = \tau(E)$ [cf. Eq. (14.29)].

We have thus found that the Bohr-Sommerfeld equation is the condition for the different oscillating terms (coming from classical trajectories repeated n times) to sum up in a coherent way.

In this discussion of the semiclassical approach we have omitted most of the technical details in order to stress the two most important points: (1) the asymptotic behavior of the level density is related to the short-time behavior of $G(x, x|t)$; (2) the oscillations in the level density for higher-dimensional systems and the Bohr-Sommerfeld condition (14.32) in one dimension have the same physical origin.

Similar results may also be obtained using the more traditional WKB method for solving the Schrödinger equation. Equation (14.32) gives the correct result only for very large n; the corrections to the leading term are given by[13]

$$N_0(E) - \frac{1}{6}\frac{d^2}{dE^2}\oint_{V(q)<E} dq\,p(q|E)^{-1/2}\left(\frac{dV}{dq}\right)^2 = n + \frac{1}{2}. \quad (14.33)$$

These corrections vanish for large n.

In many cases Eq. (14.32) is qualitatively correct for $n = 0$. A well-studied example is the quartic anharmonic oscillator,

$$\mathcal{H} = p^2 + x^4. \quad (14.34)$$

The exact ground-state energy is about 1.060, while Eq. (14.32) gives

$$E_n \sim C\left(n + \frac{1}{2}\right)^{4/3}$$

$$C = 3^{4/3} 2^{2/3} \pi^2 \Gamma\left(\frac{1}{4}\right)^{-8/3} \tag{14.35}$$

$$\Gamma\left(\frac{1}{4}\right) \simeq 3.62561 .$$

For $n = 0, 1, 2$, Eq. (14.35) is off by 20%, 5%, and 1%, respectively. Of course better results may be obtained by using the next-order corrections. In this case it is known that the following expansion holds:[14]

$$2\pi\left(n + \frac{1}{2}\right) = \tilde{E}_n\left[1 + \frac{b_1}{\tilde{E}_n^2} + \frac{b_2}{\tilde{E}_n^4} + \cdots\right] - 2(-1)^n \exp\left(-\frac{1}{2}\tilde{E}_n + \cdots\right)$$

$$\tilde{E}_n = \frac{\Gamma(\tfrac{1}{4})^2}{3}\sqrt{\frac{2}{\pi}} E_n^{3/4} \tag{14.36}$$

$$b_1 = -\frac{\pi}{3}, \qquad b_2 = \frac{11\Gamma(\tfrac{1}{4})^8}{10{,}368\,\pi^2} .$$

Many terms of the large-n expansion have been computed. The exponentially small terms are the contribution coming from complex trajectories.[15]

The semiclassical method for quantum mechanics has received a strong impetus in recent years, and a vast literature is available.

14.5. Trace identities

In writing this book I have not been able to resist the temptation to mention one of the most beautiful results of the Schrödinger equation, obtained after the second world war: the trace identities.

We consider for simplicity the one-dimensional Schrödinger equation with a nonsingular potential $V(x)$, of fast (exponential) decrease at infinity $[V(\infty) = 0]$, which for simplicity we assume to be positive. The Hamiltonian will have no discrete levels, and the spectrum is a continuum starting at $E = 0$. Very important information on the continuum spectrum is given by the phase shift $\delta(E)$ defined by the asymptotic behavior of the (un-normalized) solution of the Schrödinger equation of energy $E > 0$:

$$\psi(x|E) \simeq \begin{cases} \exp(ip_E x) & x \to -\infty \\ \exp[ip_E x + i\delta(E)] & x \to +\infty \end{cases} \tag{14.37}$$

$$p_E \equiv \sqrt{2E} .$$

14.5. Trace Identities

The trace identities[16] state that in the absence of bound states

$$\int_0^\infty dE \, E^{-s} \frac{d\delta}{dE}\bigg|_{s=-n} = 0 \quad n = 0, 1, 2, \ldots, \quad (14.38)$$

where the integral is defined as an analytic continuation in s from the region where it is convergent. The proof is very simple. We introduce at an intermediate stage the Hamiltonian $\mathcal{H}_L$ defined in the interval $(-L/2, L/2)$ with periodic boundary condition. $\mathcal{H}_L$ has only a discrete spectrum. It can be found by imposing at large L the following condition

$$\psi\left(-\frac{L}{2}\bigg|E_n\right) = \psi\left(\frac{L}{2}\bigg|E_n\right) \Rightarrow L p_{E_n} + \delta_{E_n} = 2\pi n, \quad (14.39)$$

where $\psi(x|E)$ is defined in Eq. (14.37).

We shall now derive the trace identities for $\mathcal{H}_L$ and then we shall send L to infinity. We consider for simplicity the one-dimensional case. The methods of the previous sections give $G(x, x|t)$:

$$G(x, x|t) \sim \left(\frac{1}{2\pi\hbar t}\right)^{1/2} \exp\left[-\frac{tV(x)}{\hbar}\right]\left[\sum_{k=0}^\infty C_k(x)\hbar^2 t^k\right], \quad (14.40)$$

where the C_k are polynomials in $V(q)$ and its derivatives with respect to q arising from the various commutators. In the case we consider, the integration over x cannot produce any extra power of t, so that we have[17]

$$\int_{-L/2}^{L/2} dx \, G(x, x|t) = \sum_{k=0}^\infty \frac{b_k}{t^{1/2-k}}. \quad (14.41)$$

This asymptotic behavior has a very serious consequence. We can introduce the zeta function of $\mathcal{H}_L$,

$$\zeta_L(s) = \text{Tr}[\mathcal{H}_L^{-s}] = \sum_n E_n^{-s}(L)$$

$$= \int dE \, \rho_L(E) E^{-s}$$

$$= \frac{1}{\Gamma(s)} \int dx \, \frac{dt}{t} t^s G_L(x, x|t) \equiv \frac{\mu(s)}{\Gamma(s)}. \quad (14.42)$$

For $\text{Re } s > \frac{1}{2}$ no singularities are present, while for $\text{Re } s \leq \frac{1}{2}$ we may have singularities coming from the lower integration end. Using standard techniques one finds that $\mu(s)$ is a meromorphic function having poles at $s_k = \frac{1}{2} - k$ with residuum b_k. This implies that $\mu(s)$ has no poles at negative integers; therefore

$$\zeta(s)\big|_{s=-n} = 0 \quad n = 0, 1, 2, \ldots \quad (14.43)$$

as a consequence of the poles of the Γ function. Equation (14.43) can be formally written as

$$\text{Tr}[\mathcal{H}_L^n] = 0, \tag{14.44}$$

where it is understood that Eq. (14.44) holds as an analytic continuation in n from negative to positive values. We have obtained the trace identities for a particle in a box. We should now study Eq. (14.43) in the limit $L \to \infty$.

Let us compute $\zeta(s)$ for large L. It is convenient to consider the function

$$\tilde{\zeta}(s) = \zeta(s) - \zeta_0(s) = \int_0^\infty dE \, E^{-s}[\rho(E) - \rho_0(E)]$$

$$\tilde{\zeta}(s)\big|_{s=-n} = 0 \quad n = 0, 1, 2, \ldots, \tag{14.45}$$

where ζ_0 and ρ_0 refer to the free ($V=0$) case where $\delta(E) = 0$. Now the average level spacing in p is $2\pi/L$ for the free case and $2\pi/L + 2\pi/L^2(d\delta/dp)$ for the interacting case[18] $[d\delta/dp = (d\delta/dE)(dE/dp) = \rho(d\delta/dE)]$. In the limit L going to infinity we find

$$\tilde{\zeta}(s) = \int_0^\infty \frac{1}{2\pi} E^{-s} \frac{d\delta}{dE} dE$$

$$\rho(E) - \rho_0(E) \xrightarrow[L \to \infty]{} \frac{1}{2\pi} \frac{d\delta}{dp} \frac{dp}{dE} = \frac{1}{2\pi} \frac{d\delta}{dE}. \tag{14.46}$$

If we add the contribution of the bound states which may be present if $V(x)$ is not positive definite, we get

$$\tilde{\zeta}(s) = \frac{1}{2\pi} \int_0^\infty dE \, E^{-s} \frac{d\delta}{dE} + \sum_1^N E_k^{-s}$$

$$\tilde{\zeta}(s)\big|_{s=-n} = 0 \quad n = 0, 1, 2, \ldots, \tag{14.47}$$

where N is the number of bound states. For $n = 0$, Eq. (14.47) reduces to the Levinson theorem,

$$\frac{1}{2\pi}[\delta(E)|_{E=\infty} - \delta(0)] + N = 0. \tag{14.48}$$

In this case, moreover, if $n > 0$ the integrals in Eq. (14.47) are likely to be nonconvergent, and Eq. (14.47) must be evaluated after analytic continuation in s.

Trace identities may also be derived in the case where $V(\infty) = \infty$, when their precise form depends on the large-x behavior of $V(x)$.[19] In the simplest case, e.g., $\mathcal{H}$ being given by Eq. (14.34), they are correct in the form of Eqs. (10.47) and (10.48) without modification. If the higher eigenvalues are estimated from the WKB formulae, trace identities are quite useful for finding the lower levels.

14.6. Tunneling effects

The tunneling effect is a typical quantum-mechanical phenomenon. In the simplest situation, the symmetric double-well potential of Fig. 14.3 $[V(\pm 1) = 0]$, if a particle is localized in one of the two wells with small enough energy, it will remain there forever in classical mechanics. Quantum-mechanically, however, it has a finite probability of tunneling to the other well. For real times the tunneling trajectory is classically forbidden, and the corresponding probability amplitude is exponentially small in $\hbar$.

The existence of the tunneling effect has some consequences on the structure of the low eigenvalues and eigenstates of the Hamiltonian: they can be associated in pairs $[E_n^+$ and E_n^-, $\psi_n^+(x)$ and $\psi_n^-(x)]$ whose relative energy splitting is exponentially small. More precisely, we have

$$\psi_n^+(x) = \psi_n(x) + \psi_n(-x) \qquad \psi_n^-(x) = \psi_n(x) - \psi_n(-x)$$

$$E_n^+ - E_n^- \equiv \Delta E_n \sim \exp\left(-\frac{A_n}{\hbar}\right), \tag{14.49}$$

where the function $\psi_n(x)$ is exponentially small for negative x [i.e., $\psi_n(-x)/\psi_n(x)$ is exponentially small for $x > 0$]. The consequences of this structure of the levels on the tunneling probability can be understood qualitatively by neglecting $\psi(x)$ for negative x and using only the two lower states in the representation (13.29) for Green's function. We find

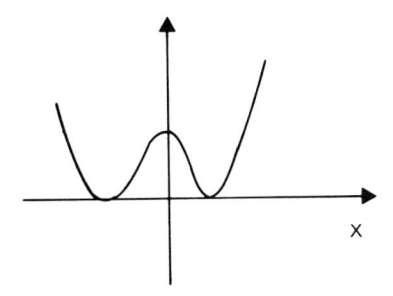

Figure 14.3. A double-well potential.

$$G(x, y|t) \cong \begin{cases} \psi_0(x)\psi_0(y)\left[1 + \exp\left(\frac{i\,\Delta E_0 t}{\hbar}\right)\right] & xy > 0 \\ \psi_0(x)\psi_0(y)\left[1 - \exp\left(\frac{i\,\Delta E_0 t}{\hbar}\right)\right] & xy < 0. \end{cases} \qquad (14.50)$$

According to these approximations the probability of remaining in the same well is $\cos^2(\Delta E_0 t/2\hbar)$, while the probability of being in the other well is $\sin^2(\Delta E_0 t/2\hbar)$; the tunneling probability P is thus proportional to $(\Delta E_0/\hbar)^{-1}$.

A direct computation of P in the real-time formalism should be carried out by deforming the functional integration path in order to pick up the contribution of the complex trajectories. Similar manipulations are normally used to extract exponentially small terms from an oscillating integral. We shall not follow this road, however, but will restrict ourselves to the computation of the exponentially small splitting between the ground state and the first excited state (ΔE_0) using the imaginary-time formalism. This exponentially small splitting implies an exponentially large coherence length $\xi = (\Delta E_0)^{-1}\hbar$ in the imaginary-time formulation, i.e., connected correlations decay like $\exp(-t/\xi)$.

We first note that the correlation function $\langle \omega(t)\omega(0)\rangle$ must go to zero when $t \to \infty$, because in one dimension no spontaneous breaking of the symmetry $\omega \to -\omega$ is possible: $\langle \omega(t)\rangle = \langle \omega(0)\rangle = 0$ (we recall that the potential is symmetric). It is also clear that in a first approximation a trajectory will spend most of its time near the bottom of one of the wells. A trajectory that crosses from one well to another must spend some time in the region where the potential is high, and its weight, which is proportional to $\exp[-1/\hbar \int dt'(\dot{\omega}^2/2) + V(\omega)]$, will be rather small. On the other hand, if we consider only trajectories that remain in the same well ($\omega(t)\omega(0) > 0$), the correlation will not go to zero. The vanishing of the correlation function $\langle \omega(t)\omega(0)\rangle$ at large t is thus connected with the existence of trajectories that go from one well to the other. The fact that we must go to exponentially large t to see a substantial decrease of the correlations is a consequence of the exponentially small weight of these trajectories.

Let us call $W(t)$ the maximum weight that a trajectory, jumping from one well to another in time t, may have [i.e., $\omega(0) = 1$, $\omega(t) = -1$]. The usual arguments suggest that, neglecting prefactors,

$$W(t) = \exp\left[-\frac{A(t)}{\hbar}\right]$$

$$A(t) = \min_{[\omega]}\left[\int_0^t dt'\left(\frac{1}{2}\dot{\omega}^2 + V(\omega)\right)\right], \qquad \omega(0) = 1, \ \omega(t) = -1.$$

(14.51)

14.6. Tunneling Effects

Postponing the computation of $W(t)$, we use only the property that, for t greater than τ [which is of the order of the inverse splitting of the ground state and the second excited level $(E_2 - E_0)^{-1}$], $A(t)$ [and consequently $W(t)$] is practically independent of t: $A(\infty) \equiv A$ and $W(\infty) \equiv W = \exp[-A/\hbar]$. Indeed, as shown in Fig. 14.4, the typical trajectory which minimizes Eq. (14.51) for large t will spend most of its time near 1 or -1 and only for a finite amount of time (around t_1) will $\omega^2(t)$ be definitely different from 1. In other words, for large t there are many trajectories that go from one well to the other; each of them has approximately the same weight and each may be labeled by the time t_1 at which the jump occurs. The total contribution of these trajectories to $G(1, -1|t)$ is

$$\int_0^t dt_1 \, W = tW. \tag{14.52}$$

There are two regimes: $tW \ll 1$, which is always the case if t is not too large, and $tW \gg 1$, which is relevant in the study of the limit t going to infinity. In this latter case we have also to consider trajectories that jump from 1 to -1 n times (see Fig. 14.5); it is easy to see that their relative weight (not normalized) is given by

$$W^n \int_0^t dt_1 \int_{t_1}^t dt_2 \cdots \int_{t_{n-1}}^t dt_n = \frac{(Wt)^n}{n!}. \tag{14.53}$$

A trajectory jumping n times may be labeled by the times $t_1, \ldots, t_n$ at which the jumps occur.

We finally conclude that the probability $P_n(t)$ of jumping n times in time t is given by

$$P_n = \exp(-Wt)\left(\frac{Wt}{n!}\right)^n, \tag{14.54}$$

i.e., by a Poisson distribution with $\langle n \rangle = Wt$.[20] The corresponding results

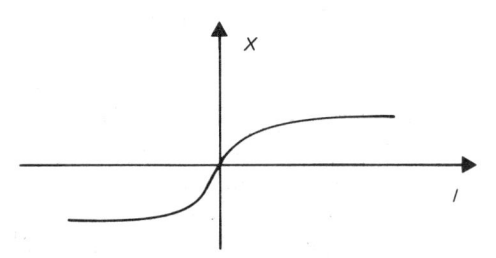

Figure 14.4. A trajectory minimizing $A(t)$.

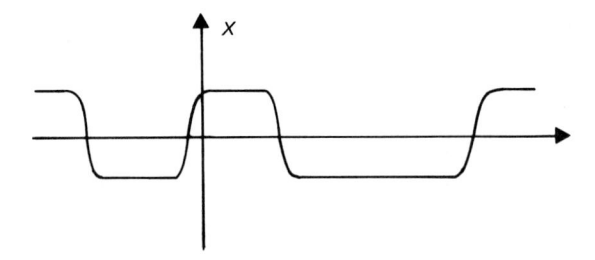

Figure 14.5. A multiple jumping trajectory.

for Green's function and the correlation functions are

$$G(1, 1|t) = \sum_{n \text{ even}} P_n(t) = \frac{1}{2}[1 + \exp(-2Wt)]$$

$$G(1, -1|t) = \sum_{n \text{ odd}} P_n(t) = \frac{1}{2}[1 - \exp(-2Wt)] \qquad (14.55)$$

$$\langle \omega(t)\omega(0)\rangle = G(1, 1|t) - G(1, -1, t) = \sum_n (-1)^n P_n(t) = \exp[-2Wt].$$

The splitting between the ground state and the first excited state is thus 2W.

We can now come back to the problem we left open, i.e., to the computation of W and A from Eq. (14.51). The usual arguments tell us that the trajectory $\omega(t')$ that minimizes A satisfies the classical equation of motion for imaginary time:

$$\frac{d^2}{dt'^2}\omega = V(\omega). \qquad (14.56)$$

We stress that the relative signs of the kinetic and potential terms at imaginary times are the opposite of the usual ones at real times, so that all the signs in front of the potential will be the opposite of the usual ones.

The quantity E (the "energy")

$$E = \frac{1}{2}\dot{\omega}^2 - V(\omega) \qquad (14.57)$$

is a constant of motion for the classical trajectory which minimizes A. For very large times t, the trajectory shown in Fig. 14.4 has a very small energy: $V(\omega(t)) = 0$ and $\dot{\omega}^2(t) \simeq 0$. In the limit $t \to \infty$ we can consider a trajectory having zero energy. Although in the general case it is necessary first to find the trajectory and later to compute the action, in this simple one-dimensional case the action can be computed without knowing the trajectory explicitly; indeed, the conservation of energy for $E = 0$ implies that

$$\frac{1}{2}\dot{\omega}^2 = V(\omega). \tag{14.58}$$

We finally obtain

$$A = \int_0^t dt'\, 2[V(\omega(t'))] = 2\int_{-1}^{1} d\omega \left(\frac{d\omega}{dt'}\right)^{-1} V(\omega)$$

$$= \int_{-1}^{1} (2V(\omega))^{1/2}\, d\omega. \tag{14.59}$$

We conclude that

$$\Delta E \simeq \exp\left[-\frac{1}{\hbar}\int_{-1}^{1}(2V(\omega))^{1/2}\,d\omega\right], \tag{14.60}$$

in agreement with a direct study of the Schrödinger equation.

This approach to the tunneling effect has the advantage (over the traditional approach) of being easily generalizable to the multidimensional case and to field theories. Indeed, one finds that even in this more general case there is a precise relation between the tunneling effect and the existence of nontrivial solutions of the classical equations of motion at imaginary times.[21]

14.7. Fermions

For completeness let us say just a few words about quantum statistics. For an N-particle wave function $\psi(x_1,\ldots,x_N)$, if all particles are different, no symmetry requirement is needed. If the particles are identical, only two possibilities are allowed in Nature: ψ symmetry (Bose statistics) and ψ antisymmetry (Fermi statistics).

In this section we want to describe some approximate methods used to evaluate the energy of the ground states of N fermions. We consider the Hamiltonian

$$\mathcal{H}_N = \frac{1}{2}\sum_{i}^{N} p_i^2 + \sum_{i}^{N} V_1(x_i) + \frac{1}{2}\sum_{ij}^{N} V_2(x_i - x_j). \tag{14.61}$$

If the fermions are not interacting among themselves ($V_2 = 0$), the ground-state wave function can be written as

$$\psi(x_1,\ldots,x_N) = \det \Psi, \tag{14.62}$$

where Ψ is an $N \times N$ matrix of elements $\Psi_{i,k} = \psi_k(x_i)$, the $\psi_k(x)$ ($k =$

$1, 2, \ldots, N$) being the eigenfunctions of the one-particle Hamiltonian

$$\mathcal{H}_1 = \frac{1}{2} p^2 + V_1(x)$$

$$\mathcal{H}_1 |\psi_k\rangle = \varepsilon_k |\psi_k\rangle, \qquad \varepsilon_k \le \varepsilon_{k+1}.$$

(14.63)

The corresponding energy of $\mathcal{H}_N$ is given by

$$E_0^N = \sum_k^N \varepsilon_k$$

$$\mathcal{H}_N |\Psi\rangle = E_0^N |\Psi\rangle.$$

(14.64)

We usually say that we fill with fermions the first N levels of $\mathcal{H}_1$. If the particles interact (i.e., $V_2 \ne 0$), the situation is more complex. One of the best simple approximations is the Hartree method.[22] Given a potential $U(x)$, we compute the wave function Ψ_U from Eq. (14.62) where the ψ_k's satisfy the equations

$$(\tfrac{1}{2} p^2 + U(x)) |\psi_k\rangle = \varepsilon_k |\psi_k\rangle.$$ (14.65)

We look for a potential U such that

$$\langle \Psi_U | \mathcal{H}_N | \Psi_U \rangle \equiv E[U]$$ (14.66)

is minimum. Obviously $E[U] \ge E$ and

$$\tilde{E} = \min_{[U]} E[U]$$ (14.67)

will be our best approximation for E.

The value of U which minimizes $E[U]$ (and which we call U_m) can be found by imposing the condition

$$E[U_m + \delta U] = E[U_m] + O((\delta U)^2)$$ (14.68)

for any δU, or equivalently $\delta E / \delta U(x) = 0$. Using the standard quantum-mechanical perturbation theory,[23] one can see that Eq. (14.68) implies that

$$U(x) = V_1(x) + \int dy \, V_2(x - y) \rho(y)$$

$$\rho(y) = \sum_k^N |\psi_k(y)|^2 \qquad \int d^D y \, \rho(y) = N$$

(14.69)

$$\left(-\frac{1}{2}\Delta + U(x)\right) \psi_k(x) = \varepsilon_k \psi_k(x),$$

14.7. Fermions

where we have used the relations

$$E[U] = \sum_{k}^{N} \varepsilon_k[U] + \int d^D x \, \rho(x)[V_1(x) - U(x)]$$

$$+ \frac{1}{2} \int d^D x \, d^D y \, \rho(x)\rho(y)V_2(x-y) \quad (14.70)$$

$$\frac{\delta}{\delta U(x)} \sum_{k}^{N} \varepsilon_k[U] = \rho(x).$$

Equation (14.69) has an obvious interpretation if we remember that $\rho(y)$ is the density of fermions at x.

The Hartree method quite often gives qualitatively and quantitatively correct results. If we use the asymptotic formulae of the previous section for estimating $\rho(y)$, we get in three dimensions the Thomas-Fermi equations

$$E[U] = \int d^D x \, C[E_F - U(x)]^{3/2} \left\{ \frac{3}{10}(E_F - U(x)) + V_1(x) \right.$$

$$\left. + \frac{1}{2} \int d^D y \, C[E_F - U(y)]^{3/2} U_2(x-y) \right\} \quad (14.71)$$

$$C = \frac{\sqrt{2}}{3\pi^2},$$

where the integral over x and y is done in the classically allowed regions $E_F - U(x) \geq 0$, $E_F - U(y) \geq 0$ and where the Fermi energy E_F and the potential U are fixed by the conditions

$$C \int dy (E_F - U(y))^{3/2} = N$$

$$U(x) = V_1(x) + \frac{C}{2} \int V_2(x,y)(E_F - U(y))^{3/2} \quad (14.72)$$

$$\rho(x) \equiv C(E_F - U(x))^{3/2}.$$

The Thomas-Fermi method is very convenient at high densities, where it gives the asymptotically correct result for $N \to \infty$ at fixed volume.

A functional integral representation for the Green's function of N fermions can be easily constructed at imaginary times,

$$G(x_1, \ldots, x_N; y_1, \ldots, y_N | t) = \sum_P (-1)^P \int d[\omega_1]_{x_1, y_{i_1(P)}}^t \cdots d[\omega_N]_{x_N, y_{i_N(P)}}^t$$

$$\cdot \exp\left\{ -\frac{1}{2} \int_0^t dt' \left[\frac{1}{2} \sum_{i}^{N} \dot\omega_i^2 + \sum_{i}^{N} V_1(\omega_i) + \frac{1}{2} \sum_{i,j} V_2(\omega_i, \omega_j) \right] \right\}, \quad (14.73)$$

where P denotes a permutation of the numbers from 1 to N ($i_k^{(p)}$, $k = 1, \ldots, N$), $(-1)^P$ is the parity of the permutation, and the sum over P runs over the $N!$ possible permutations and finally the path integral depend on P via the boundary conditions

$$\omega_k(0) = x_k \qquad \omega_k(t) = y_{i_k^{(p)}} . \qquad (14.74)$$

Indeed, Green's function of the Hamiltonian (14.61) for fermions is just the Green's function for distinguishable particles, projected onto the space of antisymmetric functions.

We observe that Eq. (14.73) is automatically antisymmetric with respect to the x's; indeed, if at a given moment the wave function is antisymmetric, it will remain antisymmetric forever.

If the two-body potential V_2 is absent, integration over the different ω's can be performed separately. One finds exactly

$$G(x_1 \cdots x_N; y_1 \cdots y_N|t) = \sum_P (-1)^P \prod_k^N G(x_k, y_{i_k}|t)$$

$$= \det(\mathcal{G}) \qquad (14.75)$$

$$G(x, y|t) = \int d[\omega]_{x,y}^t \exp\left\{-\int_0^t dt' \left[\frac{1}{2} \dot{\omega}^2(t') + V_1(\omega(t'))\right]\right\},$$

where $\mathcal{G}$ is an $N \times N$ matrix whose elements are given by

$$\mathcal{G}_{ik} = G(x_i, y_k|t) . \qquad (14.76)$$

If we recall that

$$G(x_1, \ldots, x_N; y_1, \ldots, y_N|t) = \sum_n \psi_n(x_1, \ldots, x_N) \psi_n(y_1, \ldots, y_N)$$

$$\times \exp(-E_n t) \qquad (14.77)$$

$$G(x, y|t) = \sum_n \psi_n(x) \psi_n(y) \exp[-\varepsilon_n(V_1)t] ,$$

we recover immediately Eqs. (14.62)–(14.64).

As an exercise let us rederive the Hartree approximation from functional integration. Let us suppose that the two-body potential is attractive. If we define by $V^{-1}(x, y)$ the inverse of the potential in an operator

14.7. Fermions

sense,

$$\int dz\, V(x, z)V^{-1}(z, y)\, dz = \delta^D(x - y), \tag{14.78}$$

Eq. (14.73) can be written as

$$G(x_1, \ldots, x_N; y_1, \ldots, y_N | t) = \det(V_2)^{1/2} \sum_P (-1)^P \int d[\omega]^t_{x_1 y_{i_1^{(p)}}}$$

$$\cdots d[\omega]^t_{x_N y_{i_N^{(p)}}} \int d[A] \exp\left\{ \frac{1}{2} \int\int d^D x\, d^D y\, dt\, A(x, t) A(y, t) \right.$$

$$\left. \times V_2^{-1}(x - y) - \int_0^t dt' \left[\sum_1^N \left(\frac{1}{2} \dot\omega_i^2 + V_1(\omega_i(t)) + A(\omega_i(t'), t') \right) \right] \right\}, \tag{14.79}$$

where we have used the usual Gaussian integral over the $A(x, t)$ field, which depends on space and time:

$$\det[V_2]^{1/2} \int [dA] \exp\left\{ \frac{1}{2} \int dt \int d^D x\, d^D y\, V_2^{-1}(x - y) A(x, t) A(y, t) \right.$$

$$\left. + \int dx\, dt\, J(x, t) A(x, t) \right\}$$

$$= \exp\left\{ -\frac{1}{2} \int d^D x\, d^D y\, dt\, J(x, t) J(y, t) V_2(x - y) \right\}$$

$$J(x, t) = \sum_1^N \delta(\omega_i(t) - x). \tag{14.80}$$

In this way we have reduced the two-body interaction to a one-body interaction in a random potential. The ground-state energy is given by

$$F_0 = -\lim_{t \to \infty} \frac{1}{t} \ln G(x, \ldots, x_N, y_1, \ldots, y_N | t). \tag{14.81}$$

We now try to evaluate Eqs. (14.80) and (14.81) using the method of steepest descent for the integration over A. Since we are interested in a large-t problem, we can try to maximize the integrand with a function $A(t, x)$ that does not depend on t:

$$A(x, t) = B(x). \tag{14.82}$$

If we approximate the functional integral with its value at B, we obtain, neglecting the prefactors,

$$G(x_1, \ldots, x_N, y_1, \ldots, y_N|t) \sim \exp\{-tE_0[B]\}$$

(14.83)

$$E_0[B] = -\frac{1}{2} \int d^D x \, d^D y \, V_2^{-1}(x-y) B(x) B(y) + \sum_n^N \varepsilon_n[B+V_1],$$

where the ε_n are levels in the presence of the potential $B + V_1$. G takes its minimum value when $E_0[B]$ is minimum; this happens if

$$0 = \delta E_0[B]/\delta B(x) = -\int dy \, V_2^{-1}(x-y) B(y) + \rho_{[B+V_1]}(x),$$

(14.84)

which can be rewritten as

$$B(y) = \int dx \, V_2(x-y) \rho_{[B+V_1]}(x).$$

(14.85)

If we recall Eq. (14.69), we see that Eq. (14.84) implies that $B(x) = U(x) - V_1(x)$, $U(x)$ being the solution of Eq. (14.69). The advantage of this approach is that Eq. (14.79) may be useful for evaluating numerically the corrections to the Hartree approximation.[24]

From the various examples we have seen in this chapter we should realize that the path integral formulation of quantum mechanics is a new language into which most of the old results may be translated. Whether it is more convenient to use this new language or the old traditional operator language, however, depends on the problem.

Notes for Chapter 14

1. See, for example, P. M. Morse and H. Feshbach, *Methods of Theoretical Physics*, McGraw-Hill, New York (1953), Sec. 4.6.
2. For a review see S. Albeverio and R. Høegh-Krohn, Mathematical Theory of Feynman Path Integrals, Springer, Berlin (1976) and S. Albeverio, M. Fukushima, W. Karwowski, and L. Streit, Commun. Math. Phys. *81*, 513 (1981).
3. See, for example, E. Brézin, J. C. Le Guillou, and J. Zinn-Justin, Phys. Rev. D *15*, 1558 (1977).

4. If a function $r(E)$ is analytic in the complex E plane with a cut or poles on the positive E axis and it goes like $|E|^{-\alpha}$ when $|E|$ goes to infinity (with $\alpha < 0$), we have

$$r(E) = \lim_{\varepsilon \to 0^+} \int_0^\infty \frac{1}{\pi} \frac{dE'}{E - E'} \operatorname{Im}[r(E + i\varepsilon)].$$

This relation is often called a dispersion relation. Similar relations hold if the function $r(E)$ increases not faster than a fixed power when $|E|$ goes to infinity.

5. For the application of the Baker-Hausdorf formula in this context see D. A. Kirzhnits, *The Methods of the Field Theory in the Many Body Theory*, Goratomizdat, Moscow (1963).
6. See, for example, R. Balian and C. Bloch, *Ann. Phys. (N.Y.)* 63, 592 (1970) and references therein; see also A. Voros, *Ann. Inst. H. Poincaré*, 26A, 343 (1977); B. Gramaticos and A. Voros, *Ann. Phys. (N.Y.)* 123, 359 (1979).
7. It is crucial that in Eq. (14.20) d/dE acts both on the integrand and on the integration limits, i.e., on the θ function.
8. A. Martin, *Helv. Phys. Acta* 45, 140 (1972).
9. The function $N(g)$ has been used to compute the determinant of the small fluctuation around the classical solution relevant for the evaluation of the large-order behavior of the perturbative expansion of the φ^4 interaction in three dimensions (see p. 91).
10. The interested reader can consult a vast literature on the subject; see, for example, R. Balian and C. Bloch, *Ann. Phys. (N.Y.)* 85, 514 (1974) or M. Gutzwiller, *J. Math. Phys.* 12, 343 (1971).
11. We follow closely M. C. Gutzwiller, *J. Math. Phys.* 12, 343 (1971), and R. Dashen, R. Hasslacher and A. Neveau, *Phys. Rev. D* 10, 4114 (1974), where it is possible to find the details of the calculations we sketch here. We warn the reader that we have assumed that there is only one trajectory for each energy. That would not be true in the double-well potential of the next section at low energies.
12. $\oint dq$ is the integral along a closed trajectory, and it is equal to $2 \int_{q_1}^{q_2} dq$, where q_1 and q_2 are the two turning points of the classical trajectory: In one dimension a particle traveling along a closed trajectory goes through the same point twice.
13. The traditional WKB approach is well described in N. Froman and P. O. Froman, *JWKB-Approximation, Contribution to the Theory*, North-Holland, Amsterdam (1965); see also any book on quantum mechanics. A different point of view is presented by V. P. Maslov, *Théorie des Perturbations et Méthodes Asymptotiques*, Dunod, Paris (1972); see also J. P. Eckmann and R. Sénéor, *Arch. Rational Mechanics* 61, 153 (1976).

14. R. Balian, G. Parisi and A. Voros, *Phys. Rev. Lett. 41*, 1141 (1978); *Feynman Path Integrals*, Springer Lecture Notes in Physics Vol. 106 (1979).
15. Complex trajectories are widely used in the study of scattering problems in atomic physics.
16. L. D. Faddeev, *Dokl. Akad. Nauk. 115*, 878 (1957).
17. I. M. Gelfand, *Usp. Math. Nauk. 11*, 11 (1956); *Izv. Akad. Sci., Ser. Math. 19*, 187 (1955).
18. See, for example, K. Huang, *Statistical Mechanics*, Wiley, New York (1963).
19. G. Parisi in *The Riemann Problem: Complete Integrability and Arithmetic Applications*, Lecture Notes in Mathematics No. 925, ed. by D. Chudnovsky and G. Chudnovsky, Springer-Verlag, Berlin (1982); see also A. Voros in the same volume.
20. The appearance of a Poisson distribution can be naturally understood in the framework of Nelson's formulation of quantum mechanics (see Chapter 19) as has been recently stressed by G. Jona-Lasinio, F. Martinelli, and E. Scoppola, *Phys. Rep. 77C*, 313 (1981).
21. For a review of the applications of this approach see J. Zinn-Justin in *Proceedings of the 1982 Les Houches Summer School*, ed. J. B. Zuber and R. Stora, North-Holland, Amsterdam (1984) and references therein.
22. For example, see A. L. Fetter and J. D. Walecka, *Quantum Theory of Many Particle Systems*, McGraw-Hill, New York (1971) or J. W. Negele, *Rev. Mod. Phys. 54*, 913 (1982).
23. We recall that for a single-particle Hamiltonian $\mathcal{H} = \frac{1}{2}p^2 + V(q)$ when we add a small perturbing potential δV the levels change according to $\delta E_k = \langle \psi_k | \delta V | \psi_k \rangle$; in other words, $\delta E_k / \delta V(x) = |\psi_k(x)|^2$.
24. R. Alzetta, G. Parisi, and T. Semeraro, *Nucl. Phys. B 235* [FS11], 576 (1986).

CHAPTER 15

Relativistic Quantum Field Theory

15.1. General properties

In this chapter we present some elements of relativistic quantum field theory. They should be the link between this book and the many books on quantum field theory.[1]

In the physical world, space-time is four dimensional: the zeroth direction is conventionally taken as time. It is normally assumed that physical theory is invariant under the transformation of the Lorentz group, i.e., those transformations which leave invariant the distance

$$(x-y)^2 \equiv (x_\mu - y_\mu)(x_\nu - y_\nu) g_{\mu\nu} \qquad \mu = 0, \ldots, 3, \qquad \nu = 0, \ldots, 3, \tag{15.1}$$

where the metric $g_{\mu\nu}$ is given by

$$\begin{vmatrix} -1 & 0 & 0 & 0 \\ 0 & 1 & 0 & 0 \\ 0 & 0 & 1 & 0 \\ 0 & 0 & 0 & 1 \end{vmatrix} \tag{15.2}$$

and the speed of light has been set to one by convention. In this metric x^2 is positive if x is spacelike, x^2 is negative if x is timelike. The sign convention for the metric is not the most common nowadays, but is the best for going from real time to imaginary time. Indeed, if we set

$$x_4 = ix_0 \equiv it \tag{15.3}$$

the Minkowski space here described becomes the usual Euclidean space. Therefore the imaginary-time version of a relativistic quantum theory must be invariant under the usual Euclidean group of the four-dimensional Euclidean space ($O(4)$).

283

Let us consider a concrete example. At the classical level we have a field $\varphi(x)$ (x being the set of four coordinates) defined in each point of space-time. The Lagrangian density and the total action are, respectively,

$$\mathcal{L}(x) = -\frac{1}{2}(\partial_\mu \varphi)(\partial^\mu \varphi) - \frac{m^2}{2}\varphi^2 - \frac{g}{4!}\varphi^4$$

$$S = \int d^D x\, \mathcal{L}(x).$$
(15.4)

The corresponding equations of motion are

$$\Box \varphi + m^2 \varphi + g\varphi^3 = 0$$
(15.5)

$$\Box = -\partial_\mu \partial^\mu = +\left(\frac{\partial}{\partial t}\right)^2 - \Delta = \frac{\partial^2}{\partial t^2} - \sum_1^3{}_a \frac{\partial^2}{\partial x_a^2}.$$

The theory is obviously invariant under the Lorentz group, φ being a scalar. If we want, we can go to the Hamiltonian formalism by introducing the canonical momentum of φ,

$$\Pi(x) = \dot\varphi(x).$$
(15.6)

The equal-time Poisson brackets are given by

$$\{\varphi(x), \Pi(y)\}\delta(x_0 - y_0) = \delta^4(x - y),$$
(15.7)

the total Hamiltonian and action are, respectively,

$$\mathcal{H} = \int d^3x \left[\frac{1}{2}\Pi^2(x) + \sum_1^3{}_a \frac{1}{2}(\partial_a \varphi(x))^2 + \frac{1}{2}m^2\varphi(x)^2 + \frac{1}{4!}g\varphi^4(x)\right].$$
(15.8)

$$S = \int d^4x \left[\frac{1}{2}\dot\varphi^2(x) - \sum_1^3{}_a \frac{1}{2}(\partial_a \varphi(x))^2 - \frac{1}{2}m^2\varphi^2(x) - \frac{1}{4!}g\varphi^4(x)\right].$$

In order to construct the corresponding quantum theory we can start from Eqs. (15.7) and (15.8): we declare that φ and Π are operators in a Hilbert space, whose equal-time commutation relations are given by

$$[\hat\varphi(x), \hat\Pi(y)]\delta(x_0 - y_0) = i\hbar \delta^4(x - y).$$
(15.9)

The equal-time commutation relations play the role of boundary conditions, and the time evolution is determined by the usual Heisenberg

15.1. General Properties

equation

$$\dot{\hat{A}} = -\frac{i}{\hbar}[\hat{A}, \hat{\mathcal{H}}], \qquad (15.10)$$

where $\hat{\mathcal{H}}$ is the Hamiltonian (15.8) and $\hat{\Pi}$ and $\hat{\varphi}$ are now operators in Hilbert space.

The Hamiltonian approach is not Lorentz invariant by inspection: The equal-time commutation relations have been imposed in a particular frame. However, with some additional work it is possible to prove that Eqs. (15.8) and (15.9) give rise to a Lorentz-invariant theory. In contrast, the Lorentz invariance of the theory is manifest in the Lagrangian approach: If we apply to this case the path integral formulation of quantum mechanics (see p. 237), we find that

$$\langle 0|\tau[\varphi(x_1)\cdots\varphi(x_N)]|0\rangle = \int d\mu[\varphi]\varphi(x_1)\cdots\varphi(x_N),$$

$$\int d\mu[\varphi] = 1, \qquad d\mu[\varphi] \propto d[\varphi]\exp\left\{\frac{iS[\varphi]}{\hbar}\right\}. \qquad (15.11)$$

The usual precautions must be taken when the infinite time and volume limits are taken; $|0\rangle$ is the ground state of $\mathcal{H}$ (the vacuum) and τ denotes the time-ordered product (see pp. 238–239).

In this case, as before, we can analytically continue the time-ordered products from real to imaginary times. For simplicity let us consider in detail the case of the time-ordered product of two fields. Higher-order products work in the same way. Using a complete set of intermediate states, we have for $\hbar = 1$

$$\langle 0|\varphi(t,\vec{0})\varphi(0,\vec{0})|0\rangle = \int d\rho(E)\exp(iEt),$$

$$\langle 0|\tau[\varphi(t,\vec{0})\varphi(0,\vec{0})]|0\rangle = \int d\rho(E)\exp(iE|t|), \qquad (15.12)$$

$$d\rho(E) = \langle 0|\varphi(0,\vec{0})\,d\hat{P}_E\varphi(0,\vec{0})|0\rangle,$$

where $\hat{P}_E$ is the projector on all the states of energy less than E [we recall that $\hat{\mathcal{H}} = \int d\hat{P}_E\,E$, $\exp(i\hat{\mathcal{H}}) = \int d\hat{P}_E\exp(iE)$; more generally, $f(\hat{\mathcal{H}}) = \int d\hat{P}_E\,f(E)$] and $\rho(E)$ is obviously a positive measure. [We have used the notation $(t,\vec{x})$ to indicate a point of space-time.] If we analytically continue Eq. (15.12) to imaginary times we get

$$\langle 0|\tau[\varphi(ix_4,\vec{0})\varphi(0,\vec{0})]|0\rangle = \int d\rho(E)\exp(-E|x_4|). \qquad (15.13)$$

Repeating the arguments of the previous chapters we obtain

$$\langle 0|\tau[\varphi(ix_4, \vec{0}), \varphi(0, \vec{0})]|0\rangle = \langle \varphi(x)\varphi(0)\rangle,$$

$$x \equiv (\vec{x}, x_4),$$

$$\langle g[\varphi]\rangle = \int d\mu[\varphi] g[\varphi], \tag{15.14}$$

$$\int d\mu[\varphi] = 1, \quad d\mu[\varphi] \propto d[\varphi] \exp[-A[\varphi]],$$

$$A[\varphi] = \int d^4x \left[\frac{1}{2} (\partial_\mu \varphi)^2 + \frac{m^2}{2} \varphi^2 + \frac{g}{4!} \varphi^4 \right],$$

where $g[\varphi]$ is a generic functional of φ.

We see that the correlation functions of Euclidean invariant statistical mechanics are the analytic continuation of the vacuum expectation values of the time-ordered products. All the properties of relativistic quantum field theory can be translated into corresponding properties of Euclidean statistical mechanics. For example, if ΔE is the difference in energy between the vacuum and the lowest state ($|s\rangle$) such that $\langle 0|\varphi|s\rangle \neq 0$,[2] we have $\rho(E) = 0$ for $E < \Delta E$, and $E = \Delta E$ belongs to the support of $\rho(E)$; neglecting the prefactor, we find

$$\langle \varphi(x)\varphi(y)\rangle \xrightarrow[|x-y|\to\infty]{} \exp[-\Delta E|x-y|]. \tag{15.15}$$

In other words, $(\Delta E)^{-1}$ is the coherence length of statistical mechanics, which has been denoted by ξ.

It is very easy for us to compute the correlation function of the φ fields using the perturbative expansion in powers of g. This problem has already been studied in Chapters 5, 8, and 9. After we have obtained the results in Euclidean space, the analytic continuation to Minkowsky space does not present a problem in principle, at least for scalar theories.

15.2. The Osterwalder-Schrader condition

As we have seen in a special case, we can associate Lorentz-invariant quantum theory in four dimensions, with rotationally invariant statistical mechanics. In the general case, one can prove (at least for scalar theories satisfying Bose statistics) that the time-ordered products can be continued analytically to purely imaginary times [denoted by $\langle \varphi(x_1) \cdots \varphi(x_n)\rangle$], and they can be written as

$$\langle \varphi(x_1) \cdots \varphi(x_N)\rangle = \int d\mu[\varphi] \varphi(x_1) \cdots \varphi(x_N), \tag{15.16}$$

where $d\mu[\varphi]$ is a probability measure on φ.

15.2. The Osterwalder-Schrader Condition

On the other hand, not all possible probability measures $d\mu[\varphi]$ correspond to a relativistically invariant quantum field theory. It is easy to see that we can construct probability measures for which the two-point correlation function cannot be written as

$$\langle \varphi(x)\varphi(0) \rangle = \int d\rho(E) \exp[-E|x|], \qquad (15.17)$$

or for which the spectral function $\rho(E)$ is not positive. The simplest counter example is a Gaussian distribution of φ with a covariance given by

$$\langle \varphi(x)\varphi(y) \rangle = \exp[-(x-y)^2]. \qquad (15.18)$$

The positivity of $\rho(E)$ is a direct consequence of the positivity of the scalar products $\langle n|n \rangle \geq 0$ for the states of the would-be corresponding quantum field theory.

In this situation we would like to have a general characterization of the probability measures associated with a relativistic quantum field theory. This characterization theorem exists:[3] it states that a necessary and sufficient condition (we neglect some technical conditions on the regularity of the correlation functions) for a probability measure $d\mu[\varphi]$ (invariant under rotations and translations) to be associated with a relativistic quantum field theory, is that

$$\langle g_+[\varphi]g_-[\varphi] \rangle \equiv \int d\mu[\varphi] g_+[\varphi] g_-[\varphi] \geq 0, \qquad (15.19)$$

where $g_+[\varphi]$ is a bounded smooth functional that depends only on the field φ in the half space $x_4 > 0$, and $g_-[\varphi]$ is the mirror image of $g_+[\varphi]$ defined as follows:

$$g_-[\varphi] = g_+[\varphi^M]$$
$$\varphi^M(\vec{x}, x_4) = \varphi(\vec{x}, -x_4). \qquad (15.20)$$

Obviously g_- depends only on the values of φ in the region $x_4 < 0$. It is evident that the positivity condition is very powerful. As an example we consider the case in which $g_+[\varphi] = \int_0^\infty g(x_4)\varphi(\vec{0}, x_4)$, $g(0) = 0$. The positivity condition Eq. (15.19) implies that

$$\langle g_+[\varphi]g_-[\varphi] \rangle = \int_0^\infty dx_4 \int_0^\infty dx_4' \, g(x_4)g(x_4')C(x_4 + x_4') \geq 0$$
$$C(x_4 + x_4') = \langle \varphi(\vec{0}, x_4)\varphi(\vec{0}, -x_4') \rangle \qquad (15.21)$$

for any bounded smooth function g.

The theorem on positive functions referred to above tells us that the condition (15.21) is equivalent to

$$C(x_4) = \int d\rho(E) \exp[-Ex_4], \qquad x_4 \geq 0, \qquad (15.22)$$

where $d\rho(E)$ is a positive measure.[4] The proof of this theorem [Eq. (15.19)] is rather complicated and will not be reported here. It is normally referred to as the Osterwalder-Schrader (OS) theorem. The OS theorem is rather useful because it is in the form of inequalities. Indeed, if we consider the regularized Euclidean theory on a lattice (which is not rotationally invariant) we can sometimes easily prove the OS inequalities [Eq. (15.19)] using the transfer metric technique, as we shall see. In the limit in which the lattice spacing a goes to zero, the inequalities must remain valid in the continuum limit if they were satisfied on the lattice. This means that if we can prove that the lattice regularized theory has a limit when $a \to 0$ and that the limit is rotationally invariant, there is a corresponding relativistic quantum field theory, provided that the OS inequalities are satisfied on the lattice.

How to prove the OS inequalities for a lattice theory? We consider the case in which the interaction is of nearest-neighbor form. In this case we can introduce a transfer matrix T such that

$$\langle \varphi(\vec{0}, k_4) \varphi(\vec{0}, 0) \rangle = \langle \psi_0 | \hat{\varphi}(\vec{0}) \hat{T}^{k_4} \hat{\varphi}(\vec{0}) | \psi_0 \rangle$$

$$= \sum_{n}^{\infty} |\langle \psi_0 | \hat{\varphi}(\vec{0}) | \psi_n \rangle|^2 (t_n)^{k_4}$$

$$= \sum_n c_n^2 \exp[k_4 \ln(t_n)], \qquad k_4 \geq 0 \quad (15.23)$$

where we indicate a point on the four-dimensional lattice by $(\vec{k}, k_4)$. $\vec{k}$ denotes the first three coordinates, and $|\psi_0\rangle$ is the eigenstate with largest eigenvalue of $\hat{T}$. If the transfer matrix $\hat{T}$ has only positive eigenvalues ($\ln t_n$ are real) the two-point correlation function has the needed form [see Eq. (15.17)].

The lattice equivalent of the Osterwalder-Schrader condition may be obtained by defining the reflection with respect to a hyperplane which does not intersect the lattice (to avoid problems at $k_4 = 0$):

$$\varphi^M(\vec{k}, k_4) = \varphi^M(\vec{k}, -1 - k_4)$$

$$\langle g_+[\varphi] g_+[\varphi^M] \rangle \geq 0 \qquad (15.24)$$

$$\frac{\partial g_+[\varphi]}{\partial \varphi(\vec{k}, k_4)} = 0, \qquad \text{if } k_4 < 0.$$

15.2. The Osterwalder-Schrader Condition

It is instructive to compute the equivalent of Eq. (15.21) on the lattice. We obtain

$$\sum_{0}^{\infty}{}_{k_4} \sum_{0}^{\infty}{}_{k_4'} g(k_4)g(k_4')\langle \varphi(\vec{0}, k_4)\varphi(\vec{0}, -k_4 - 1)\rangle$$

$$= \sum_{0}^{\infty}{}_{k_4} \sum_{0}^{\infty}{}_{k_4'} g(k_4)g(k_4')\langle \psi_0|\hat{\varphi}(\vec{0})\hat{T}^{k_4}\hat{T}\hat{T}^{k_4'}\hat{\varphi}(\vec{0})|\psi_0\rangle$$

$$= \langle \psi_0|\hat{\varphi}(\vec{0})\hat{T}_g\hat{T}\hat{T}_g\hat{\varphi}(\vec{0})|\psi_0\rangle = \langle \psi_g|\hat{T}|\psi_g\rangle \geq 0$$

(15.25)

$$\hat{T}_g = \sum_{0}^{\infty}{}_{k_4} g(k_4)\hat{T}^{k_4}, \quad |\psi\rangle = \hat{T}_g\hat{\varphi}(\vec{0})|\psi_0\rangle,$$

where the last inequality holds if $\hat{T}$ is a positive operator.

The argument may be extended to higher-order correlations: the OS conditions are valid if $\hat{T}$ is positive (the positivity of $\hat{T}$ can be easily checked in many cases). It is easy to understand why $\hat{T}$ must be positive. If that happens we can write

$$\hat{T}^n = \exp[-n\hat{\mathcal{H}}_L], \qquad (15.26)$$

where $\hat{\mathcal{H}}_L$ is a Hermitian operator that will become the Hamiltonian after the rotation from real to imaginary time. It is interesting to note that we can associate a theory defined on a $d + 1$ dimensional lattice with a theory defined on a d-dimensional lattice and on continuous time (imaginary or real), if $\hat{T}$ is positive; for example, the two-field correlation function would be defined in Euclidean space as

$$\langle \varphi(\vec{k}, t)\varphi(\vec{0}, 0)\rangle = \langle \psi_0|\hat{\varphi}(\vec{k}) \exp(-t\hat{\mathcal{H}}_L)\hat{\varphi}(\vec{0})|\psi_0\rangle$$

$$= \langle \psi_0|\hat{\varphi}(\vec{k})\hat{T}^t\hat{\varphi}(\vec{0})|\psi_0\rangle. \qquad (15.27)$$

For integer t Eq. (15.27) reproduces the usual result [cf. Eq. (15.23)]. It naturally gives a correlation function that interpolates between the known results at integer t.[5]

It is important to understand that the crucial condition is the existence and the self-adjointness of the transfer matrix $\hat{T}$, while the positivity condition is not crucial; indeed $\hat{T}^2$ is always positive, independently of the sign of $\hat{T}$. If $\hat{T}$ is a symmetric matrix, the correlation functions at even distance satisfy all the needed positivity requirements.

15.3. The particle interpretation

We can now ask the question, which physical system does the quantum Hamiltonian $\mathcal{H}$ [Eq. (15.2)] describe? Now $\mathcal{H}$ is the Hamiltonian of a quantized field theory, and we know from the early days of quantum mechanics that the eigenstates of $\mathcal{H}$ are single-particle or many-particle states, the particles being the quanta of the field.

Let us see what happens in the Gaussian model for $g = 0$. The two-point function in momentum space is given by

$$G(p_0, \vec{p}) \equiv \int d^3t\, dx_0 \exp[i\vec{p} \cdot \vec{x} - ip_0 x_0] \langle 0|\tau[\varphi(x_0, \vec{x}_0)\varphi(0, \vec{0})]|0\rangle$$

$$= \frac{1}{-p_0^2 + |\vec{p}|^2 + m^2}, \qquad (15.28)$$

which implies the existence of a state whose energy depends on the momentum $\vec{p}$ as

$$E(\vec{p}) = \sqrt{m^2 + \vec{p}^2} \Rightarrow E^2(\vec{p}) = |\vec{p}|^2 + m^2, \qquad (15.29)$$

as can be seen from the position of the pole in p_0 of $G(p_0, \vec{p})$. Equation (15.29) is the usual result for a single-particle state of mass m. The Gaussian model is a generalized harmonic oscillator. If there are two states of energy E_1 and E_2, respectively (the ground-state energy being zero by convention), there is also another state of energy $E_1 + E_2$ (this statement may be checked by computing the many-field correlation functions (see Sec. 13.5). In this way the n-particle states have a total energy exactly equal to the sum of the energies of the single particles; no interaction is present and the theory is said to be free.

The particle interpretation of relativistic quantum field theory finds its natural setting in the formalism of second quantization, where the field $\hat{\varphi}$ is decomposed into the sum of creations and annihilations of single-particle states. If the model is not Gaussian, interactions between particles are present. In such a case we have the possible formation of bound states (if the interaction is attractive), scattering between the particles, and particle creation (two incoming and four outgoing particles). All the information needed to compare the theory with real experiments can be extracted from the knowledge of the appropriate time-ordered products; for example, the scattering amplitude for two-particle elastic collisions is given by the four-point amputated time-ordered product evaluated when the square of the external momenta $(-p_{0,i}^2 + |\vec{p}_i|^2)$ is equal to $-m^2$, m being the physical mass of the particles.

This subject is quite broad; the interested reader can find a complete treatment in any of the many books dedicated to relativistic quantum field theory.

Notes for Chapter 15

1. Many books could be added to those already cited; let me mention only two among the most recent: P. Ramond, *Field Theory: a modern primer*, Benjamin, and T. D. Lee, *Particle Physics and Introduction to Field Theory*, Harwood Academic Publisher, New York (1981).
2. In the terminology of relativistic quantum field theory, $(\Delta E)^{1/2}$ is normally called the mass gap. Indeed, if (as we shall later see) the state is a single-particle state, its mass is equal to $(\Delta E)^{1/2}$.
3. The original results are contained in K. Osterwalder and R. Schrader, *Commun. Math. Phys.* **31**, 83 (1973) and **42**, 281 (1976); for a recent review see, for example, K. Osterwalder in *Gauge theories: fundamental interactions and rigorous results*, P. Dita, V. Georgescu, and R. Purice, eds., Birkhäuser, Boston (1982).
4. It is evident that Eq. (15.22) implies Eq. (15.21):

$$\int_0^\infty dx_4\, g(x_4) \int_0^\infty dx_4'\, g(x_4') \int dE\, \rho(E) \exp[-E(x_4 + x_4')]$$

$$= \int dE\, \rho(E)(g_E)^2$$

where $g_E = \int_0^\infty dx_4 \exp(-Ex_4) g(x_4)$. The inverse statement has a more complex proof which can be found in M. Reed and B. Simon, *Methods of modern mathematical physics IV, analysis of operators*, Academic Press, New York (1978), p. 75. We notice here that if Eq. (15.22) holds with a function $\rho(E)$ that is negative for some E, it is reasonable to suppose that we can find a function $g(E)$ such that $\int dE\, \rho(E)(g_E)^2$ is negative.
5. The positivity of the transfer matrix plays a crucial role in establishing the connection between Euclidean statistical theory and relativistic quantum mechanics; it is very important to note that the argument on p. 233, note 6, can be used to prove that the more complex transfer matrix (12.62) is also a positive operator.

CHAPTER 16

Particle-Field Duality

16.1. The free case

From the early days of quantum mechanics it has been known that the same quantum system could manifest itself under different conditions as particles or as waves. This fact is not a paradox: the system behaves in a unique way (particles or waves) only in the two extreme classical limits. The real description of the system is a quantum one and not a classical one. We have just seen that the same quantum theory (i.e., quantum relativistic field theory) can be considered either as a theory of interacting particles or as a field theory. We have also seen that in a quantum system the probability amplitude can be written as the sum of all possible trajectories of the corresponding classical system. In order to formulate the particle-wave duality in the most complete form we should be able to write the same quantum amplitude both as an integral over field configurations (as we have already done) and as an integral over trajectories of particles in space-time (as we shall do in this chapter). In this way neither of the two classical limits is treated on a privileged footing.

Let us begin by analyzing the free case. We have already seen [Eqs. (13.17) and (5.7)] that

$$G(x) = \langle \varphi(x)\varphi(0) \rangle = \langle x| \frac{1}{-\Delta + m^2} |0\rangle$$

$$= \int_0^\infty dt \langle x| \exp[-t(-\Delta + m^2)]|0\rangle$$

$$= \int_0^\infty dt \, d[\omega]_{0,x}^t \exp\left[-m^2 t - \int_0^t \frac{\dot{\omega}(t')^2}{2} dt'\right]$$

$$\equiv \int_0^\infty dt \, dP_{0,x}^t[\omega]. \qquad (16.1)$$

16.1. The Free Case

In other words, the free correlation function can be written as an integral over all the trajectories going from 0 to x with the appropriate measure $dP^t_{0,x}[\omega]$. These trajectories can be interpreted as the trajectories of a real particle going from 0 to x. Similar formulae hold for higher-order Green's functions, e.g.,

$$\langle \varphi(x_1)\varphi(x_2)\varphi(x_3)\varphi(x_4)\rangle = \int_0^\infty dt_1\, dt_2 \{dP^{t_1}_{x_1,x_2}[\omega_1] dP^{t_2}_{x_3,x_4}[\omega_2]$$

$$+ dP^{t_1}_{x_1,x_3}[\omega_1]\, dP^{t_2}_{x_2,x_4}[\omega_2]$$

$$+ dP^{t_1}_{x_1,x_4}[\omega_1]\, dP^{t_2}_{x_2,x_3}[\omega_2]\}. \quad (16.2)$$

In this way we have succeeded in writing the quantum amplitudes as an integral over trajectories of particles; unfortunately, Eq. (16.1) is not illuminating for the classical limit because the parameter t which parameterizes the trajectory does not have a clear classical meaning.[1] On the other hand, the final measure $dP[\omega] = \int_0^\infty dt\, P^t_{0,x}[\omega]$ depends only on the trajectory ω; in order to obtain the explicit expression for such a measure we should integrate over all possible parametrizations of the trajectory which lead to the same trajectory as a geometrical object in space. A classical interpretation of the measure $dP[\omega]$ can be obtained (without performing this painful job) by studying the same problem on a lattice of spacing (a). In this case we know that G has a representation in terms of random walks,

$$G(k) \equiv \langle \varphi(k)\varphi(0)\rangle = C \sum_\omega \exp[-Al[\omega]]$$

$$C = (2D)^{-1} a^{2-D} \quad (16.3)$$

$$A = -\frac{1}{a}\ln(q), \quad q = \frac{1}{2D}(1 - a^2 m^2/2D),$$

where the sum is done over all paths (walks) going from 0 to k [cf. Eqs. (4.25)–(4.28)] and $l[\omega]$ is the length of the path ω in lattice units (a). Here the trajectories receive a weight proportional to the exponential of their length. This result is highly interesting because we know that the classical action for a free particle is

$$S = m \int dl = ml, \quad (16.4)$$

where l is the length of the classical trajectory. Unfortunately A [defined in Eq. (16.3)] is divergent when the lattice spacing (a) goes to zero, so that Eq. (16.3) cannot be considered literally the quantum version of

(16.4). Independently of the precise identification of the classical theory for free particles corresponding to Eq. (16.1),[2] we have succeeded in our aim in the free case. We shall see now what happens in the interacting case.

16.2. The φ^4 interaction

We shall now consider only the simplest case, the φ^4 interaction. Although it is not strictly necessary, it is convenient to introduce an auxiliary $\sigma(x)$ field in order to write

$$\exp\left[-\frac{g}{24}\int d^D x\, \varphi^4(x)\right] \propto \int d[\sigma]\exp\left\{-\int d^D x\left[\frac{6\sigma^2(x)}{g} - i\varphi^2(x)\sigma(x)\right]\right\},$$
(16.5)

where the proportionality constant is φ independent. We thus find that the partition function Z and the correlation function $\langle \varphi(x)\varphi(0)\rangle$ can be written, respectively, as

$$Z = \int d[\sigma][\det(-\Delta + m^2 + i\sigma)]^{-z}\exp\left[-\int d^D x\, \frac{6\sigma^2(x)}{g}\right]$$

$$\langle \varphi(y)\varphi(0)\rangle = \frac{1}{Z}\int d[\sigma][\det(-\Delta + m^2 + i\sigma)]^{-z}$$
(16.6)
$$\times \exp\left[-\int d^D x\, \frac{6\sigma^2(x)}{g}\right] G(y, 0|\sigma)$$

$$(-\Delta_y + m^2 + i\sigma)G(y, 0|\sigma) = \delta^D(y) \qquad z = \frac{1}{2},$$

where we have used the integral representation (16.5) and the resulting Gaussian integral over the φ field. A similar formula holds in a theory with N component fields [cf. Eq. (10.37)] with the main difference that $z = N/2$. Let us see the consequences of Eq. (16.6) first for $z = 0$. In this case the partition function Z is one. Let us study the correlation function $G(x)$. Equation (16.1) can be modified as follows:

$$G(x, 0|\sigma) = \langle x|(-\Delta + m^2 + i\sigma)^{-1}|0\rangle$$

$$= \int_0^\infty dt\, \langle x|\exp[-t(-\Delta + m^2 + i\sigma)]|0\rangle$$

$$= \int_0^\infty dt\, d[\omega]_{0,x}^t \exp\left\{-m^2 t - \int_0^t dt'\left[\frac{\dot{\omega}^2}{2} + i\sigma(\omega(t'))\right]\right\}, \quad (16.7)$$

where we have used the standard Feynman path representation for the

16.2. The φ^4 Interaction

kernel of the Schrödinger equation at imaginary time. We can now integrate again over the σ variables. We finally obtain

$$\langle \varphi(x)\varphi(0)\rangle = \int_0^\infty dt\, d[\omega]_{0,x}^t \exp\left\{-\int_0^t dt'\, \frac{\dot\omega^2}{2} - tm^2\right.$$
$$\left. - \frac{g}{24} \int_0^t d\tau_1 \int_0^t d\tau_2\, \delta^D[\omega(\tau_1) - \omega(\tau_2)]\right\}. \qquad (16.8)$$

In the same approximation ($z = 0$) we obtain for the four-point function

$$\langle \varphi(x_1)\varphi(x_2)\varphi(x_3)\varphi(x_4)\rangle = \int dt_1\, dt_2\, d[\omega_1]_{x_1 x_2}^{t_1} d[\omega_2]_{x_3 x_4}^{t_2} \exp\left\{-m^2(t_1 + t_2)\right.$$
$$- \int_0^{t_1} dt'_1\, \frac{\dot\omega_1^2}{2} - \int_0^{t_2} dt'_2\, \frac{\dot\omega_2^2}{2}$$
$$- \frac{g}{24}\left[\int_0^{t_1} d\tau_1 \int_0^{t_1} d\tau_2\left\{\delta^D[\omega_1(\tau_1) - \omega_1(\tau_2)]\right.\right.$$
$$+ \int_0^{t_2} d\tau_1 \int_0^{t_2} d\tau_2\, \delta^D[\omega_2(\tau_1) - \omega_2(\tau_2)]$$
$$\left.\left.\left. + 2\int_0^{t_1} d\tau_1 \int_0^{t_2} d\tau_2\, \delta^D[\omega_1(\tau_1) - \omega_2(\tau_2)]\right]\right\}\right.$$
$$+ (x_2 \leftrightarrow x_3) + (x_2 \leftrightarrow x_4). \qquad (16.9)$$

Equations (16.8) and (16.9) are the same as in the free case, with the crucial difference that now we have an extra weight for each time that the trajectories intersect. Equation (16.8) can be classically interpreted as the sum over all the trajectories of a particle with a repulsive self-interaction with zero range (δ function potential).

Let us now see the effect of the determinant. We can use the identities

$$\det^{-z}(-\Delta + m^2 + i\sigma) = \exp[-z\, \text{Tr}\ln(-\Delta + m^2 + i\sigma)]$$
$$= \sum_k \frac{(-z)^k}{k!} [\text{Tr}\ln(-\Delta + m^2 + i\sigma)]^k$$
$$= \sum_k \frac{z^k}{k!} \prod_1^k \left[\int_0^\infty \frac{dt_l}{t_l} \int d^D x_l \int d[\omega_l]_{x_l,x_l}^{t_l}\right]$$
$$\times \exp\left\{-\sum_1^k \int_0^{t_l} d\tau_l \left[\frac{1}{2}\dot\omega_l^2 + m^2 + i\sigma(\omega_l(\tau_l))\right]\right\}$$

$$\ln A = -\int_0^\infty dt\, \exp(-tA)/t \qquad (16.10)$$

$$\text{Tr}\ln(-\Delta + m^2 + i\sigma) = \int_0^\infty dt/t \cdot d^D x\langle x|\exp[-t(-\Delta + m^2 + i\sigma)]|x\rangle,$$

where we have neglected possible divergences at $t = 0$. We finally obtain that the partition function Z is given by

$$Z = \sum_{0\ k}^{\infty} \frac{z^k}{k!} \prod_1^k{}_l \left[\int_0^\infty \frac{dt_l}{t_l} \int d^D x_l \int d[\omega_l]_{x_l,x_l}^{t_l} \right]$$

$$\cdot \exp\left\{ -\sum_1^k{}_l \left[\int_0^{t_l} d\tau_l \left(\frac{\dot\omega_l^2}{2} + m^2 \right) \right] \right.$$

$$\left. - \frac{g}{24} \sum_1^k{}_l \sum_1^k{}_j \left[\int_0^{t_l} d\tau_l \int_0^{t_j} d\tau_j\, \delta[\omega(\tau_l) - \omega(\tau_j)] \right] \right\}. \quad (16.11)$$

Expression (16.11) is formally divergent, but this divergence may be eliminated by using the usual techniques.[3] Here we shall not pay too much attention to the problems connected with the divergences, but rather try to underline the physical picture.

Using the same approach, we find that the correlation function is given by

$$\langle \varphi(x)\varphi(0) \rangle = \frac{1}{Z} \sum_{0\ k}^{\infty} \frac{z^k}{k!} \prod_1^k{}_l \left[\int_0^\infty \frac{dt_l}{t_l} \int d^D x_l \int d[\omega_l]_{x_l,x_l}^{t_l} \right]$$

$$\cdot \int_0^\infty dt_0\, d[\omega_0]_{0,x}^{t_0} \cdot \exp\left\{ -\sum_0^k{}_l \int_0^{t_l} d\tau_l \left(\frac{\dot\omega_l^2}{2} + m^2 \right) \right.$$

$$\left. - \frac{g}{24} \sum_0^k{}_l \sum_0^k{}_j \int_0^{t_l} d\tau_l \int_0^{t_j} d\tau_j\, \delta[\omega(\tau_l) - \omega(\tau_j)] \right\}. \quad (16.12)$$

A similar formula holds for the higher-order correlation functions. In other words, we have a kind of gas of closed trajectories in a grand canonical ensemble, where the total number of closed trajectories is not fixed, z playing the role of the fugacity. Z becomes the partition function of this gas of self-repulsive trajectories, and the correlation function is the expectation value of an open trajectory embedded in this gas.

Qualitatively we could say that, when we construct a relativistic quantum theory of self-repulsive particles, the total number of particles is not fixed, but we have the possibility of particle creation. Moreover, if there is only one particle in the initial or final state, quantum fluctuations may induce the creation and destruction of particles from the vacuum at intermediate times. These vacuum fluctuations may be considered as closed trajectories in space-time, and they correspond to the closed trajectories of Eqs. (16.11) and (16.12).[4] We have thus reached our aim of obtaining a representation in terms of particle trajectories of the correlation functions. The particle-wave duality is implemented in its most complete form.

16.3. Rigorous considerations

The results of the previous subsection have laid the groundwork for some very important rigorous considerations; we shall now describe the main results.[5] We first observe that it would be very easy (but wrong) to conclude that $\langle \varphi(x_1)\varphi(x_2)\varphi(x_3)\varphi(x_4)\rangle_c$ must be zero as soon as $D>2$. Let us consider the case $z=0$, which is not qualitatively different from the general case (as long as $z \geq 0$), but has the advantage of having simpler expressions. Combining Eqs. (16.7) and (16.8), we get[6]

$$\langle \varphi(x_1)\varphi(x_2)\varphi(x_3)\varphi(x_4)\rangle_c = \int_0^\infty dt_1 \int_0^\infty dt_2\, dP_g[\omega_1]_{x_1 x_2}^{t_1}$$

$$\cdot dP_g[\omega_2]_{x_3 x_4}^{t_2} \exp\left[-\frac{g}{12}\int_0^{t_1} d\tau_1\right.$$

$$\left.\times \int_0^{t_2} d\tau_2\, \delta(\omega_1(\tau_1) - \omega_2(\tau_2))\right]$$

$$\leq \int_0^\infty dt_1 \int_0^\infty dt_2\, dP_g[\omega_1]_{x_1 x_2}^{t_1}\, dP_g[\omega_2]_{x_3 x_4}^{t_2}$$

$$= \langle \varphi(x_1)\varphi(x_2)\rangle\langle \varphi(x_3)\varphi(x_4)\rangle$$

$$dP_g[\omega]_{x,y}^t \equiv d[\omega]_{x,y}^t \cdot \exp\left\{-\int_0^t dt'\left[\frac{\dot\omega^2}{2} + m^2\right]\right.$$

$$\left. -\frac{g}{24}\int_0^t d\tau_1 \int_0^t d\tau_2\, \delta[\omega(\tau_1) - \omega(\tau_2)]\right\}.$$

(16.13)

Since the exponential of a negative number is less than 1, the connected four-field correlation function must be negative. The argument of the exponent (or by consequence the integrand) will be different from zero only if the two trajectories ω_1 and ω_2 cross. Now ω_1 and ω_2 are two random trajectories, extracted with the probability distribution (16.13). Let us assume for the moment that the trajectories ω's are smooth (wrong assumption!); in more than two dimensions, two generic smooth trajectories never intersect. The case in which the ω's intersect (and in which the argument of the integrand is different from zero) will have zero measure: the integral will therefore automatically be zero. However, we have seen that the trajectories are not smooth, and the argument presented here does not work. A careful study of the trajectories that dominate the functional integral tells us that two generic trajectories will intersect as long as $D<4$, and no intersection will be present when $D>4$.[7]

The simplest way to obtain this result rigorously consists in starting from the lattice regularized theory with a lattice spacing a and proving that for any value of g the connected correlation function is smaller than a^{D-4}. As a consequence the connected correlation functions (for more than two fields) will be zero in the limit $a \to 0$, if $D > 4$, independently of the value of g. It is possible to arrive at these conclusions not only for $z = 0$ but for any positive value of z. This recently proved theorem states that if $D > 4$, only the free theory may be obtained in the limit $a \to 0$; a similar result (whose proof is still lacking) is supposed to be valid in four dimensions. These conclusions are the same as in Chapter 9.

Notes for Chapter 16

1. Sometimes t is called the proper time: in Minkowski space the saddle-point trajectory for going from 0 to x^2 ($x^2 < 0$, i.e., timelike) is given by $\omega_\mu(t') = x_\mu t'/t$; the corresponding action (as function of t) is stationary for $t = \sqrt{-x^2/m^2}$. In the classical limit, at the saddle point mt is the time in the reference system of the moving particle.
2. A very interesting possibility consists in modifying the classical Lagrangian by introducing the proper time at the classical level in such a way that the solutions of the equation of motion are not changed. The quantization of this new theory leads directly to Eq. (16.1); see R. Casalbuoni, *Riv. Nuovo Cimento A 33*, 336 (1976).
3. Equation (16.11) as it stands is also divergent in the free theory; however, this divergence may be easily eliminated by replacing $\int_0^\infty dt_I$ with $\int_{1/\Lambda^2}^\infty dt_I$ and by studying the ratio Z/Z_F before taking the limit $\Lambda^2 \to \infty$. Of course, the usual divergences of field theory are still present. They can be eliminated in $D < 4$ by adding counterterms proportional to the mass. At an intermediate stage it is convenient to fully regularize the theory by substituting for $\delta[\omega(\tau_i) - \omega(\tau_j)]$ a smoothed δ function, e.g., $\delta_\Lambda(x) = (2\pi\Lambda^2)^{-D/2} \exp(-x^2/2\Lambda^2))$.
4. The results presented in this section are mainly due to K. Symanzik in *Local Quantum Theory*, ed. Jost, Academic Press, New York (1969). He has also stressed the possibility of writing the Euclidean quantum field theory as the statistical mechanics of a gas of closed trajectories.
5. Only recently was it discovered that Symanzik's ideas presented in the previous subsection could be used to obtain rigorous results; recent

general papers on this subject are: D. Brydges, in *Gauge theories: fundamental interactions and rigorous results* ed. by P. Dita, V. Georgescu, and R. Purice, Birkhäuser, Boston (1982); C. Aragao de Carvalho, S. Caracciolo, and J. Fröhlich, Nucl. Phys. B *125* [FS7], 209 (1983); M. Aizenman, in *Scaling and Self-Similarity in Physics*, ed. by J. Fröhlich, Birkhäuser, Boston (1983).

6. It is understood that Eq. (16.13) must be symmetrized with respect to x_1, x_2, x_3, and x_4, as is done in Eq. (16.2).
7. This argument is very easy to understand in a nonrigorous way: see, for example, P. De Gennes, *J. Phys. (Paris) Lett.* *31*, 16 (1977); G. Parisi, *Phys. Lett. B 81*, 357 (1979).

CHAPTER 17

Time-Dependent Correlations

17.1. Quantum-statistical mechanics

Up to now we have studied classical statistical mechanics. Quantum statistical mechanics can be approached in a similar way: one postulates that at thermal equilibrium the expectation value of an operator A and the partition function Z are given by

$$\langle A \rangle = \frac{\text{Tr}\{A \exp[-\mathcal{H}\beta]\}}{\text{Tr}\{\exp[-\mathcal{H}\beta]\}}$$

$$Z = \text{Tr}\{\exp[-\mathcal{H}\beta]\} \,, \quad (17.1)$$

where $\mathcal{H}$ is the Hamiltonian. If the levels of $\mathcal{H}$ are discrete, the system stays at the nth level with energy E_n, with a probability proportional to $\exp(-E_n \beta)$. Equation (17.1) is the generalization of the Boltzmann-Gibbs distribution [Eq. (1.5)] to the quantum case; indeed, in the classical limit the former reduces to the latter. In the simplest case, where

$$\mathcal{H} = \sum_{\nu}^{D} \frac{\partial_\nu^2}{2} + V(q) \,,$$

using the results of Chapter 14, we recover the classical formulae in the limit of small $\hbar$ or small β:

$$\langle A \rangle \simeq \frac{1}{(2\pi\hbar)^D} \int d^D q \, d^D p \, \exp[-\beta \mathcal{H}(q, p)] \frac{A(q)}{Z}$$

$$Z \simeq \frac{1}{(2\pi\hbar)^D} \int d^D q \, d^D p \, \exp[-\beta \mathcal{H}(q, p)] \,, \quad (17.2)$$

with the only difference being that Z (and consequently the entropy) has an absolute normalization for continuous system: the nondegeneracy of the ground state tells us that the zero-temperature entropy is zero.[1]

17.1. Quantum-Statistical Mechanics

Equation (17.1) can be easily written using the path integral formulation:[2]

$$\langle A \rangle = \int dP[\omega] A(\omega(0))/Z$$

$$dP[\omega] = \int d^D x \, d[\omega]_{x,x}^{\beta\hbar} \exp\left\{ -\frac{1}{\hbar} \int_0^{\beta\hbar} d\tau \left[\sum_{1}^{D} \frac{\dot{\omega}_\nu^2}{2} + V(\omega(\tau)) \right] \right\} \quad (17.3)$$

$$\int dP[\omega] = Z,$$

where the sum is done over all the trajectories defined on the imaginary time interval $0 - \beta\hbar$ with periodic boundary conditions

$$\omega_\nu(0) = \omega_\nu(\beta\hbar). \quad (17.4)$$

In the same way, the statistical mechanics of a quantum field theory at finite temperature may be constructed by considering only field configurations defined in a strip of size $\beta\hbar$ with periodic boundary conditions, i.e., $\varphi(\vec{x}, 0) = \varphi(\vec{x}, \beta\hbar)$, where $\vec{x}$ denotes the space coordinates and $\varphi(\vec{x}, t)$ is defined only for $0 \leq t \leq \beta\hbar$.

Up to now we have considered only the expectation values of observables at the same time; it is also possible to study what happens if we measure observables at different times on the same system at thermal equilibrium. For example, in a classical theory we can consider

$$C_{AB}(t) \equiv \int dP_{\text{eq}}(C) A(C(t)) B(C) \equiv \langle A(t) B(0) \rangle, \quad (17.5)$$

where C is the generic configuration of the system, P_{eq} is the equilibrium probability distribution, and $C(t)$ is the configuration of the system at the time t under the condition $C(0) = C$ [obviously $C(t)$ is a function of C]. In other words, the configurations at time 0 are supposed to follow the equilibrium distribution, and the configurations at time t are computed as the time evolutions of the configurations at time 0; from the definition of equilibrium the configurations at time t have the same probability distribution as the configurations at time 0.

These time-dependent correlations give us information on the time evolution of the system at equilibrium; it is not very difficult to measure them experimentally in real materials. Their study is very important for the comparison of experimental data with the theory.

Time-dependent correlations can also be introduced in the quantum case: Eq. (17.5) becomes now

$$C_{AB}(t) = \text{Tr}\{\exp[-\beta\mathcal{H}]A(t)B(0)\}Z \equiv \langle A(t)B(0)\rangle$$

$$A(t) = \exp\left[\frac{i\mathcal{H}t}{\hbar}\right]A(0)\exp\left[-\frac{i\mathcal{H}t}{\hbar}\right], \quad (17.6)$$

[cf. Eq. (13.3)]. In the quantum case, operators do not commute, and $\langle B(t)A(0)\rangle \neq C_{AB}(t)$. Sometimes it is convenient to consider the function $C_{\{A,B\}}(t) \equiv \langle A(t)B(0) + B(0)A(t)\rangle$ in which the order of A and B does not matter.[3]

In the quantum case the correlation functions satisfy very interesting identities, whose proof is relatively simple. We first notice that

$$\langle A(t)B(0)\rangle = \sum_{n,m} \frac{\langle n|\exp[-\beta\mathcal{H}]|n\rangle\langle n|A(t)|m\rangle\langle m|B(0)|n\rangle}{Z}$$

$$= \sum_{n,m} \frac{\langle n|A|m\rangle\langle m|B|n\rangle \exp\left[-i\frac{t}{\hbar}(E_m - E_n) - \beta E_n\right]}{Z}$$

$$\mathcal{H}|n\rangle = E_n|n\rangle. \quad (17.7)$$

$C_{AB}(t)$ is thus an analytic function of t which can be extended at imaginary times. The following remarkable condition holds:

$$\langle A(t)B(0)\rangle = \langle B(0)A(t + i\beta\hbar)\rangle = \langle B(-t - i\beta\hbar)A(0)\rangle$$

$$C_{AB}(t) = C_{BA}(-t - i\beta\hbar), \quad (17.8)$$

as can be explicitly checked in Eq. (17.7). The condition (17.8), often denoted as the KMS condition,[4] is rather strong. Indeed, let us suppose that the equilibrium expectation values are given by

$$\langle A\rangle = \text{Tr}(\rho A)$$

$$\langle A(t)B(0)\rangle = \text{Tr}[\rho A(t)B(0)] \quad (17.9)$$

$$\text{Tr}[\rho] = 1,$$

where ρ is a given positive operator. The KMS condition implies that $\rho \propto \exp[-\beta\mathcal{H}]$. Indeed,

$$\langle B(0)A(t + i\beta\hbar)\rangle = \text{Tr}[\rho B(0)A(t + i\hbar\beta)]$$

$$= \text{Tr}[\exp(-\beta\mathcal{H}) \cdot A(t) \cdot \exp(\beta\mathcal{H}) \cdot \rho \cdot B(0)]. \quad (17.10)$$

17.2. The Quantum Fluctuation-Dissipation Theorem 303

The last expression must be equal to $\mathrm{Tr}[\rho A(t)B(0)]$. This is possible only if $A(t)$ commutes with $\exp[\beta\mathcal{H}]\cdot\rho$ for any $A(t)$; thus $\exp[\beta\mathcal{H}]\rho$ must be a pure number, i.e., $\rho \propto \exp[-\beta\mathcal{H}]$.

It is interesting to note that the KMS condition implies that $C_{\{AB\}}(t) = C_{\{AB\}}(t+i\hbar\beta)$, i.e., the correlation function of the anticommutator is periodic with period $i\hbar\beta$.

Using the formalism of Chapter 13, it is easy to see that at imaginary times $C_{AB}(-it)$ has a simple functional representation for $0 \le t \le \beta\hbar$:

$$C_{AB}(-it) = C_{BA}(-i(\hbar\beta - t))$$

$$= \int \frac{dP[\omega] A[\omega(t)] B[\omega(0)]}{Z}, \qquad (17.11)$$

where $P[\omega]$ is defined in Eq. (17.3).

This could be the beginning of a book dedicated to quantum statistical mechanics; here, for reasons of space, we must leave this interesting subject practically unexplored.

17.2. The quantum fluctuation-dissipation theorem

If we have a quantum system at equilibrium at temperature β for a given Hamiltonian $\mathcal{H}$, the time evolution is such that the system will remain at equilibrium forever. If we apply an external force to the system, i.e., by adding to the Hamiltonian an external perturbation $f(t)B(t)$ ($f(t) = 0$, $t < 0$) for positive time, the expectation value of A will depend on the time:[5]

$$\langle A(t)\rangle = \frac{\mathrm{Tr}[\exp[-\beta\mathcal{H}]U_t[f]A(0)(U_t[f])^{-1}]}{Z}, \qquad (17.12)$$

where the unitary operator $U_t[f]$ satisfies the differential equation $-i\hbar(\partial/\partial t)U_t[f] = [\mathcal{H} + f(t)B(t)]U_t[f]$, whose solution is given by

$$U_t[f] = U(t) + \frac{i}{\hbar}\int_0^t d\tau\, U(\tau)f(\tau)B(\tau)U(t-\tau) - \frac{1}{\hbar^2}\int_0^t d\tau_1 \int_0^{\tau_1} d\tau_2\, U(\tau_2)f(\tau_2)$$

$$\times B(\tau_2)U(\tau_1-\tau_2)f(\tau_1)B(\tau_1)U(t-\tau_1) + O(f^3)$$

$$U(\tau) \equiv \exp\left[\frac{i\tau\mathcal{H}}{\hbar}\right]. \qquad (17.13)$$

We can write Eq. (17.13) in a compact way by using the time-ordered symbol τ [cf. Eq. (13.24)]:

$$U_t[f] = \tau\left[\exp\left(\frac{i}{\hbar}\int_0^t d\tau(\mathcal{H} + f(\tau)B(\tau))\right)\right]. \quad (17.14)$$

Equation (17.14) is a short notation for (17.13). At the first order in f we have

$$\langle A(t)\rangle = \langle A(0)\rangle + \int_0^t R_{AB}(t-\tau)f(\tau)\,d\tau + O(f^2)$$

$$R_{AB}(t-\tau) = -\frac{i}{\hbar}\langle [A(t), B(\tau)]\rangle \quad (17.15)$$

$$= -\frac{i}{\hbar}\text{Tr}\{\exp[-\beta\mathcal{H}][A(t), B(\tau)]\}/Z \quad t > \tau,$$

where $R_{AB}(t)$ is by definition the time-dependent linear response function.

The fluctuation-dissipation theorem gives a relation between the response function (i.e., the expectation value of the commutator) and the correlation function. This can be seen in the following way. We define the correlation functions τ in Fourier space

$$\langle A(t)B(0)\rangle = \frac{1}{2\pi}\int_{-\infty}^{+\infty} d\omega\,\exp[it\omega]\tilde{C}_{AB}(\omega)$$

$$\langle B(t)A(0)\rangle = \frac{1}{2\pi}\int_{-\infty}^{+\infty} d\omega\,\exp[it\omega]\tilde{C}_{BA}(\omega). \quad (17.16)$$

Now $\langle A(t)B(0)\rangle = \langle B(0)A(t+i\beta\hbar)\rangle = \langle B(-t-i\beta\hbar)A(0)\rangle$, the KMS condition, becomes

$$\tilde{C}_{AB}(\omega) = \tilde{C}_{BA}(-\omega)\exp[\beta\hbar\omega]. \quad (17.17)$$

Equation (17.17) implies that the Fourier transform of the commutator $\tilde{C}_{[A,B]}(\omega)$ is given by

$$\tilde{C}_{[AB]}(\omega) = \tilde{C}_{AB}(\omega) - \tilde{C}_{BA}(-\omega) = \tilde{C}_{AB}(\omega)[1 - \exp(-\beta\hbar\omega)]$$

$$C_{[AB]}(t) \equiv \langle [A(t), B(0)]\rangle = \langle A(t)B(0) - B(-t)A(0)\rangle \quad (17.18)$$

$$= \frac{1}{2\pi}\int_{-\infty}^{+\infty} d\omega\,\exp[it\omega]\tilde{C}_{[AB]}(\omega).$$

17.2. The Quantum Fluctuation-Dissipation Theorem

It is convenient to introduce the Fourier transform of the anticommutator $\tilde{C}_{\{AB\}}(\omega)$. We easily find that

$$\tilde{C}_{\{AB\}}(\omega) = [1 + \exp(-\beta\hbar\omega)]\tilde{C}_{AB}(\omega) \qquad (17.19)$$

and therefore

$$\tilde{C}_{[AB]}(\omega) = 2\tanh\left[\frac{1}{2}\beta\hbar\omega\right]\tilde{C}_{\{AB\}}(\omega). \qquad (17.20)$$

Combining Eqs. (17.20) and (17.15), we get, for $t > 0$,

$$R_{AB}(t) = -\frac{1}{2\pi}\int_{-\infty}^{+\infty} d\omega \exp(i\omega t)\frac{2}{\hbar}\tanh\left(\frac{\omega\hbar\beta}{2}\right)C_{\{AB\}}(\omega). \qquad (17.21)$$

The response function is thus fixed by the expectation value of the anticommutator. It is interesting to obtain the result for the static response $R^s_{AB} = \int_0^\infty dt\, R_{AB}(t)$, i.e., the response to a time-independent perturbation. We find[6]

$$R^s_{AB} = -\frac{i}{\hbar}\int_0^\infty [\langle A(t)B(0)\rangle - \langle B(0)A(t)\rangle]\, dt$$

$$= -\frac{i}{\hbar}\int_0^\infty [C_{AB}(t) - C_{AB}(t + i\beta\hbar)]\, dt = -\frac{i}{\hbar}\left[\int_0^\infty + \int_{\infty+i\beta\hbar}^{i\beta\hbar}\right]C_{AB}(t)\, dt$$

$$= -\frac{i}{\hbar}\int_0^{i\beta\hbar} C_{AB}(t)\, dt = -\frac{1}{\hbar}\left\langle A(0)\int_0^{\beta\hbar} B(it)\, dt\right\rangle, \qquad (17.22)$$

where we have assumed that the integral $\int_\infty^{\infty+i\beta\hbar} dt\, C_{AB}(t)$ is zero [i.e., $C_{AB}(t)$ goes to zero at infinite time]. We have deformed the integration path in the complex t plane from 0 to infinity and from infinity to $i\beta\hbar$, producing a new integration path going directly from 0 to $i\beta\hbar$.

It is also possible to derive Eq. (17.22) in the static approach. It is evident that for a time-independent perturbation B

$$R^s_{AB} = \lim_{\varepsilon\to 0}\frac{\text{Tr}\{[\exp(-\beta(\mathcal{H} + \varepsilon B)) - \exp(-\beta\mathcal{H})]A\}}{\varepsilon\,\text{Tr}\{\exp(-\beta\mathcal{H})\}}. \qquad (17.23)$$

Again using Eqs. (17.14) and (17.15) in the case of a time-independent perturbation, we find the useful operator identity

$$\exp(-E - \varepsilon F) = \exp(-E)\tau\left[\exp-\int_0^1 \varepsilon F(t')\,dt'\right]$$

$$= \exp(-E) - \varepsilon \int_0^1 dt_1 \exp(-E(1-t_1))F\exp(-Et_1)$$

$$+ \varepsilon^2 \int_0^1 dt_1 \int_0^{t_1} dt_2 \exp(-E(1-t_1))F\exp(-E(t_2-t_1))$$

$$\times F\exp(-Et_2) + O(\varepsilon^3), \tag{17.24}$$

where $F(t') = \exp(Et')F\exp(-Et')$.

By applying Eq. (17.24) to Eq. (17.23), we find again the result of Eq. (17.22), i.e.,

$$R^s_{AB} = -\beta \int_0^1 dt_1 \frac{\text{Tr}[\exp(-\beta\mathcal{H}(1-t_1))B\exp(-\beta\mathcal{H}t_1)A]}{Z}$$

$$= -\frac{1}{\hbar}\int_0^{\beta\hbar} dt_1\,\text{Tr}[\exp(-\beta\mathcal{H})B(it)A(0)]$$

$$= -\frac{1}{\hbar}\int_0^{\beta\hbar} dt\,\langle B(it)A(0)\rangle. \tag{17.25}$$

It is also evident that Eq. (17.22) has an instantaneous derivation if we start from Eq. (17.3).

Equation (17.20) is interesting in Fourier space: if we use the integral representation of the θ step function, $[\theta(t) = \int_{-\infty}^{+\infty}(\exp i\omega t)/2\pi(\omega - i\varepsilon)$ for $\varepsilon \to 0^+]$, and the convolution theorem, we get

$$\tilde{R}_{AB}(\omega) = \frac{1}{\pi\hbar}\int_{-\infty}^{+\infty}\frac{d\omega'}{\omega-\omega'-i\varepsilon}\tanh\left(\frac{\omega'\hbar\beta}{2}\right)\tilde{C}_{\{A,B\}}(\omega')$$

$$\tilde{R}_{A,B}(\omega) = \int_0^\infty dt\,R(t)\exp(i\omega t) \tag{17.26}$$

$$\tilde{C}_{\{A,B\}}(\omega) = \int_{-\infty}^{+\infty} dt\,C_{\{A,B\}}(t)\exp(i\omega t).$$

In other words, $\tilde{R}_{AB}(\omega)$ is an analytic function in the upper complex plane, with a cut or poles on the real axis, the discontinuity on the axis being $(1/\pi\hbar)\tanh(\omega\hbar\beta/2)\tilde{C}_{\{A,B\}}(\omega)$.

17.3. The classical KMS condition

We have already remarked that in the classical theory we can consider time-dependent correlation functions. Indeed, the probability distribution

17.3. The Classical KMS Condition

at zero time is known, and we need only to use the classical equation of motion,

$$\dot{q} = \frac{\partial \mathcal{H}}{\partial p} = \{q, \mathcal{H}\}_p$$

$$\dot{p} = -\frac{\partial \mathcal{H}}{\partial q} = \{p, \mathcal{H}\}_p \quad (17.27)$$

$$\dot{A} = \{A, \mathcal{H}\}_p,$$

where the bracket denotes the Poisson bracket (not the anticommutator!):

$$\{A, B\}_p = \frac{\partial A}{\partial q}\frac{\partial B}{\partial p} - \frac{\partial A}{\partial p}\frac{\partial B}{\partial q}. \quad (17.28)$$

We recall that in the associated quantum theory the quantum commutator becomes the classical Poisson bracket when $\hbar \to 0$: $i/\hbar[A, B] \to \{A, B\}_p$. If we consider the correlation function of two observables A and B, we find that

$$\frac{d}{dt_1}\langle A(t_1)B(t_2)\rangle\big|_{t_1=0} \equiv \langle\{A(0), \mathcal{H}\}_p B(t_2)\rangle$$

$$= -\frac{1}{\beta}\langle\{A(0), B(t_2)\}_p\rangle. \quad (17.29)$$

The last equality may be proved by noticing that[7]

$$\langle\{A(0), \mathcal{H}\}_p B(t_2)\rangle = \frac{1}{Z}\int dq\, dp \left[\frac{\partial A}{\partial q}\frac{\partial \mathcal{H}}{\partial p} - \frac{\partial A}{\partial p}\frac{\partial \mathcal{H}}{\partial q}\right]$$

$$\times B(t_2)\exp[-\beta\mathcal{H}(q, p)]$$

$$= \frac{1}{Z}\frac{1}{\beta}\int dq\, dp\left[\frac{\partial A}{\partial q}\frac{\partial B(t_2)}{\partial p} - \frac{\partial A}{\partial p}\frac{\partial B(t_2)}{\partial q}\right]$$

$$\times \exp[-\beta\mathcal{H}(q, p)]. \quad (17.30)$$

Integrating by parts, we have obtained the needed result. Equation (17.29) is called the classical KMS condition.[8] Indeed for $\hbar$ small we can replace $A(t + i\hbar\beta)$ in Eq. (17.8) with $A(t) + \dot{A}(t)i\hbar\beta$; in this way we obtain

$$\left\langle \frac{1}{2}\{\dot{A}(t)B(0)\}\right\rangle \simeq \frac{i}{\hbar\beta}\langle[A(t), B(0)]\rangle, \quad (17.31)$$

which in the limit $\hbar \to 0$ becomes Eq. (17.29) for $t_2 = -t$. A similar result can be obtained for the response function. Indeed, an explicit computation (which we omit) or the simple use of the quantum formula for $\hbar \to 0$ tell us that

$$R_{AB}(t) = \langle \{A(t), B(0)\}_p \rangle = \beta \langle \dot{A}(t) B(0) \rangle \qquad t > 0$$
$$R_{AB}(t) = 0 \qquad t < 0. \tag{17.32}$$

Equation (17.32) is clearly the classical limit of Eq. (17.21): in the limit $\hbar \to 0$, $i \tanh((\hbar \omega \beta)/2)/\hbar$ becomes $(i\beta/2)\omega$ and $i\omega$ is the time derivative.

Equation (17.32) is the time-dependent classical fluctuation-dissipation theorem; the static theorem can be easily obtained:

$$R^s_{AB} \equiv \int_0^\infty R_{AB}(t)\, dt = \beta \int_0^\infty \langle \dot{A}(t) B(0) \rangle\, dt$$
$$= -\beta \langle A(0) B(0) \rangle. \tag{17.33}$$

We have thus recovered the result of the static linear response theory [cf. Eq. (2.12)]. As we have seen for time-dependent perturbations in the framework of the linear response approximation, the response of the system can be extracted from the knowledge of the appropriate expectation values. Equation (17.32) is also simple in Fourier space. We obtain

$$\tilde{R}_{AB}(\omega) = \int_0^{+\infty} dt \exp(-i\omega t) R_{AB}(t)$$
$$= \beta \int_{-\infty}^{+\infty} \frac{d\omega' \omega'}{\omega - \omega' - i\varepsilon} \tilde{C}_{AB}(\omega') \tag{17.34}$$
$$= -\beta C_{AB}(0) + \omega \int_{-\infty}^{+\infty} \frac{d\omega'}{(\omega - \omega' - i\varepsilon)} \tilde{C}_{AB}(\omega')$$

$$\tilde{C}_{AB}(\omega) = \int_{-\infty}^{+\infty} dt \exp(-i\omega t) \langle A(t) B(0) \rangle,$$

where $i\varepsilon$ is infinitesimal ($\varepsilon > 0$). Equation (17.34) at $\omega = 0$ reduces to Eq. (17.33).

The classical KMS condition is very strong. Indeed, for a finite system, if $d\mu(C)$ is a probability distribution over the configuration space and the classical KMS condition is satisfied, i.e.,

$$\int d\mu(C) \{A(0), \mathcal{H}\}_p B = -\frac{1}{\beta} \int d\mu(C) \{A, B\}_p, \tag{17.35}$$

17.4. A classical example

In order to give to the reader some feeling for the problems that may arise in the actual evaluation of time-dependent correlation functions, we shall study an explicit case, the free classical field theory. A field $\varphi(\vec{x}, t)$ and a momentum $\Pi(\vec{x}, t)$ are defined at each point of the space-time. They satisfy the classical Poisson brackets (see Chapter 15). The Hamiltonian and the equations of motion are, respectively,

$$\mathcal{H} = \int d^3x \left[\frac{1}{2} \sum_a^3 (\partial_a \varphi)^2 + \frac{m^2}{2} \varphi^2 + \frac{1}{2} \Pi^2 \right]$$

$$\dot{\varphi} = \Pi \qquad \dot{\Pi} = -(-\Delta + m^2)\varphi$$

$$\Delta = \sum_a^3 \partial_a^2 \,. \tag{17.36}$$

If we introduce the Fourier transforms of φ and Π,

$$\tilde{\varphi}(p, t) = \int d^3x \, \exp(ipx)\varphi(x, t)$$

$$\tilde{\Pi}(p, t) = \int d^3x \, \exp(ipx)\Pi(x, t) \,, \tag{17.37}$$

the solution of Eq. (17.36) can be written as

$$\tilde{\varphi}(p, t) = \cos[\omega(p)t]\tilde{\varphi}(p, 0) + \frac{1}{\omega(p)} \sin[\omega(p)t]\tilde{\Pi}(p, 0)$$

$$\tilde{\Pi}(p, t) = -\omega(p) \sin[\omega(p)t]\tilde{\varphi}(p, 0) + \cos[\omega(p)t]\tilde{\Pi}(p, 0) \tag{17.38}$$

$$\omega(p) = (p^2 + m^2)^{1/2} \,.$$

We notice that $\omega^2(p)\tilde{\varphi}^2(p, t) + \tilde{\Pi}^2(p, t)$ does not depend on t and it is a constant of the motion for any p. The correlation functions in momentum space are given by

$$\langle \tilde{\varphi}(p_1, t)\tilde{\varphi}(p_2)\rangle = \cos[\omega(p_1)t]\cos[\omega(p_2)t]\langle \tilde{\varphi}(p_1, 0)\tilde{\varphi}(p_2, 0)\rangle$$
$$+ \cos[\omega(p_1)t]\frac{\sin[\omega(p_2)t]}{\omega[p_2]}\langle \tilde{\varphi}(p_1, 0)\tilde{\Pi}(p_2, 0)\rangle$$
$$+ \cdots. \qquad (17.39)$$

We can now use the information we have on the distribution at $t = 0$, which is supposed to be the equilibrium one:

$$\langle \tilde{\Pi}(p_1, 0)\tilde{\Pi}(p_2, 0)\rangle = \frac{\delta(p_1 + p_2)}{\beta}$$

$$\langle \tilde{\Pi}(p_1, 0)\tilde{\varphi}(p_2, 0)\rangle = 0 \qquad (17.40)$$

$$\langle \tilde{\varphi}(p_1, 0)\tilde{\varphi}(p_2, 0)\rangle = \frac{\delta(p_1 + p_2)}{[\beta\omega^2(p_1)]}.$$

We finally obtain

$$\langle \tilde{\varphi}(p_1, t)\tilde{\varphi}(p_2, 0)\rangle = \frac{\delta(p_1 + p_2)}{\beta\omega^2(p_1)}\cos[\omega(p)t]$$

$$\langle \tilde{\varphi}(p_1, t)\tilde{\Pi}(p_2, 0)\rangle = \frac{\delta(p_1 + p_2)}{\beta\omega(p_1)}\sin[\omega(p)t] \qquad (17.41)$$

$$\langle \tilde{\Pi}(p_1, t)\tilde{\Pi}(p_2, 0)\rangle = \frac{\delta(p_1 + p_2)}{\beta}\cos[\omega(p)t].$$

We could obtain the same result just by observing that the equations of motion for φ imply that

$$(\Box + m^2)\langle \varphi(x, t)\varphi(0, 0)\rangle = 0 \qquad (17.42)$$
$$\left(-\frac{d^2}{dt^2} + p_1^2 + m^2\right)\langle \tilde{\varphi}(p_1, t)\tilde{\varphi}(p_2, 0)\rangle = 0.$$

The solution of Eq. (17.41) is fixed by the boundary conditions in Eq. (17.40). Going back to position space we find

$$\langle \varphi(x, t)\varphi(0, 0)\rangle = \frac{1}{(2\pi)^3}\int \frac{d^3p \exp(-ipx)}{\beta(p^2 + m^2)}\cos[(p^2 + m^2)^{1/2}t]$$

$$\langle \Pi(x, t)\Pi(0, 0)\rangle = \frac{1}{(2\pi)^3}\int \frac{d^3p}{\beta}\exp(-ipx)\cos[(p^2 + m^2)^{1/2}t]$$

$$\langle \Pi(x, t)\varphi(0, 0)\rangle = \frac{1}{(2\pi)^3}\int \frac{d^3p \exp(-ipx)}{\beta(p^2 + m^2)^{1/2}}\sin[(p^2 + m^2)^{1/2}t].$$
$$(17.43)$$

17.5. A Quantum Example

For large times, when the integral is dominated by the region around $p = 0$, we obtain [cf. Eq. (14.4)]

$$\langle \varphi(x, t)\varphi(0, 0)\rangle \sim \frac{1}{\beta m^2} \left(\frac{m}{2\pi t}\right)^{3/2} \exp\left[\frac{ix^2 m}{2t}\right]$$

$$\langle \Pi(x, t)\Pi(0, 0)\rangle \sim \frac{1}{\beta} \left(\frac{m^2}{2\pi t}\right)^{3/2} \exp\left[\frac{ix^2 m}{2t}\right] \qquad (17.44)$$

$$\langle \Pi(x, t)\varphi(0, 0)\rangle \sim O(t^{-5/2}).$$

The correlation functions have a $t^{-3/2}$ tail when the time becomes large.

If we add a term $(g/24) \int d^3x \, \varphi^4(x)$ in the Hamiltonian, a perturbative expansion in powers of g can be constructed. For example, we can use the equation

$$\left\langle \left(-\Box + m^2 + \frac{g}{6} \varphi^2(x, t)\right)\varphi(x, t)\varphi(0, 0)\right\rangle = 0 \qquad (17.45)$$

for the two-field correlation function and similar equations for the higher-order correlation functions. These equations, supplemented by the boundary conditions at $t = 0$, make it possible for us to compute the time-dependent correlation function recursively. With a certain amount of additional work, it is possible to organize the perturbative expansion in the form of diagrams; a detailed study of this interesting field would take us too far from the aim of this book.[9]

17.5. A quantum example

We consider the anharmonic oscillator, whose quantum Hamiltonian is given by

$$\mathcal{H} = \frac{1}{2} p^2 + \frac{1}{2} m^2 q^2 + \frac{1}{4!} gq^4. \qquad (17.46)$$

For positive m^2 and nonnegative g, the levels of $\mathcal{H}$ are discrete: The partition function can be computed using Eq. (17.1).

At $g = 0$ the levels are given (for $\hbar = 1$) by

$$E_n = \left(n + \frac{1}{2}\right)m. \qquad (17.47)$$

We consequently find

$$Z_0 \equiv Z(g)|_{\hat{g}=0} = \sum_0^\infty {}_n \exp\left[-\beta\left(n + \frac{1}{2}\right)m\right] = \frac{1}{\sinh\left(\frac{\beta m}{2}\right)}. \qquad (17.48)$$

For small positive g, the energy levels can be computed by using the usual perturbative expansion of nonrelativistic quantum mechanics: one easily finds

$$E_n = \left(n + \frac{1}{2}\right)m + \frac{g}{16m^2}\left(n^2 + n + \frac{1}{2}\right) + O(g^2)$$

$$Z(g) = Z_0\left\{1 - \frac{g\beta}{32m^2}\left[\tanh\left(\frac{\beta m}{2}\right)\right]^{-2} + O(g^2)\right\}.$$

(17.49)

Here we want to show how to recover these results using Eq. (17.3). In this case we have

$$Z(g) = \int d[q] \exp\left\{-\int_0^\beta dt \left[\frac{1}{2}\dot{q}^2 + \frac{m^2}{2}q^2 + \frac{g}{4!}q^4\right]\right\}. \quad (17.50)$$

Let us start at $g = 0$. We first compute the derivative of the partition function with respect to m^2; the absolute normalization factors of functional integrals are usually difficult to evaluate.

The two-q correlation function satisfies at $g = 0$ the following differential equation:

$$G(t) = \langle q(t)q(0) \rangle, \quad -\frac{d^2}{dt^2}G(t) + m^2 G(t) = \sum_{-\infty}^{+\infty} \delta(n\beta - t), \quad (17.51)$$

where $G(t)$ is a periodic function of period β. At $\beta = \infty$, we already know that

$$G(t) = \frac{1}{2m}\exp[-m|t|]. \quad (17.52)$$

At $\beta \neq \infty$, we find that

$$G(t) = \frac{1}{2m}\sum_{-\infty}^{\infty} \exp[-m|t - n\beta|]$$

$$= \frac{\exp(-mt) + \exp(-\beta m + mt)}{2m[1 - \exp(-\beta m)]} \quad \text{for } 0 \leq t \leq \beta. \quad (17.53)$$

Alternatively we could impose the periodicity condition by writing

$$G(t) = \sum_{-\infty}^{+\infty} \exp(itl2\pi/\beta)\tilde{G}(k_l), \quad k_l = 2\pi l/\beta. \quad (17.54)$$

The differential equation (17.51) implies that

17.5. A Quantum Example

$$\tilde{G}(k) = \frac{2\pi}{\beta} \frac{1}{k^2 + m^2}. \tag{17.55}$$

We easily recover Eq. (17.53) by using the Poisson formula (see p. 101, note 2). It is evident that

$$\frac{d}{dm}\ln Z_0 = -\beta m \langle q^2 \rangle = -\frac{\beta}{2}\sum_{-\infty}^{+\infty} \exp(-\beta m|n|) = -\frac{\beta}{2}\left[\tanh\left(\frac{\beta m}{2}\right)\right]^{-1}. \tag{17.56}$$

By integrating Eq. (17.56) we find

$$Z_0 = \frac{a(\beta)}{\sinh(\beta m/2)}, \tag{17.57}$$

where an ambiguity remains in the β-dependent term $a(\beta)$. Using the same arguments as in Sec. 14.6, we find that $a(\beta) = \frac{1}{2}$.

The perturbative expansion can be easily obtained by expanding the exponential in the functional integrals in powers of g. After some easy work we find

$$\ln Z(g) = \ln Z_0 - \frac{g\beta}{4!}\langle q^4 \rangle + \frac{1}{2}\frac{g^2\beta}{(4!)^2}\int_0^\beta dt \langle q^4(t) q^4(0) \rangle_c + O(g^3)$$

$$= \ln Z_0 - \frac{g\beta}{8}G^2(0) + \frac{1}{2}\frac{g^2\beta}{(4!)^2}\int_0^\beta dt\{4! G^4(t) + 72 G^2(0) G^2(t)\}$$

$$+ O(g^3). \tag{17.58}$$

We have thus recovered Eq. (17.49). We leave to the careful reader the task of verifying the correctness of Eq. (17.58) by using the usual perturbative expansion for the spectrum.

Similar results can be obtained for a Hamiltonian of a quantum field; in the free case we have

$$\mathcal{H} = \int_{0\le \bar{x} \le L} d^3x \left\{ \frac{1}{2}\Pi^2(x) + \frac{1}{2}[\partial_\nu \varphi(x)]^2 + \frac{1}{2}m^2\varphi^2(x) \right\}, \tag{17.59}$$

where periodic boundary conditions have been used. We find that the free energy density and the partition function are given, respectively, by

$$f(\beta) = (L^3\beta)^{-1}\sum_l \ln\left\{2\sinh\left[\frac{\beta}{2}(k_l^2 + m^2)^{1/2}\right]\right\}, \quad k_l^2 = (2\pi)^2 l^2/L^2$$
$$Z = \exp[-L^3\beta f(\beta)], \tag{17.60}$$

where the sum runs over vectors l of integer coordinates.

In the infinite-volume limit we find that the free-energy density is given by

$$f(\beta) - f(\infty) = \frac{\beta^{-1}}{(2\pi)^3} \int d^3k \ln[1 - \exp[-\beta(k^2 + m^2)^{1/2}]\} , \quad (17.61)$$

which in the limit $m^2 = 0$ reduces to the well-known Planck formula.

Notes for Chapter 17

1. The zero value of the entropy for a quantum system is a statement that is strictly true for a finite system; it is generally believed, but as far as I know has never been proven, that for any "reasonable" quantum Hamiltonian the limits $T \to 0$ and $V \to \infty$ commute for the entropy density, so that the infinite-volume entropy density goes to zero at zero temperature, i.e., the third law of thermodynamics is true. This statement is highly nontrivial insofar as there are plenty of discrete classical systems (e.g., the antiferromagnetic Ising model on a triangular lattice) that have a nonzero zero-temperature entropy.
2. See, for example, R. P. Feynman, *Statistical Mechanics*, W. A. Benjamin, New York (1972).
3. Many different types of correlation functions (advanced, retarded, time-ordered) can be considered; a nice compact formalism is described in Zhou Guang-zhao, Su Zho-lin, Hao Bai-lin, Yu Lu, *Comm. Theor. Phys.* **1**, 245 (1982) and references therein.
4. See, for example, L. P. Kadanoff and G. Baym, *Quantum Statistical Mechanics*, W. A. Benjamin, New York (1962).
5. A more complete treatment can be found in many places, e.g. L. D. Landau and E. M. Lifshitz, *Statistical Physics*, Pergamon, Oxford (1968); see also K. Kubo in *The many body problem*, Benjamin (1969).
6. For simplicity we assume that the expectation values of the A and B operators are zero; otherwise we would have to consider connected correlation functions in the rest of the chapter.
7. The Poisson brackets between an operator A at time 0 and an operator B at time t are defined by noticing that the canonical coordinates at time t are a function of the canonical coordinates at time 0; in other words, $B(t)$ may be considered as a very complicated function of the

canonical coordinates at time zero. The definition of the Poisson brackets does not depend on the reference time: the time evolution of the canonical coordinates is a canonical transformation, and the Poisson brackets are invariant under canonical transformations.
8. G. Gallavotti and E. Verboven, *Nuovo Cimento B 28*, 274 (1975).
9. There are many books on this subject. A well-known example is E. Prigogine, *Non-Equilibrium Statistical Mechanics*, Interscience, New York (1962).

CHAPTER 18

The Approach to Equilibrium

18.1. The microcanonical ensemble

In the first chapter of this book we presented the hypothesis that the long-time behavior of a large system is described by the canonical probability distribution. In the rest of the book we have studied the properties of such a distribution; now we shall say a bit more about this hypothesis to justify it. The natural setting for such a discussion would be a study of the time evolution of an infinite system. Unfortunately, to study an infinite system in a rigorous way is not simple: the very existence of a solution of the equations of motion for the infinite system cannot be proved without problems. For example, we can construct for an assembly of hard spheres an apparently innocent looking initial condition such that, after a cascade of collisions, a sphere is pushed at infinite velocity in a finite amount of time (what will happen later is unclear!). It is evident that such initial conditions are pathological, i.e., we must fix the positions and velocities of an infinite number of particles with infinite precision, but the very existence of these pathologies requires that we devote a certain amount of work to excluding their relevance for a generic "reasonable" initial configuration. More precisely, we need to prove that inside a given class of initial conditions the time evolution will be well defined for most of the configurations, and those configurations which produce pathologies have zero measure. Unfortunately a rigorous proof of such an important theorem is still lacking, although some partial results have been obtained.[1] Let us forget these difficulties and go on with the heuristic physically motivated approach.

The traditional argument for justifying the canonical distribution is the following: we consider a system with a finite number of degrees of freedom in a box, with a given Hamiltonian H. In such a situation the energy is conserved. If we call $C(t)$ the position of the system in phase space $[C = (\vec{q}, \vec{p})]$, at time t, $C(t)$ will move on the constant energy surface $\Sigma(E)$, E being the initial energy. If there are other constants of

18.1. The Microcanonical Ensemble

motion, $C(t)$ will be confined to only a small part of the energy surface Σ; if not, we may assume that $C(t)$ will pass through nearly all points on Σ; if that happens the system is said to be ergodic. More precisely, a system is ergodic if

$$\lim_{t\to\infty} \frac{1}{t} \int_0^t dt'\, g(C(t')) = \frac{\int dC\, \delta[H(C) - E] g(C)}{\int dC\, \delta[H(C) - E]}. \tag{18.1}$$

The time average of an ergodic system is given by the microcanonical ensemble

$$dP(C) \propto dC\, \delta[H(C) - E]. \tag{18.2}$$

Depending on the form of H, one can construct either nonergodic or ergodic systems. To find nonergodic systems is very simple; it is sufficient to have extra conservation laws. Noninteracting particles in a box with reflecting walls are a prototype of a nonergodic system, the speed of each particle being a constant of motion. [The Hamiltonian (17.36) is also nonergodic.] To prove that a Hamiltonian system is ergodic is normally far from being trivial; the most famous results are due to Sinai, who proved that a few particles in a reflecting box with a hard-sphere interaction (like billiard balls) constitute an ergodic system. Actually in this case one can prove more: if two systems are in two very close initial states C_1 and C_2, after a sufficiently large time the systems will be in two completely different states with no apparent correlation. The time evolution is said to be mixing. Mixing is a stronger property than ergodicity (indeed it implies ergodicity).[2]

If we cut the discussion by assuming that Eq. (18.2) is correct, we still must go from the microcanonical distribution to the canonical one [i.e., Eq. (1.10)]. This can be done for a very large system if we concentrate our attention on a subsystem. The traditional proof runs as follows. We first suppose that the Hamiltonian can be decomposed as (see p. 3)

$$H = H_1(C_1) + H_2(C_2) + \varepsilon H_{12}(C_1, C_2), \tag{18.3}$$

where the interaction Hamiltonian is very small but sufficient to have the system at equilibrium. The microcanonical distribution, neglecting H_{12}, is

$$dP(C_1, C_2) \propto \delta(H(C_1) + H(C_2) - E)\, dC_1\, dC_2, \tag{18.4}$$

which restricted to C_1 (i.e., after integration over C_2) becomes

$$dP(C_1) \propto W[E - H(C_1)] \, dC_1$$

$$\int dP(C_1) = 1 \,, \qquad (18.5)$$

where

$$W(E_2) = \int dC_2 \, \delta(E_2 - H_2(C_2)) \,. \qquad (18.6)$$

Now if the second system is very large (e.g., it describes N degrees of freedom), we expect that the volume $W(E_2)$ of the fixed energy surface increases as

$$W(E_2) \propto \exp\left[Ng\left(\frac{E_2}{N}\right) \right] \,, \qquad (18.7)$$

when N goes to infinity at fixed ratio $E_2/N \equiv W$.

This behavior of $W(E_2)$ is crucial in the rest of the argument: it may be justified either by dividing system 2 into a large number of weakly interacting subsystems or by noticing that the partition function of system 2 can be written, if (18.7) holds, as

$$Z_2(\beta) \equiv \int dE_2 \, \exp(-\beta E_2) W(E_2) \simeq \int du \, \exp[+N(-\beta u + g(u))] \,. \qquad (18.8)$$

The integral can be done by using the method of steepest descents; we find that

$$-\frac{1}{\beta N} \ln Z_2(\beta) = u_\beta - \frac{1}{\beta} g(\varepsilon_\beta) \,, \qquad \left.\frac{dg}{du}\right|_{u=u_\beta} = \beta \,. \qquad (18.9)$$

In other words, Eq. (18.7) is essentially equivalent to the statement that the free-energy density of the canonical distribution exists in the infinite-volume limit; this should be true for all good Hamiltonians, but it is easy to find counterexamples (e.g., N particles interacting with an attractive potential). If we compare Eq. (18.7) with Eq. (1.19) we see that $g(u)$ is the entropy density of the canonical distribution.

If we substitute (18.7) in Eq. (18.5), we obtain

$$dP(C_1) \propto dC_1 \, \exp\left[Ng\left(u - \frac{H_1(C_1)}{N}\right) \right]$$

$$\simeq dC_1 \, \exp[-g'(u) H_1(C_1)] \cdot \exp[Ng(u)] \,. \qquad (18.10)$$

18.1. The Microcanonical Ensemble

The small system has a canonical distribution with

$$\beta = \frac{dg}{du}. \tag{18.11}$$

The value of β obtained in this way is the same for any of the subsystems of the big system, and β^{-1} can be identified with the temperature. The consistency of these results may be checked by observing that the well known relation for the entropy as a function of the temperature, $ds/dT = 1/T(du/dT)$, implies that

$$\frac{ds}{du} = \frac{1}{T}, \tag{18.12}$$

in agreement with Eq. (18.11).

The main assumption here has been Eq. (18.7). If we add the extra hypothesis that the expectation values of local observables are smooth functions of the total energy, the argument may be simplified. As an exercise let us check that the classical KMS condition is satisfied in the microcanonical ensemble in the infinite-volume limit under this hypothesis.

An elementary computation shows that

$$\langle \{A, \mathcal{H}\}_P B \rangle_E = \int_0^E dE' \langle \{A, B\}_P \rangle_{E'} \frac{W(E')}{W(E)},$$

$$W(E) \equiv \int dC\, \delta(\mathcal{H} - E), \tag{18.13}$$

where $\langle\ \rangle_{E'}$ denotes the microcanonical average with energy E'; indeed we have

$$\sum_i \int dC \left[\frac{\partial A}{\partial q_i} \frac{\partial \mathcal{H}}{\partial p_i} - \frac{\partial A}{\partial p_i} \frac{\partial \mathcal{H}}{\partial q_i} \right] B \delta(H - E)$$

$$= \sum_i \int dC\, B \left[\frac{\partial A}{\partial q_i} \frac{\partial}{\partial p_i} - \frac{\partial A}{\partial p_i} \frac{\partial}{\partial q_i} \right] \theta(H - E). \tag{18.14}$$

If we integrate Eq. (18.14) by parts and divide the result by the normalization factor $W(E)$, we find Eq. (18.13).

As we have already remarked, $W(E)$ is supposed to be a rapidly increasing function of E so that the integral will be dominated by the integration region E' near E if the expectation values $\langle\ \rangle_E$ are slow functions of E, as is likely if we consider functions A and B of only a finite (i.e., small with respect to N) number of variables. Under these hypotheses we get the KMS condition:

$$\langle \{A, H\}_p B \rangle_E \simeq \beta(E) \langle \{A, B\}_p \rangle$$

$$\beta(E) = \int_0^E dE' \, \frac{W(E')}{W(E)} \, .$$

(18.15)

If Eq. (18.7) is assumed, we recover Eq. (18.9).

We conclude that if the system is large and ergodic, the canonical distribution for subsystems is correct: for a generic initial condition the large-time probability distribution of its subsystems will be given by the canonical one. Indeed for practically all nonpathological Hamiltonians (such that the canonical distribution is well defined in the infinite-volume limit), the previous considerations on the equivalence of the canonical and microcanonical ensembles can be rigorously justified.

At this point we should say a few words on the problem of irreversibility, i.e., how the irreversible behavior seen at the macroscopic level is compatible with the microscopic reversibility of the equations of motion. The ergodicity property of the system does not imply any irreversible behavior. The irreversibility shows up if we divide the phase space of the system into two parts—a small one which we call "ordered" and a large one which we call "disordered"; without breaking the time-reversal symmetry, an ergodic system in an "ordered" state at time zero will soon become "disordered" and will become ordered again only after a very large time. If, however, the system becomes very large and the fraction of phase space that we call ordered is exponentially small compared with the volume, we expect that the time required to come back to the ordered state will become exponentially large compared with the volume, so that it can be neglected for practical purposes.

A more precise statement can be made by introducing the entropy of the system at a given time. This concept seems at first sight self-contradictory, because the entropy is normally defined for a probability distribution and not for a single copy of the system. The entropy of a single configuration can be defined for a very large system. We consider for simplicity a discrete system like the Ising model (the extension to a continuous model is possible by dividing the phase space into discrete cells). To be definite, we start from a two-dimensional Ising model with $N \times N$ sites. We now consider the N^2 different squares of side L which can be constructed: In each square there are L^2 spins. The corresponding number of states of each square is 2^{L^2}. If $P^L(i)$ ($i = 1, 2^{L^2}$) is the probability for the occurrence of each of these states in any of these squares during the time evolution of the system; we could define a corresponding entropy for spins of the square,

$$S_L = -\frac{1}{L^2} \sum_{i=1}^{2^{L^2}} P^L(i) \ln P^L(i) \, .$$

(18.16)

It is clear that when $L \to \infty$, S_L must go to the correct entropy density for the spin. In order to estimate correctly the entropy from a given sample of M configurations of the squares, we must have $M \gg 2^{L^2}$. This condition is realized for the squares obtained by a single configuration of the system if $N^2 = M \gg 2^{L^2}$. This means that in a system of side 10, S_2 can be reasonably estimated for a single configuration; in a system of side 100 we can also estimate S_3 and so on. In this way we have constructed approximate entropies for a finite system. The ordered states of the previous discussion can now be characterized by small entropy, while the disordered state will be characterized by a higher entropy. If we consider a very large ergodic system (so that we can reasonably define an entropy S_L for high L), we expect that the generic trajectory will be most of the time in a region of high entropy, while it will spend only an exponentially small fraction of the time in the region of low entropy. If the initial configuration has been chosen to be a highly unlikely one, with a small entropy, we expect to have near the initial time a sharp increase in the entropy up to the moment the system reaches equilibrium. In a nearly infinite system the entropy should never decrease substantially (it should oscillate around its maximum).

It should be clear that we do not have at our disposal the mathematical tools to substantiate this simple physical picture. Most of the work done up to now consists in derivations of approximate (or exact) evolution equations (which generalizes the Boltzmann equation), the so-called master equations, which should describe the behavior of the system far from equilibrium. These equations under suitable conditions, show irreversible behavior, in the sense mentioned above.

18.2. The KAM theorem

We have seen that ergodicity seems to play a crucial role in our understanding of the approach to equilibrium and to the microcanonical distribution. Sinai's theorem on hard spheres cannot be generalized too much: there is an opposite theorem of Kolmogoroff, Arnold, and Moser, which, roughly speaking, says that if the Hamiltonian H_0 has some integrals of motion, beyond the energy, the Hamiltonian

$$H = H_0 + gH_1 \tag{18.17}$$

will not be ergodic for most of the g, if $|g|$ is sufficiently small and H_1 is a smooth perturbation. In other words, if we substitute for the hard-sphere potential a small repulsive (or attractive) potential, the system will no longer be ergodic for small enough but nonzero potential. We could thus say that we need a strong perturbation of the original integrable Hamiltonian H_0 to produce an ergodic behavior.[3]

Let us consider the following model Hamiltonian defined by

$$H = \sum_{k}^{L} \left[\frac{(\varphi_k - \varphi_{k+1})^2}{2} + \frac{m^2 \varphi_k^2}{2} + \frac{g}{24} \varphi_k^4 + \frac{1}{2} \Pi_k^2 \right]$$

$$H_1 = \sum_{k}^{L} \frac{1}{24} \varphi_k^4$$

(18.18)

for a chain of length L with periodic boundary conditions. The corresponding equations of motion are

$$\dot{\varphi}_k = \Pi_k$$

$$\dot{\Pi}_k = -\left(m^2 + \frac{g}{6} \varphi_k^2\right)\varphi_k - 2\varphi_k + \varphi_{k-1} + \varphi_{k+1}.$$

(18.19)

If $g = 0$, the explicit solution of the equation of motion is given by

$$\varphi_k(t) = \sum_{n}^{L} \exp[ip_n k]\tilde{\varphi}(p_n, t) \qquad p_n = \frac{n}{L} 2\pi$$

$$\Pi_k(t) = \sum_{n}^{L} \exp[ip_n k]\tilde{\Pi}(p_n, t)$$

$$\tilde{\varphi}(p_n, t) = \tilde{\varphi}(p_n, 0) \cos[\omega(p_n)t] + \tilde{\Pi}(p_n, 0) \frac{\sin[\omega(p_n)t]}{\omega(p_n)}$$

(18.20)

$$\tilde{\Pi}(p_n, t) = -\tilde{\varphi}(p_n, 0)\omega(p_n) \sin[\omega(p_n)t] + \tilde{\Pi}(p_n, 0) \cos[\omega(p_n)t]$$

$$\omega(p_n) = [m^2 + 2 - 2\cos(p_n)]^{1/2}.$$

The quantities $\omega^2(p_n)\tilde{\varphi}(p_n, t)^2 + \tilde{\Pi}^2(p_n, t)$ are constants of motion. The system thus has $2L$ degrees of freedom and L constants of motion and is thus quite far from being ergodic.

The KAM theorem tells us that for a given initial configuration for g small enough the system will be far from being ergodic, while for g high enough the system is expected to be ergodic. This qualitative analysis is confirmed by the data coming from numerical simulations. It turns out that for a given initial configuration we can define with small ambiguity an ergodicity threshold g_T; for $g < g_T$ the system will not be ergodic for most of the g's, and for $g > g_T$ the system will most likely be ergodic. The ergodicity threshold seems to depend mainly on the energy.[4] If $g_T \to 0$, when L goes to infinity at fixed energy density E/L, the use of the canonical distribution in the infinite volume can be justified; if g_T remains distant from zero, the situation is more complex. It is also possible that

18.2. The KAM Theorem

the existence of a fixed number of globally conserved quantities does not destroy the canonical distribution at infinite volume. The question is quite difficult to decide, either analytically or by doing numerical simulations for large values of L. We argue here that the system will actually go to the canonical distribution (and g_T to zero), if the volume is large enough, but the time needed may be very large.[5]

We first note that in the infinite volume many simplifications are possible even in the free case. If we look at the distribution probability of φ_k ($dP(\varphi_k)$) in a finite-length chain, $dP(\varphi_k)$ will have a rather complicated shape depending crucially on the initial data; on the other hand, in the infinite-volume limit one obtains a Gaussian distribution,

$$dP(\varphi_k) = \left(\frac{1}{2\pi\langle\varphi^2\rangle}\right)^{1/2} \exp[-\tfrac{1}{2}\varphi_k^2/\langle\varphi^2\rangle]\, d\varphi_k, \qquad (18.21)$$

where $\langle\varphi^2\rangle$ depends on the initial data. This striking result is rather easy to prove. If we look at Eq. (18.20), we see that $\varphi_k^2(t)$ is written as the sum of terms with different phases. If t is large enough, $\varphi_k(t)$ will be the sum of L terms with random phase ($\exp[i\omega(p_n)t]$ becomes a random number) and it will have a Gaussian distribution as a consequence of the central limit theorem.

This result can be easily understood. At $g=0$, H describes the motion of waves of different frequency; the value of $\varphi_k(t)$ can be written as the sum over waves that come from different points of the space. If there were no long-range correlations in the initial state, $\varphi_k(t)$ can be written as the sum of independent contributions; consequently we have a Gaussian distribution. Now we have already seen that in order to get beyond the region of validity of the KAM theorem the perturbation must be large. The relative weight of H_0 and H_1 is given, roughly speaking, by

$$\frac{H_1}{H_0} \sim \frac{(g/6)\varphi_k^2}{m^2+\omega^2(p_n)} \sim O\left(\frac{g\varphi_k^2}{m^2+1}\right). \qquad (18.22)$$

In a finite system, the distribution of φ_k is bounded and the ratio H_1/H_0 will be small for small g. On the other hand, in the infinite-volume limit φ_k^2 has an unbounded distribution at $g=0$; the probability that the ratio in (18.22) will be greater than 1 can be estimated to be

$$P \simeq \exp\left[-\frac{m^2+1}{g\langle\varphi^2\rangle}\right]. \qquad (18.23)$$

In other words, for any arbitrarily small value of g, the perturbation will become arbitrarily large if we wait a sufficiently long time. These events with large $g\varphi^2$ should play the same rôle as collisions between hard spheres in destroying the ergodicity of the system and pushing it toward the microcanonical or canonical distribution. The time needed for having such an event will be proportional to $P^{-1} \sim \exp[(m^2+1)/g\langle\varphi^2\rangle]$. The prediction of this qualitative analysis is that the time needed to reach equilibrium will become, in the infinite system, exponentially large $[O(\exp(A/g))]$ for small g, suggesting that much care should be used in the analysis of computer data for small g and not too large times. The situation is not yet fully clarified, and much work is still needed in this field.

Notes for Chapter 18

1. O. E. Lanford, *Commun. Math. Phys.* **11**, 257 (1969); Ya. G. Sinai, *Theor. Math. Phys.* **11**, 248 (1972); C. Marchioro, A. Pellegrinotti, and E. Presutti, *Commun. Math. Phys.* **40**, 195 (1975).
2. For a precise definition of these very important concepts the reader is referred to two very nice review articles: J. L. Lebowitz and O. Penrose, *Phys. Today* **26** (Feb.), 23 (1973); J. L. Lebowitz in *Statistical mechanics, new concepts, new problems, new applications*, ed. by S. A. Rice et al., The University Press, Chicago (1972).
3. A careful and interesting presentation of the KAM theory can be found in G. Gallavotti in *Scaling and Self-Similarity in Physics*, ed. by J. Frohlich, Birkhauser, Boston (1983).
4. G. Bellettin, L. Galgani, and J. Strelcyn, *Phys. Rev. A* **19**, 2338 (1976).
5. This point has been discussed in F. Fucito, E. Marinari, F. Marchesoni, G. Parisi, L. Peliti, S. Ruffo, and A. Vulpiani, *J. Phys. (Paris)* **43**, 707 (1982).

CHAPTER 19

The Stochastic Approach

19.1. Brownian motion

In the preceding chapter we considered the dynamics of an isolated Hamiltonian system. The opposite situation is given by a system that remains in strong thermal contact with another system at thermal equilibrium. The prototype of such behavior is provided by small particles that interact among themselves with a given force (which may be electrical, magnetic, gravitational, etc., in origin) suspended in a highly viscous liquid (i.e., oil) at a given temperature. In the high-viscosity limit the equation of motion of the ith particle is

$$\frac{dx_i}{dt} = \frac{1}{\eta} F_i, \qquad (19.1)$$

where η is proportional to the viscosity. In many cases the force F_i is given by

$$F_i = -\frac{\partial}{\partial x_i} V(x), \qquad (19.2)$$

where $V(x)$ is the potential energy of the configuration; for example,

$$V(x) = \sum_{i}^{N} V_1(x_i) + \sum_{i,j} \frac{1}{2} V_2(x_i, x_j), \qquad (19.3)$$

where we have assumed the existence of only one- and two-body forces (V_1 may be the gravitational force). It is clear that Eq. (19.1) cannot be correct and something is missing. It implies that

$$\lim_{t \to \infty} x_i(t) = x_i^{(m)}, \qquad (19.4)$$

where $x_i^{(m)}$ are the coordinates of a minimum of $V(x)$, while we know that at large times the distribution probability of the x_i's should be the canonical one,

$$dP(x) \propto \exp[-\beta V(x)]\, dx_i\,, \tag{19.5}$$

where $(k\beta)^{-1}$ is the temperature of the oil. The missing term in Eq. (19.1) is the force acting on the particles due to the collisions of the molecules of oil: they produce an extra random force $b_i(t)$, which should be added to (19.2). We have

$$\dot{x}_i = -\frac{1}{\eta}\frac{\partial V}{\partial x_i} + b_i(t) \equiv \frac{1}{\eta} F_i(x) + b_i(t)\,. \tag{19.6}$$

This new force prevents the particles from remaining at the minimum position $(x^{(m)})$, and it is the origin of the Brownian motion. Let us study in detail the nature of this force. It is convenient to consider the quantities

$$B_i^\varepsilon(t) = \int_t^{t+\varepsilon} b_i(t')\, dt'\,, \tag{19.7}$$

which are the total forces due to collisions with the molecules in a time interval ε. In the simplest case, each collision gives a contribution to the force, and contributions coming from different collisions are uncorrelated: the forces acting on the same particle at different times (or on different particles at the same time) are practically uncorrelated. We can thus write

$$\begin{aligned}\overline{B_i^\varepsilon(t_1)B_i^\varepsilon(t_2)} &= 0\,, \quad |t_1 - t_2| > \varepsilon \\ \overline{B_i^\varepsilon(t_1)B_j^\varepsilon(t_2)}\,, &= 0\,, \quad i \neq j\,,\end{aligned} \tag{19.8}$$

where the bar denotes the average over many repeated experiments: i.e., we have M identical copies of the same system (or a single system on which M measurements are taken at widely separated time intervals) and we average the $B_i^\varepsilon(t)$'s over these M replicas. The bar denotes the average when M goes to infinity. If the number of collisions in a time ε is quite large, the $B_i^\varepsilon(t)$ will be Gaussian-distributed variables with a variance

$$\overline{(B_i^\varepsilon(t))^2} \equiv 2A\varepsilon\,. \tag{19.9}$$

In the physical world Eq. (19.9) holds only for not too small ε; for practical reasons it is convenient to suppose that Eq. (19.9) is correct for any ε arbitrarily small. If this happens, the correlations of the $b_i(t)$ are

19.1. Brownian Motion

given by

$$\overline{b_i(t_1)b_j(t_2)} = 2\delta_{ij}\delta(t_1 - t_2)A, \qquad (19.10)$$

the $b_i(t)$'s being Gaussian-distributed variables with zero mean. Equation (19.10) implies that the probability distribution of the $b_i(t)$'s is

$$dP[b_i] \propto \prod_1^N d[b_i] \exp\left[-\frac{1}{2}\sum_1^N \int \frac{b_i^2(t)}{2A}\right]. \qquad (19.11)$$

If the probability distribution of the b's is given by Eqs. (19.10) and (19.11), Eq. (19.6) is the so-called Langevin equation,[1] which is usually written by mathematicians as

$$dx_i = \frac{1}{\eta} F_i \, dt + dB_i(t)$$

$$\langle dB_i(t) \, dB_j(t') \rangle = 2A\delta_{ij} \, dt. \qquad (19.12)$$

Insofar as the $b_i(t)$ are not good smooth functions [as can be seen by Eq. (19.10)], Eq. (19.6) may be ambiguous; we shall follow the procedure of substituting for (19.6) the regularized forward equation

$$x_i(t_{n+1}) = x_i(t_n) + \frac{\varepsilon}{\eta} F_i(x(t_n)) + B_i^\varepsilon(t_n) \qquad (19.13)$$

and only at the end perform the limit $\varepsilon \to 0$.[2] It is impossible to solve Eq. (19.6) in the usual sense, i.e., to find the trajectory $x_i(t)$, because the term $b_i(t)$ on the r.h.s. of (19.6) is not known—only its probability distribution is known. Equation (19.6) may be solved only in a probabilistic sense. We must find $x_i(t)$ as an explicit functional of the noise $b_i(t)$ and compute the probability distribution of the x_i's induced by the probability distribution of the b's; in order to be more precise we define $P(x \mid t)$ by

$$\int dx \, P(x \mid t) g(x) = \int dP[b] g(x^{[b]}(t)) \equiv \overline{g(x^{[b]}(t))}, \qquad (19.14)$$

where g is a generic function and $x^{[b]}(t)$ is the b-dependent solution of Eq. (19.6). We could also write

$$P(x \mid t) = \left[\int dP[b] \prod_1^N \delta(x_i - x_i^{[b]})\right]. \qquad (19.15)$$

Of course, the probability $P(x \mid t)$ will depend on the initial condition at $t = 0$.

Let us work out the case $V = 0$ here; the general case will be studied in the next section. The differential equation (19.6) can be easily integrated:

$$x_i(t) = x_i(0) + \int_0^t b_i(t')\, dt' . \tag{19.16}$$

$x_i(t)$ is the sum of Gaussian variables. It has a Gaussian distribution given by

$$P(x_i, t) = \prod_i^N \left(\frac{1}{4\pi tA}\right)^{1/2} \exp\left[-\frac{(x_i - x(0))^2}{4tA}\right], \tag{19.17}$$

as can be checked by computing the variance

$$\overline{(x_i(t) - x_i(0))^2} = \int_0^t dt_1 \int_0^t dt_2 \, \overline{b_i(t_1) b_i(t_2)}$$

$$= 2tA . \tag{19.18}$$

The probability P satisfies the differential equation

$$\frac{\partial P}{\partial t} = A \sum_i^N \left(\frac{\partial}{\partial x_i}\right)^2 P , \tag{19.19}$$

with the boundary condition

$$P(x \mid 0) = \prod_i \delta(x_i - x_i(0)) . \tag{19.20}$$

Equation (19.19) can also be derived from Eq. (19.13). We have

$$P(x \mid t_{n+1}) = \overline{P(x + B^\varepsilon t_n \mid t_n)}$$

$$\simeq P(x \mid t_n) + \sum_i^N \overline{B_i^\varepsilon(t_n)} \frac{\partial P}{\partial x_i} + \frac{1}{2} \sum_i^N \sum_j^N \overline{B_i^\varepsilon(t) B_j^\varepsilon(t)} \frac{\partial^2 P}{\partial x_i \partial x_j}$$

$$+ O((B^\varepsilon)^3) , \tag{19.21}$$

which implies, after the averages over the B's,

$$\frac{P(x \mid t_{n+1}) - P(x, t_n)}{\varepsilon} = A \sum_i^N \frac{\partial^2 P}{\partial x_i^2} + O(\varepsilon^{1/2}) . \tag{19.22}$$

In other words, B^ε is not of order ε but of order $\varepsilon^{1/2}$, so that one must be quite careful in evaluating the derivatives [$(dx)^2$ is of order dt].

The trajectory $x(t)$ is essentially a random walk defined on the continuum. The typical distance from the origin increases like $t^{1/2}$.

19.2. The Langevin and Fokker-Planck equations

We now study the Langevin equation (19.12). We want to find the associated differential equation for the probability, the so-called Fokker-Planck (FP) equation. If the F_i's are zero, we get Eq. (19.19); if the b_i's are zero ($A=0$), we have the "continuity" equation,

$$\frac{\partial P}{\partial t} = -\sum_{i}^{N}\frac{\partial}{\partial x_i}\left[\frac{F_i}{\eta}P\right], \qquad (19.23)$$

which is the N-dimensional generalization of the usual continuity equation $\dot{\rho} = -\text{div}(\vec{v}\rho)$. If both terms are present we obtain

$$\frac{\partial P}{\partial t} = \sum_{i}\left\{-\frac{\partial}{\partial x_i}\left(\frac{F_i}{\eta}P\right) + A\frac{\partial^2}{\partial x_i^2}P\right\}, \qquad (19.24)$$

as the reader can easily check;[3] the r.h.s. of (19.24) is just the linear combination of the r.h.s. of Eqs. (19.19) and (19.23). We expect that at very large times the probability distribution P becomes the canonical distribution; this may be possible only if the r.h.s. of (19.24) is zero when $P \sim \exp[-\beta V]$. If equilibrium is asymptotically reached at large times $\dot{P}$ must be zero. We easily obtain

$$\sum_{i}^{N}\left\{\frac{\partial}{\partial x_i}\left[-\frac{F_i}{\eta} + A\frac{\partial}{\partial x_i}\right]\exp[-\beta V(x)]\right\}$$

$$= \sum_{i}^{N}\left\{\frac{\partial}{\partial x_i}\left[\left(-\frac{F_i}{\eta} - A\beta\frac{\partial V}{\partial x_i}\right)\exp[-\beta V(x)]\right]\right\}. \qquad (19.25)$$

We find that Eq. (19.25) is zero (we recall that $F_i = -\partial V/\partial x_i$) if

$$\eta A \beta = 1. \qquad (19.26)$$

Equation (19.26) is essentially the celebrated Einstein relation between the constants of the Brownian motion and the viscosity. In the following, for simplicity we shall set $A = \eta = 1$ and absorb β in the definition of V, so that the FP equation becomes

$$\frac{\partial P}{\partial t} = \sum_{i}^{N}\left\{\frac{\partial}{\partial x_i}\left(\frac{\partial V}{\partial x_i}P\right) + \frac{\partial^2}{\partial x_i^2}P\right\}$$

$$= \sum_{i}^{N}\left\{\frac{\partial^2 V}{\partial x_i^2}P + \frac{\partial V}{\partial x_i}\frac{\partial P}{\partial x_i} + \frac{\partial^2 P}{\partial x_i^2}\right\}. \qquad (19.27)$$

Now we must prove that the solution of Eq. (19.27) actually goes to the canonical distribution for large times (this may happen only if V goes rapidly enough to infinity that $\exp(-V)$ is a normalizable function). The condition (19.26) implies only that the canonical distribution is a stationary distribution. Fortunately here the approach to equilibrium is completely under control. Equation (19.27) is quite similar to the Schrödinger equation for imaginary time, apart from the presence of a term proportional to $(\partial V/\partial x_i)(\partial P/\partial x_i)$. This term may be eliminated by introducing the function ρ defined by

$$P(x\mid t) = \psi_0(x)\rho(x\mid t)$$
$$\psi_0(x) \propto \exp[-\tfrac{1}{2} V(x)] \,.$$
(19.28)

The function $\rho(x\mid t)$ satisfies the differential equation

$$\frac{\partial \rho}{\partial t} = \left[\sum_{1}^{N} \frac{\partial^2}{\partial x_i^2} - U_{FP}(x)\right]\rho \equiv -\hat{\mathcal{H}}_{FP}\rho$$

$$U_{FP}(x) = \sum_{1}^{N}{}_i \left[\frac{1}{4}\left(\frac{\partial V}{\partial x_i}\right)^2 - \frac{1}{2}\frac{\partial^2 V}{\partial x_i^2}\right].$$
(19.29)

$\hat{\mathcal{H}}_{FP}$ is now a Schrödinger-type operator. If $V(x)$ goes rapidly to infinity, at infinity $U_{FP}(x)$ also goes to infinity and $\hat{\mathcal{H}}$ has a discrete spectrum. It is easy to check that $\psi_0(x)$ is an eigenfunction of $\hat{\mathcal{H}}$,

$$\hat{\mathcal{H}}|\psi_0\rangle = 0$$
(19.30)

with zero eigenvalue.

Now $\psi_0(x)$ is a non-negative function; consequently it must be the ground state of $\hat{\mathcal{H}}$: All other eigenvalues E_n ($\hat{\mathcal{H}}|\psi_n\rangle = E_n|\psi_n\rangle$) are positive. The solution of Eq. (19.27) is given by

$$\rho(x\mid t) = \sum_n \psi_n(x) c_n \exp[-E_n t]$$
$$c_n = \int dx\, \psi_n(x)\rho(x\mid 0) \,,$$
(19.31)

or in Dirac's notation

$$|\rho(t)\rangle = \sum_n \exp(-E_n t)\langle \psi_n\mid \rho\rangle |\psi_n\rangle \,.$$
(19.32)

When t goes to infinity all terms (but $n = 0$) go exponentially to zero: We

19.2. The Langevin and Fokker-Planck Equations

obtain

$$\rho(x|t) \xrightarrow[t \to \infty]{} \psi_0(x) c_0$$

$$c_0 = \int dy\, \psi_0(y) \rho(y|0) = \int dy\, P(y|0) = 1 \qquad (19.33)$$

$$P(x|t) \xrightarrow[t \to \infty]{} \psi_0^2(x),$$

where we have used the property $\int dy\, P(y|0) = 1$ [the probability distribution $P(y|0)$ at $t = 0$ is normalized]. It is also easy to see that the normalization of the probability is conserved by Eq. (19.27),

$$\int dx\, P(x|t) = \int dx\, \psi_0(x) \rho(x|t) = c_0 = 1, \qquad (19.34)$$

where we have again used the orthonormality condition of the eigenstates of $\hat{\mathcal{H}}$:

$$\langle \psi_n | \psi_{n'} \rangle = \int dx\, \psi_n(x) \psi_{n'}(x) = \delta_{n,n'}. \qquad (19.35)$$

We have proven that the probability distribution generated by the Langevin equation (19.12) goes to the canonical distribution. Therefore we have

$$\langle g(x) \rangle = \lim_{t \to \infty} \overline{g[x(t)]}, \qquad (19.36)$$

where the brackets denote the average over the canonical distribution. However, Eq. (19.36) is a statement that involves the average over many trajectories. We may ask what happens if we pick a typical trajectory and we consider

$$\lim_{t \to \infty} \frac{1}{t} \int_0^t f[x(t')]\, dt'. \qquad (19.37)$$

We argue now that the expression in Eq. (19.37) is equal, with probability 1, to $\langle f \rangle$. Indeed, we can divide the time interval t into N intervals of length $t_0 \simeq t^{1/2}$ with $N \simeq t^{1/2}$. When t goes to infinity, both t_0 and N also go to infinity; the expression in Eq. (19.37) can be considered as the average of N independent trajectories of length t_0, and because of Eq. (19.36) it must be equal to $\langle f \rangle$.

These considerations may be sharpened if we notice that

$$\overline{\frac{1}{t} \int_0^t g[x(t')]\, dt'} \simeq \langle g \rangle \qquad (19.38)$$

for large t; this equation is an evident consequence of Eq. (19.36). It is not too difficult to estimate the variance of the quantity defined in Eq. (19.37),

$$\overline{\left(\frac{1}{t}\int_0^t dt'\, g(x(t')) - \frac{1}{t}\int_0^t g(x(t'))\, dt'\right)^2}$$
$$= \frac{1}{t^2}\int_0^t dt_1 \int_0^t dt_2\, [\overline{g(x(t_1))g(x(t_2))} - \overline{g(x(t_1))}\,\overline{g(x(t_2))}]. \quad (19.39)$$

The integral in Eq. (19.39) is of order $1/t$. Indeed, for large t_1 and $t_2 - t_1$,

$$\overline{g(x(t_1))g(x(t_2))} = \sum_n |\langle \psi_0|\hat{g}|\psi_n\rangle|^2 \exp[-|t_1 - t_2|E_n]_{(t_1-t_2)\to\infty}$$
$$\sim |\langle \psi_0|\hat{g}|\psi_0\rangle|^2 = \bar{g}^2 = \langle g\rangle^2. \quad (19.40)$$

More precisely, Eq. (19.39) is for large t given by G/t, where

$$G = \sum_1^\infty \frac{2|\langle \psi_0|\hat{g}|\psi_n\rangle|^2}{E_n}. \quad (19.41)$$

A similar argument for the higher powers of Eq. (19.38) (which we leave to the reader as an exercise) shows that

$$\langle g\rangle - \frac{1}{t}\int_0^t g(x(t'))\, dt' \quad (19.42)$$

has for large times a Gaussian probability distribution with variance G/t, so that (19.42) goes to zero with probability 1. The reader should notice that while in the case of the ensemble distribution the approach to equilibrium was exponential, in the case of a single trajectory the expression in (19.37) differs from the asymptotic value by a random number proportional to $(G/t)^{-1/2}$.

19.3. An example

As usual, our favorite example will be a scalar field theory, the Hamiltonian H being given by Eq. (17.36). The number of degrees of freedom is now infinite; the associated Langevin equation is

$$\frac{\partial}{\partial t}\varphi(x,t) = -\left(-\Delta_x + m^2 + \frac{g}{6}\varphi^2(x,t)\right)\varphi(x,t) + b(x,t), \quad (19.43)$$

where the $b(x,t)$ are random Gaussian variables with variance

19.3. An Example

$$\overline{b(x, t)b(x', t')} = 2\delta(x - x')\delta(t - t'). \quad (19.44)$$

Equations (19.43) and (19.44) can be obtained by applying the formalism of the previous subsection to the lattice regularized theory and sending the lattice spacing to zero. Of course in two or more dimensions ultraviolet divergences are present if $g \neq 0$. In order to avoid this kind of problem here we consider only the one-dimensional case (or add a cutoff in the Hamiltonian). We stress that the time we introduce here has nothing to do with the time of continuation to Minkowsky space, but for a four-dimensional theory it is a fifth-dimensional time. If $g = 0$, Eq. (19.43) can be easily solved:

$$\varphi(x, t) = \int_0^t dt' \int d^D y \, \mathcal{D}(x - y, t - t') b(y')$$

$$\frac{\partial}{\partial t} \mathcal{D}(x, t) = -(-\Delta + m^2) \mathcal{D}(x, t)$$

$$\mathcal{D}(x, 0) = \delta^D(x) \quad (19.45)$$

$$\mathcal{D}(x, t) = \frac{1}{(2\pi)^D} \int d^D p \, \exp[-t(p^2 + m^2) - ipx]$$

$$= \frac{1}{(4\pi t)^{D/2}} \exp\left(-\frac{x^2}{4t} - m^2 t\right),$$

where for simplicity we have assumed that

$$\varphi(x, 0) = 0. \quad (19.46)$$

The equal-time correlation function can be easily computed in momentum space,

$$G(p, t) = \int dx \, \overline{\varphi(x, t)\varphi(0, t)} \exp(ipx)$$

$$= \frac{1}{p^2 + m^2} \{1 - \exp[-2t(p^2 + m^2)]\}, \quad (19.47)$$

where we have used the relation

$$\overline{\varphi(x, t)\varphi(0, t)} = \int_0^t dt_1 \int_0^t dt_2 \int d^D y_1 \int d^D y_2$$

$$\times \mathcal{D}(x - y_1, t - t_1) \mathcal{D}(-y_2, t - t_2) \overline{b(y_1, t_1) b(y_2, t_2)}$$

$$= \int_0^t dt_1 \int d^D y_1 \, 2\mathcal{D}(x - y_1, t - t_1) \mathcal{D}(-y_1, t - t_1). \quad (19.48)$$

As usual $\varphi(x, t)$ is the sum of Gaussian variables, and it is Gaussian distributed. If g is different from zero, a perturbative expansion in powers of g can be constructed. Equation (19.42) can be written as

$$\varphi(x, t) = \int_0^t dt' \int d^D y \, \mathcal{D}(x - y, t - t') \\ \times \left[b(y, t') - \frac{g}{6} \varphi^3(y, t') \right]. \quad (19.49)$$

In this way at a finite order in g, φ is a polynomial of the b's; the average over the b's can be easily taken using Eq. (19.44). A diagrammatic representation of this expansion can be obtained. With a certain amount of work it is possible to check explicitly at the diagrammatic level that

$$\overline{\varphi(x, t)\varphi(0, t)} \xrightarrow[t \to \infty]{} \langle \varphi(x)\varphi(0) \rangle, \quad (19.50)$$

as a power expansion in g, as it should be from general considerations.[4]

Equation (19.43) is often used in the framework of statistical mechanics in order to compute the time-dependent correlation functions of ferromagnets near the critical point. It is possible to define a dynamical exponent which tells us how slowly the correlation functions go to equilibrium near the critical temperature. The interested reader can find a vast literature on this subject.[5]

19.4. Nelson's quantum mechanics

In this section we present in a modified and simplified version some of Nelson's ideas on the stochastic formulation of quantum mechanics:[6] We limit ourselves to the study of the quantum theory at imaginary time of a single particle in a potential $V(x)$.[7]

The correlation functions of the position $\omega(t)$ (which are the imaginary-time continuation of expectation values of the time-ordered products in the ground state) can be written for $\hbar = 1$ as [cf. Eqs. (13.21) and (13.26)]

$$\langle \omega(t_1) \cdots \omega(t_n) \rangle = \frac{\int d[\omega]\omega(t_1) \cdots \omega(t_n) \exp(-S[\omega])}{\int d[\omega] \exp[-S[\omega]]}$$

$$= \langle \psi_0 | \hat{q} \exp[-(t_2 - t_1)\hat{\mathcal{H}}] \hat{q} \cdots \hat{q}$$

$$\times \exp(-(t_n - t_{n-1})\hat{\mathcal{H}} | \psi_0 \rangle, \quad t_k > t_{k-1}. \quad (19.51)$$

Nelson's modified proposal is to write them as

19.4. Nelson's Quantum Mechanics

$$\langle \omega(t_1) \cdots \omega(t_n) \rangle = \overline{x(t_1) \cdots x(t_n)}, \qquad (19.52)$$

where $x(t)$ satisfies a Langevin-type equation:

$$\dot{x}(t) = f(x) + \eta(t), \qquad f(x) \equiv -\frac{dU}{dx}, \qquad \overline{\eta(t)\eta(t')} = \delta(t-t'), \qquad (19.53)$$

for $-\infty < t < +\infty$.

Which U must be taken? If we limit ourselves to the equal-time expectation values we see that

$$\overline{(x(t))^n} = \frac{\int dx \exp(-U(x)) x^n}{\int dx \exp(-U(x))}$$

$$\langle \omega^n(t) \rangle = \langle \psi_0 | \hat{q}^n | \psi_0 \rangle = \int dx \, \psi_0^2(x) x^n \qquad \left(\int \psi_0^2(x) = 1 \right). \qquad (19.54)$$

Equation (19.52) may be correct only if

$$\psi_0^2(x) \propto \exp(-U(x)), \qquad U(x) = -2 \ln \psi_0(x) + \text{constant}. \qquad (19.55)$$

The potential U is fixed by the equal-time expectation values. The real surprise is that with this choice of U, Eq. (19.51) is also correct at unequal times. Indeed, if Eq. (19.55) is satisfied, the associated Fokker-Planck Hamiltonian [Eq. (19.29)] is given by[8]

$$H_{\text{PF}} = -\left(\frac{d}{dx}\right)^2 + U_{\text{FP}}(x), \qquad U_{\text{FP}}(x) = \frac{1}{4}\left(\frac{d}{dx} U\right)^2 - \frac{1}{2} \frac{d^2}{dx^2} U$$

$$= -\frac{d\psi_0(x)/dx^2}{\psi_0(x)} = V(x) - E_0. \qquad (19.56)$$

Using Eq. (19.31), we obtained the announced result.

We see that "by miracle" the imaginary-time quantum evolution may be reformulated in the language of stochastic differential equations. Nelson's fundamental paper contains more: the same results can be obtained for the real-time evolution, where the form of the drift f is fixed by the correspondence principle which requires on the average that

$$a = -\frac{dV}{dx} \qquad a = \text{``}\frac{d^2}{dt^2}\text{''} x, \qquad (19.57)$$

with a suitable definition of the second derivative "d^2/dt^2" adapted to nondifferentiable trajectories of the type arising from Eq. (19.52), where the forward and backward velocities (see footnote 3) are different. Nelson's quantum mechanics has recently been recast in the form of a variational principle similar to that used in classical mechanics.

On the interpretation of Nelson's fundamental observation, the physical community is split. Some consider it a beautiful mathematical tool, while others think that Nelson's formulation of quantum mechanics is more fundamental than the other formulations and that it may be the starting point for a new mechanics of which quantum mechanics will be only a particular case.[9] Time will be the judge.

Notes for Chapter 19

1. A very nice book, which collects many original and important contributions is *Selected papers on noise and stochastic processes*, ed. by N. Wax, Dover Publications, Inc., New York (1954); more recent (and more mathematically minded) contributions can be found in "New Stochastic Methods in Physics", ed. by C. DeWitt-Morette and K. D. Elworthy, *Phys. Rep.* 77, n.3, and references therein.

2. The other alternative would be to use the mathematical apparatus that has been developed over the last fifty years to study stochastic differential equations like (19.12), in particular the Ito differential calculus. While the introduction of a calculus is definitely advantageous in the long run, Eq. (19.13) is sufficient for the few applications we study here.

3. It is interesting that although the trajectories are not differentiable, it is possible to define an average forward and backward velocity at point x at time t,

$$V^+ = \lim_{\varepsilon \to 0} \overline{\frac{x(t+\varepsilon) - x(t)}{\varepsilon}}, \qquad V^- = \lim_{\varepsilon \to 0} \overline{\frac{x(t) - x(t-\varepsilon)}{\varepsilon}},$$

where the average is taken over all the trajectories arriving at point x at time t. While from Eq. (19.13) it is evident that $V^+ = \lim_{\varepsilon \to 0} \overline{(F_i(x) + \varepsilon B_n^\varepsilon)} = F_i(x)$, the evaluation of the backward velocity is more involved: the probability of arriving at point x at time t_n depends on the probability distribution of the x'_i at the time t_{n-1} and on B_n^ε. An explicit computation shows that $V_i^- = V_i^+ - (2A/\rho)(\partial \rho/\partial x_i)$. Stated simply, the extra term in V_i^- tells us that it is more likely that the particle will arrive at point x coming from a region of high probability (high ρ) than from a region of low probability. It is remarkable that if we define a mean velocity as $V_i^M = (V_i^+ + V_i^-)/2$, Eq. (19.24) becomes $\partial \rho / \partial t = -\Sigma_i \, \partial(V_i^M \rho)$, i.e., the usual continuity equation.

4. C. De Dominicis, *Lett. Nuovo Cimento* 12, 567 (1975); G. Parisi and Wu Yongshi, *Scientia Sinica* 24, 483 (1981); W. Grimus and H. Hüffel, *Z. Phys.* C 18, 129 (1983).
5. A nice review paper is P. C. Hohenberg and B. Halperin, *Rev. Mod. Phys.* 49, 435 (1977).
6. E. Nelson, *Phys. Rev.* 150, 1079 (1966) and *Dynamical Theory of the Brownian Motion*, Princeton University Press, Princeton (1967); F. Guerra, *Phys. Rep.* 77, 263 (1981); F. Guerra and L. M. Morato, *Phys. Rev.* D 27 1774 (1983).
7. Nelson's quantum mechanics is normally formulated at real time; its imaginary-time version is discussed in F. Guerra and P. Ruggero, *Phys. Rev. Lett.* 31, 1022 (1973); G. Jona Lasinio, F. Martinelli, and E. Scoppola, *Phys. Rep.* C 77, 313 (1981).
8. We have used the Schrödinger equation $[(-d^2/dx^2)\psi_0(x) + V(x)]\psi_0(x) = E_0 \psi_0(x)$ for the ground-state wave function.
9. E. Nelson in *Mathematical Physics VII*, ed. by W. E. Brittin *et al.*, North-Holland, Amsterdam (1984).

CHAPTER 20

Computer Simulation

20.1. Molecular dynamics

Up to now in this book we have looked at analytic methods, especially perturbative expansions. However, there are many systems for which the perturbative expansion is very cumbersome to obtain, or for which, for various reasons, the perturbative expansion may be poorly convergent. It may thus be quite useful to have a technique for computing numerically the free energy and the correlation functions, not in the infinite-volume limit (which would be impossible), but in a finite volume, where the number of degrees of freedom retained may be relatively high (e.g., 10^3).[1] This possibility is now offered by present-day technology: chips with a memory of 10^5 words or with a computational power of 10^7 floating-point operations (with 32 bits) are available to us at the price of a few hundred dollars; on larger computers we may have $O(10^6)$ words of memory with $O(10^8)$ floating-point operations (with 64 bits) per second (e.g. Cray-1). Large-scale computations which involve up to 10^{13} floating-point operations are not uncommon in the literature.

The availability of this impressive computational power opens new possibilities which were closed to us in the past. Let us consider a typical example: a system of N particles in a box with reflecting walls, interacting with a given potential

$$H = \sum_{i}^{N} \frac{1}{2} p_i^2 + V(x). \qquad (20.1)$$

If we want to compute the internal energy E as function of the temperature we have two ways, which we hope are equivalent.

(A) We used the statistical mechanics formulation, which states that

$$E = \langle H \rangle = \frac{\int dx\, dp\, \exp[-\beta H] H}{\int dx\, dp\, \exp[-\beta H]}. \qquad (20.2)$$

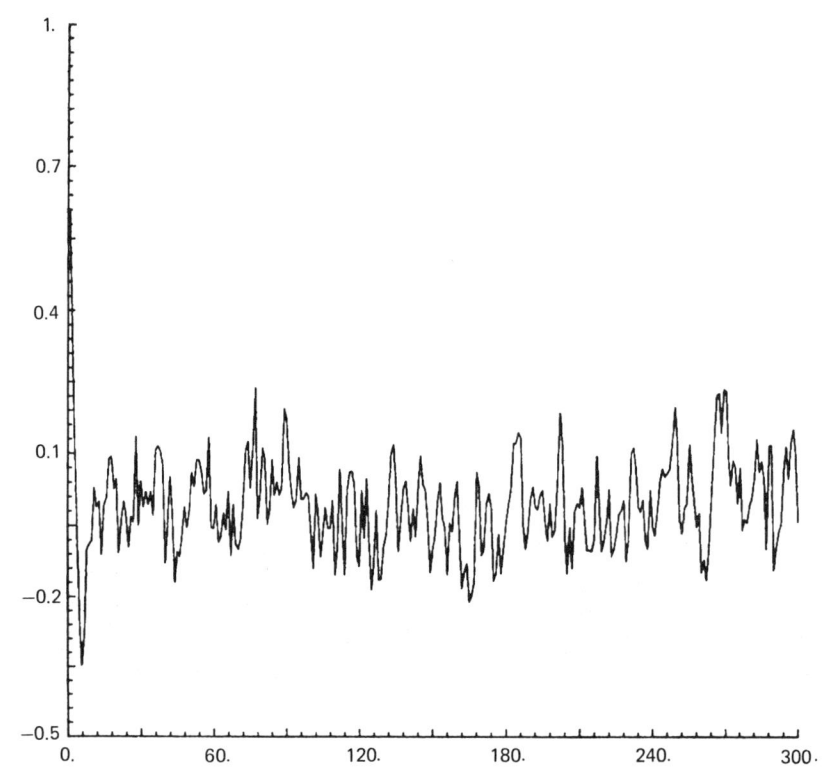

Figure 20.1. The first 300 Monte Carlo sweeps for a two-dimensional Ising model on a 30 × 30 lattice with periodic boundary conditions; in the initial state all spins are equal to 1. The magnetization is plotted as a function of the number of sweeps at $\beta = 0.3$. The value of the instantaneous magnetization density oscillates around zero: the oscillations are rather fast.

(B) We start with an arbitrary configuration at $t = 0$, with given energy E; we solve the equations of motion

$$\dot{x}_i = p_i, \qquad \dot{p}_i = -\frac{\partial V}{\partial x_i} \qquad (20.3)$$

up to a large time t_0. The temperature T is computed using the relation

$$\langle p_i^2 \rangle = (1/2)kT \qquad (20.4)$$

and doing the approximation

$$\frac{1}{N}\sum_i^N \langle p_i^2 \rangle \simeq \frac{1}{t_0} \int_0^{t_0} dt \left[\frac{1}{N} \sum_i p_i^2(t) \right], \qquad (20.5)$$

which is justified for large times t_0.

The physical formulation is clearly (B), while (A) is obviously much better for doing analytical computations, as we have already seen.

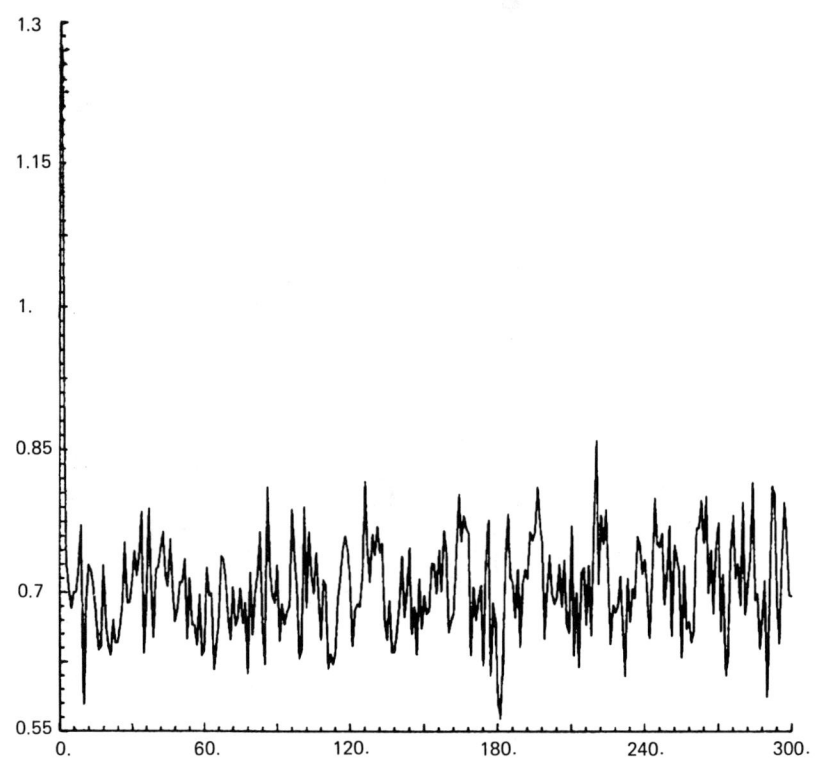

Figure 20.2. As in Fig. 20.1 minus the internal energy at $\beta = 0.3$.

However, if we try to evaluate Eqs. (20.2) and (20.5) numerically by brute force, Eq. (20.5) is much simpler. The integral over the x's in Eq. (20.2) (the p integral is trivial) is an N-dimensional integral. If such an integral is approximately done on a mesh of n points in x space, Eq. 20.2 becomes a sum over n^N terms. In the very rough approximation $n = 2$ we still have 2^N terms, which, for $N \sim 10^3$ or more, is beyond the computational power of any present or future computers. In contrast, if the differential equations (20.5) are replaced by a finite difference equation in times (ε being the spacing in time), the number of operations needed to evaluate (20.5) is of order $N^2 t_0/\varepsilon$, which is much smaller than in the previous case and (if N is not too much larger than 10^3) well within our computer capacities, provided that the time t_0 needed to reach equilibrium is not terribly large. From the point of view of numerical simulations we could say that the work of the founding fathers of statistical mechanics has been to transform a simple problem (B) into a terrible mess (A). Of course we have learned a lot in this process.

Technique (B) has the advantage of allowing the computation of both static and time-dependent correlations. It is very often used in the study of real gases (liquid and solids) with realistic potentials like the Lennard-Jones potential in the monoatomic case,[2] and it is normally referred to as

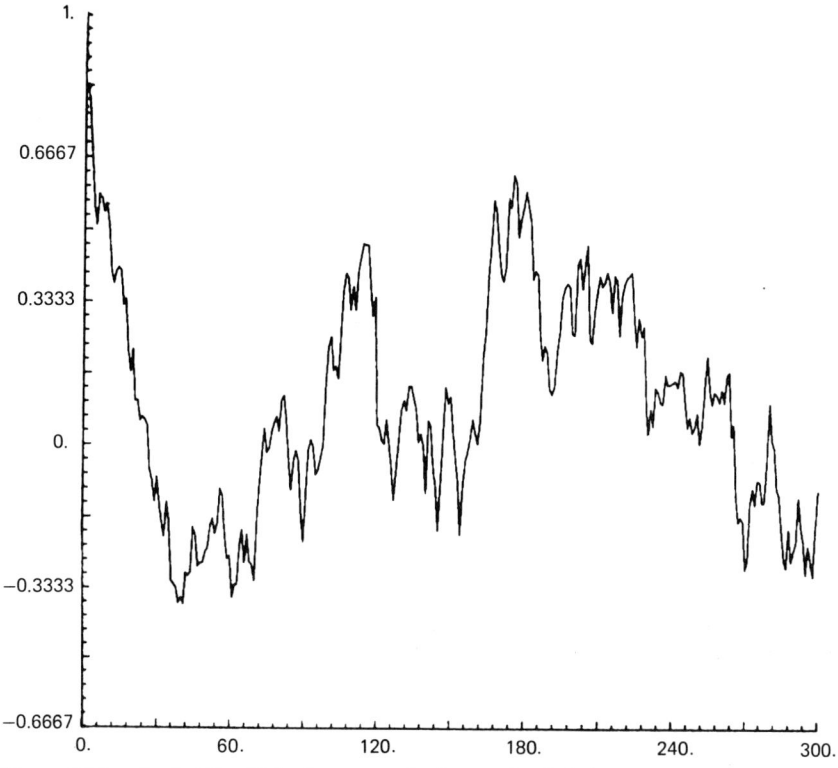

Figure 20.3. As in Fig. 20.1, the magnetization at $\beta = 0.4$; here we are very near the critical temperature and the oscillations in the magnetization are slow and wide.

molecular dynamics.[3] In other cases (like spin models) or Euclidean versions of relativistic quantum field theories it is not so interesting to compute realistic correlations at different times, but only the equal-time equilibrium correlations are needed; here the choice of the dynamics is arbitrary, provided that the large-time probability distribution is the canonical one. Various possibilities are open: e.g., we could use a generator of random numbers[4] to find a particular solution of the Langevin equation [(19.6) or (19.13)] to compute the equilibrium expectation values using Eq. (19.25); the most popular technique, which can also be applied to discrete systems, is the Monte Carlo method, which will be the subject of the next section.

20.2. The Monte Carlo method

We now describe a general method for computing the equilibrium statistical expectation values. This method makes heavy use of random numbers, so it is called the Monte Carlo method[5] (as we shall see, it can also be called the importance sampling method). The strategy is the

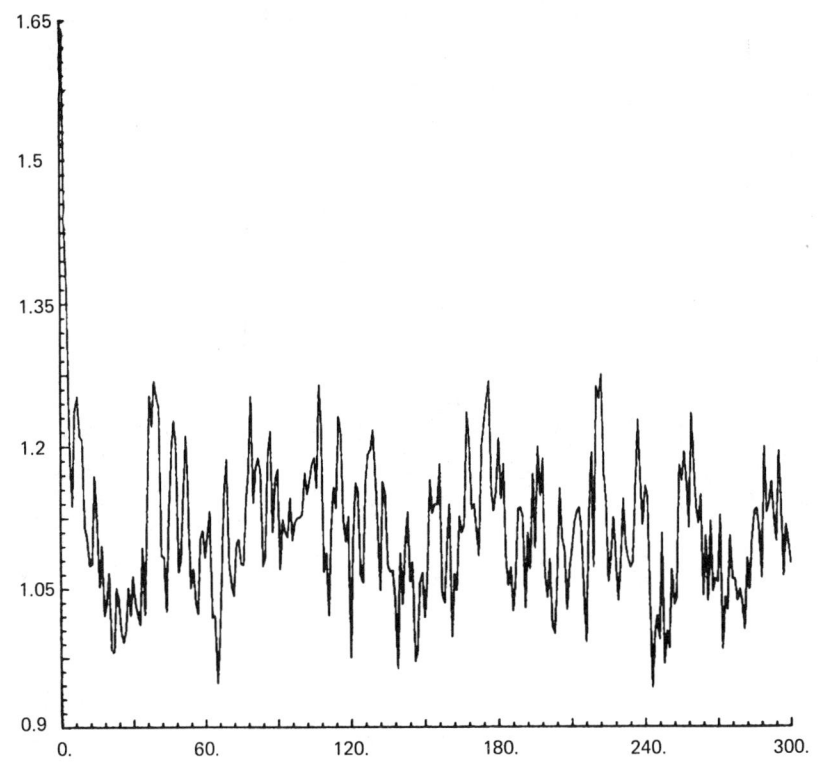

Figure 20.4. As in Fig. 20.1 minus the internal energy at $\beta = 0.4$; here the fluctuations are wider and slower than in Fig. 20.2, but the effect of the transition is not so evident as in Fig. 20.3.

following: if the equilibrium distribution is $\exp[-\beta H(C)]$, where C denotes the configuration of the system, we would like to construct an algorithm to generate a sequence C_n of configurations such that

$$\lim_{N \to \infty} \frac{1}{N} \sum_{n}^{N} A(C_n) = \int d\mu(C) A(C),$$

$$d\mu(C) \propto \exp[-\beta H(C)], \quad \int d\mu(C) = 1.$$

(20.6)

The Newton and Langevin equations [Eqs. (19.6) and (20.1), respectively] provide explicit examples of such an algorithm, the former being deterministic, the latter probabilistic. The algorithms we shall consider are such that the probability of having a given configuration C_n depends only on C_{n-1} and not on the previous history of the system. These algorithms are the discrete equivalences of first-order differential equations: the sequences generated are usually referred to (in the mathematical literature) as Markov chains. These algorithms are completely described by the transition probability TP (C, C'), which tells us the

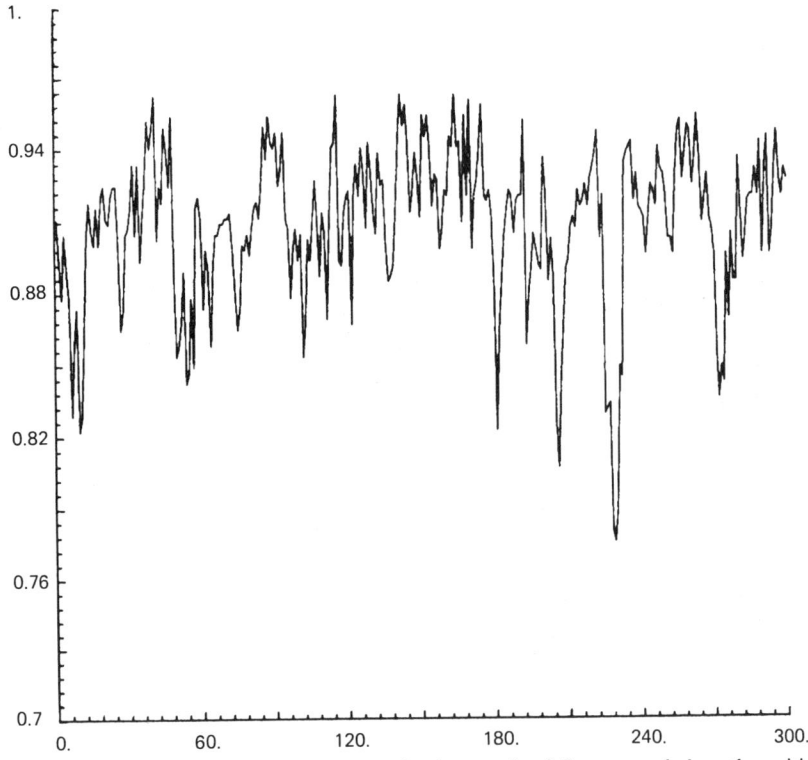

Figure 20.5. As in Fig. 20.1, the magnetization at $\beta = 0.5$; we are below the critical temperature and a spontaneous magnetization is observed.

probability of having $C_{n+1} = C'$ if $C_n = C$.

If the configuration space does not factorize into two or more disconnected parts, i.e., for any C and C' we can find a finite sequence C_i, $i = 1, \ldots, m+1$ such that

$$\text{TP}(C_i, C_{i+1}) \neq 0 \quad i = 1, \ldots, m$$
$$C_1 = C, \quad C_{m+1} = C', \quad (20.7)$$

the algorithm is said to be ergodic. If $\text{TP}(C, C')$ satisfies the so-called detailed balance condition,

$$\text{TP}(C, C') \exp[-\beta H(C)] = \text{TP}(C', C) \exp[-\beta H(C')], \quad (20.8)$$

and the transformation is ergodic, then Eq. (20.6) holds for this algorithm. Before sketching the proof of this statement, we only observe that it is evident that $P_{\text{eq}} \sim \exp[-\beta H(C)]$ is a stationary probability distribution, i.e., if C_n is distributed according to P_{eq}, C_{n+1} also has the same probability distribution, as can be seen by integrating both sides of Eq. (20.8) over C.

If $\rho_n(C)$ is the probability of the system's staying in configuration C

343

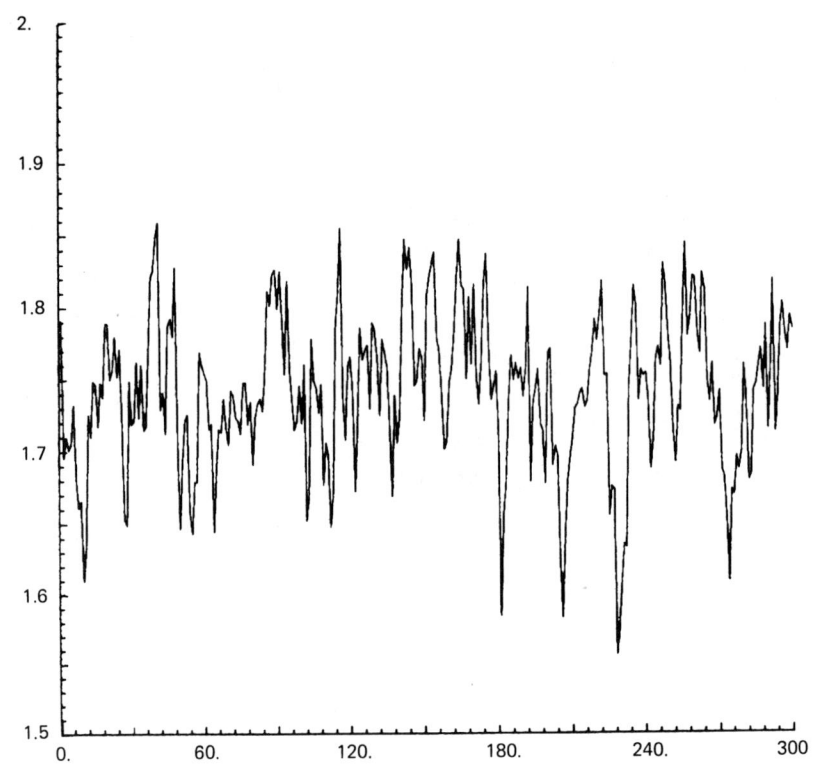

Figure 20.6. As in Fig. 20.1 minus the internal energy at $\beta = 0.5$; the reader may notice that the downward spikes of the magnetization and of the internal energy are clearly correlated.

after n steps [$\rho_0(C)$ is the initial probability distribution], arguments very similar to those employed in Chapter 12 tell us that

$$\rho_n(C) = \hat{T}^n \rho_0(C), \tag{20.9}$$

where $\hat{T}$ is an integral operator acting on the space of states ($\hat{T}$ is a finite-dimensional operator if the space of states is finite), e.g.,

$$\rho_1(C) \equiv \hat{T}\rho_0(C) = \int dC_0 \, \mathrm{TP}(C_0, C_1)\rho(C_0). \tag{20.10}$$

While the transfer matrix is a self-adjoint operator, here $\hat{T}$ does not have this property; fortunately if we define

$$u_n(C) = \rho_n(C) \exp\left[\frac{\beta}{2} H(C)\right], \tag{20.11}$$

(as we have done in Chapter 19), Eq. (20.9) is equivalent to

$$u_{n+1}(C) = \int dC_n \, R(C_n, C_{n+1}) u_n(C_n), \tag{20.12}$$

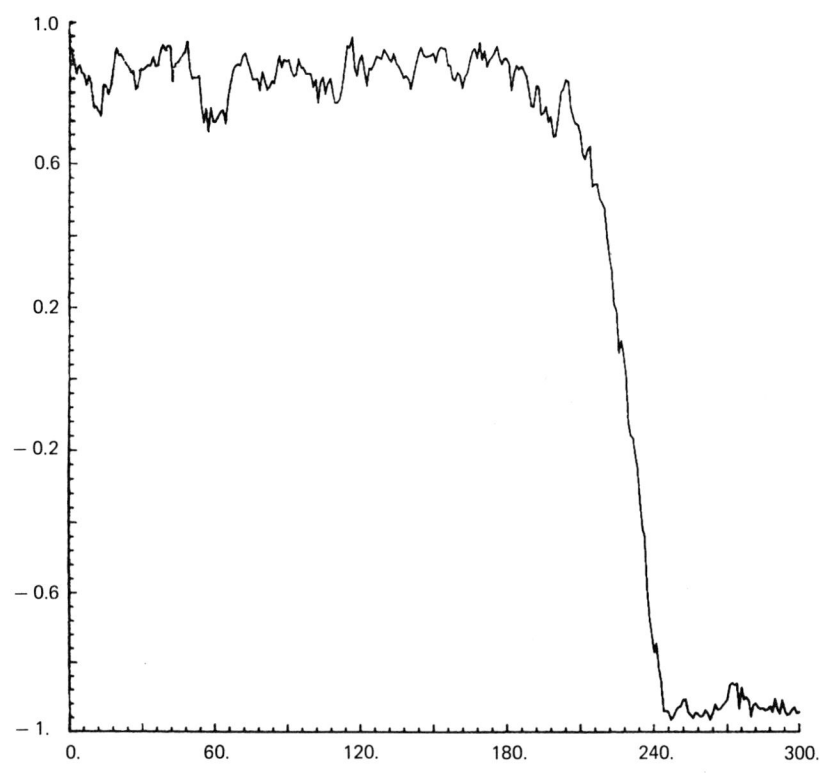

Figure 20.7. As in Fig. 20.1, the magnetization at $\beta = 0.5$, $h = -0.077$ (in the previous cases we had $h = 0$): We see a metastable state decaying into a stable state with negative magnetization.

which can also be written as

$$u_{n+1}(C) = \hat{R}u_n(C) = \hat{R}^{n+1}u_0(C). \tag{20.13}$$

The detailed balance condition (Eq. 20.8) is equivalent to

$$R(C, C') = R(C', C). \tag{20.14}$$

The operator $\hat{R}$ is therefore self-adjoint; when n goes to infinity, $\hat{R}^n$ projects on the states with maximum eigenvalue i.e., 1. The positivity of the matrix elements of $\hat{R}$ and the ergodicity condition (20.7) can be used to show that there is only one eigenvector of R with eigenvalue 1. This eigenvector is obviously[6] $u_{eq}(C) \equiv \exp(-(\beta/2)H(C))$.

This formulation's advantage is that it is very easy to construct algorithms that satisfy condition (20.8). This is done normally in two steps; given a configuration C_n, one extracts a new configuration C_T with a random algorithm characterized by a symmetric transition probability TS,

$$\text{TS}(C_n, C_T) = \text{TS}(C_T, C_n). \tag{20.15}$$

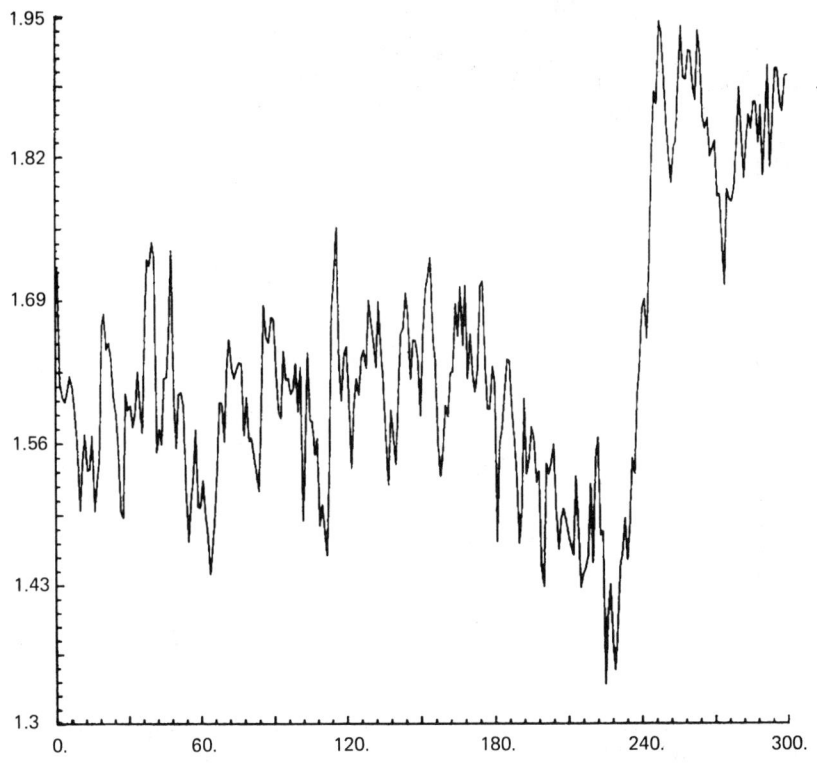

Figure 20.8. As in Fig. 20.1 minus the internal energy at $\beta = 0.5$, $h = -0.077$: The internal energy is lower in the stable state than in the metastable state. Just before the formation of the stable phase we observe an increase of the internal energy, as expected.

In the Ising case the configuration C_T could be the configuration C_n in which a random chosen spin has been flipped. After the ergodic algorithm corresponding to TS has been constructed (TS looks like a random walk in configuration space and it is the same for any choice of the Hamiltonian H), it is quite easy to write the algorithm for TP: we compute $H(C_n) - H(C_T) \equiv \Delta H$, we extract a uniformly distributed random number r in the interval $0 = 1$, and we set

$$C_{n+1} = C_n \quad \text{if } \exp[\beta \, \Delta H] < r$$
$$C_{n+1} = C_T \quad \text{if } \exp[\beta \, \Delta H] > r. \tag{20.16}$$

It is easy to see that this algorithm satisfies the detailed balance conditions. In this way we can generate a sequence of configurations from which we compute equilibrium properties.

In most cases it is not convenient to implement the detailed balance condition, but it is enough to impose only the balance condition,[7] which is sufficient to assure that the probability distribution goes to the canonical

20.2. The Monte Carlo Method

one after many iterations. This happens, for example, if the spins to be flipped are not taken randomly, but are chosen in a sequential way.

The Monte Carlo method is a very powerful investigative tool, especially at the heuristic level, i.e., helping us to know what should be proved using analytical tools.

In Table 20.1 we show a typical Fortran subroutine which performs a Monte Carlo cycle for a two-dimensional Ising model with periodic boundary conditions in a lattice of size $N \times N$ with nearest-neighbor interaction at temperature T with a magnetic field H (a Monte Carlo cycle consisting in applying algorithm (20.10) to each spin of the lattice in sequential order). We warn the reader that we have not attempted to produce an optimized code, and the program in Table 20.1 is relatively slow. For large n the upgrading takes about 150 µs per spin updating on a Vax 11/780; using multispin coding one can go down to 20 µs per spin updating on the same machine. Some applications of this program are shown in Figs. 20.1–8.

Table 20.1 A Fortran program for Monte Carlo simulations of the two-dimensional Ising model

```
C Auxiliary vectors forward and backward are used to impose periodic
C boundary conditions.
C Each spin may take only values plus or minus 1.
         PARAMETER maximal_side=100
         INTEGER forward(maximal_side),backward(maximal_side)
         COMMON /boundaries/ forward,backward
         INTEGER side,spin(maximal_side,maximal_side)
         INTEGER random_seed
         REAL magnetic_field,magnetization_density

         CALL read_input(side,number_of_iterations,beta,magnetic_field)
         CALL get_random_seed(random_seed)
         CALL compute_backward_and_forward(side)
         CALL set_spin_to_1(spin,side)
         DO 1 iteration=1,number_of_iterations
            CALL one_Monte_Carlo_cycle(spin,side,random_seed,
       1    beta,magnetic_field,energy_density,magnetization_density)
            CALL write_output(iteration,energy_density,magnetization_
       1    density)
1        CONTINUE
         END

C ****************************************************************

         SUBROUTINE compute_backward_and_forward(side)
         PARAMETER maximal_side=100
         INTEGER forward(maximal_side),backward(maximal_side)
         COMMON /boundaries/ forward,backward
         INTEGER position,side
```

Table 20.1 (continued)

```
            DO 1 position=1,side
              forward(position)=MOD(position,side)+1
              backward(position)=MOD((position-2+side),side)+1
1           CONTINUE
            RETURN
            END
C  ****************************************************************
            SUBROUTINE set_spin_to_1(spin,side)
            PARAMETER maximal_side=100
            INTEGER side,spin(maximal_side,maximal_side)
            INTEGER x_position,y_position

            DO 1 x_position=1,side
              DO 2 y_position=1,side
                spin(x_position,y_position)=1
2           CONTINUE
1           CONTINUE
            RETURN
            END
C  ****************************************************************
            SUBROUTINE one_Monte_Carlo_cycle(spin,side,random_seed,
1           beta,magnetic_field,energy_density,magnetization_density)
            PARAMETER maximal_side=100
            INTEGER forward(maximal_side),backward(maximal_side)
            COMMON /boundaries/ forward,backward
            INTEGER side,random_seed,spin(maximal_side,maximal_side)
            REAL magnetic_field,magnetization_density
            INTEGER current_spin,sum_of_the_neighbours
            INTEGER x_position,y_position

            total_magnetization=0
            total_energy=0
            DO 5 x_position=1,side
              DO 6 y_position=1,side
                current_spin=spin(x_position,y_position)
                sum_of_the_neighbours=
1               spin(x_position,forward (y_position))+
2               spin(x_position,backward(y_position))+
3               spin(forward (x_position),y_position)+
4               spin(backward(x_position),y_position)
                effective_force=sum_of_the_neighbours+magnetic_field
                IF ( EXP(-beta*effective_force*current_spin*2.)
1               .GT. RAN(random_seed) ) THEN
                   new_spin=-current_spin
                ELSE
                   new_spin=current_spin
                END IF
                spin(x_position,y_position)=new_spin
                total_magnetization=total_magnetization+new_spin
                total_energy=total_energy+
1               (.5*sum_of_the_neighbours+magnetic_field)*new_spin
```

20.2. The Monte Carlo Method

Table 20.1 (continued)

```
C please note the factor .5 multiplying sum_of_the_neighbours
6          CONTINUE
5          CONTINUE
           magnetization_density=total_magnetization/FLOAT(side**2)
           energy_density=total_energy/FLOAT(side**2)
           RETURN
           END
C ****************************************************************
           SUBROUTINE read_input(side,number_of_iterations,beta,magnetic_field)
           REAL magnetic_field
           INTEGER side
           PARAMETER maximal_side=100

1          WRITE(6,*) 'Which is the length of the side ?
    1          (Please less then',maximal_side,')'
           READ(5,*) side
           IF (side .GT. maximal_side .OR. side .LT. 1) THEN
                   WRITE (6,*) 'The value of side is not good'
                   go to 1
           ENDIF
           WRITE(6,*) 'How many iterations?'
           READ(5,*) number_of_iterations
           WRITE(6,*) 'Which is the value of beta ?'
           READ(5,*) beta
           WRITE(6,*) 'Which is the value of the magnetic field?'
           READ(5,*) magnetic_field
           RETURN
           END
C ****************************************************************
           SUBROUTINE get_random_seed(random_seed)
           INTEGER random_seed

           WRITE(6,*)'Which is the random seed?'
           WRITE(6,*) 'Please insert a positive odd number of 7-8 digits'
           READ(5,*) random_seed
           RETURN
           END
C ****************************************************************
           SUBROUTINE write_output(iteration,energy_density,magnetization_
    1                 density)
           REAL magnetization_density

           WRITE (6,*) 'Iteration=        ',iteration
           WRITE (6,*) 'Energy density =       ', energy_density
           WRITE (6,*) 'Magnetization density=      ',magnetization_density
           RETURN
           END
```

Notes for Chapter 20

1. At the critical temperature finite-volume corrections are exponentially small (in most cases), so that a not too large volume may be enough. For computing the critical possible properties it may be possible to use a combination of the renormalization group ideas of the Chapter 7 and computer simulations; see, for example, R. Swendson, *Phys. Rev. Letter*, 52, 1155 (1984) and references therein.
2. For a review of molecular dynamics, see, for example, U. G. Hover, *Ann. Rev. Phys. Chem.* 34, 103 (1983) and *Phys. Today*, 37, 44 (1984).
3. There is a catch: if the system is mixing and two trajectories which are close at the initial time separate exponentially as the time increases, i.e. if $x_1(0) - x_2(0) = \delta$, $x_1(t) - x_2(t) \sim \delta \exp(\alpha t)$, $\alpha > 0$, until $\delta \exp(\alpha t) = O(1)$, the presence of truncation errors forbids the computer to follow a given trajectory for a long time. However, it is believed that the time averages along a trajectory do not depend too much on the very precise choice of the initial point, at least for the systems we consider here.
4. Although digital computers are deterministic machines, it is possible to find deterministic algorithms which "fake" for all practical purposes a random process; for a discussion see D. E. Knouth, *The Art of Computer Programming*, Addison-Wesley Publishing Co., Reading (1973).
5. For a review of Monte Carlo methods see *Monte Carlo Methods*, ed. by K. Binder, Springer-Verlag, Berlin, (1979); see also G. Parisi in *Les Houches Session XLIII 1984*, ed. by K. Osterwalder and R. Stora. North-Holland, Amsterdam, (1986).
6. There is a small gap in the argument: $\hat{R}$ may have a negative eigenvalue equal to -1; in this case $u_n(C)$ would have two different limits when n goes to infinity, one for even n, the other for odd n, neither of them being $u_{eq}(C)$. In order to exclude this unpleasant situation it is handy to consider $\hat{R}^2$, whose eigenvalues are obviously nonnegative; if the ergodicity condition (20.7) holds when restricted to even $m+1$, $u_{2n}(C)$ [and by consequence $u_{2n+1}(C)$] goes to the equilibrium distribution $u_{eq}(C)$ when n goes to infinity.
7. The balance condition requires only that

$$\exp[-\beta H(C)] \equiv \sum_{C'} \text{TP}(C, C') \exp[-\beta H(C)]$$
$$= \sum_{C'} \text{TP}(C', C) \exp[-\beta H(C')].$$

The balance condition is a weaker condition than the detailed balance: It implies only that the equilibrium probability is a stationary probability.

Index

Absolute temperature, 7–12, 319
Analytic continuation, 202
Anharmonic oscillator, 311–313
Antisymmetry, 278
Appel's comparison theorem, 60, 204
Asymptotic freedom, 174, 197
Attraction basin, 118, 119
Auxiliary field, 294
BHP theorem, 155, 165
Backward velocity, 336, n. 3
Baker-Hausdorf relation, 262
Balance condition, 342, 350, n. 7
Bernoulli numbers, 247
Bohr-Sommerfeld equation, 267
Boltzmann distribution, 2, 300
Bond moving, 132–133
Borel sum, 149, 168, 169
Bose statistic, 275
Brillouin zone, 34, 200
Broken symmetry, 13–17, 114, 183
Callan-Symanzik equation, 152, 168, 172
Canonical transformation, 248
Carlson theorem, 179, n. 6
Casimir effect, 253
Central limit theorem, 113, 323
Characteristic curves, 152, 172
Classical action, 235, 259
Classical equations of motion, 258, 260, 310
Classical field, 284, 309
Classical Hamiltonian, 1
Classical limit, 235, 257–282, 300
Classical trajectories, 265
Co-dimension, 119
Computers, 338

Conformal mapping, 89
Connected correlations, 53
Conservation of energy, 274
Conservation laws, 317, 322
Continuity equation, 329, 336, n. 3
Convexity, 131
Correlation functions, 33–39, 50, 224, 238–240
Correlation length, 36, 121, 225, 286
Counterterms, 73, 167
Critical exponents, 30, 40, 60, 110, 123, 130, 142, 146–150, 175–176, 198
Critical exponents in four dimensions, 173, 174
Critical point, 152, 153, 159
Critical surface, 120
Cutoff, 69, 166, 168, 176–178
Decimation, 132
Degeneration of the eigenvalues, 223
Degree of divergence, 84, 165
Detailed balance condition, 341
Determinant, 246–248, 249–253, 260
Diagrams, 52, 58, 61, 77–80, 107, 109, 212
Diagrams (two loops), 92–100, 146
Diffraction, 254, n. 6
Dimensional counting, 74
Discrete spectrum of Hamiltonian, 260
Disorder, 320
Dispersion relation, 261
Dobrushin-Lanford-Ruelle equations, 27
Droplet, 203
Einstein relation, 329
Entropy, 4, 320

Equal-time commutation relations, 241, 284
Equal-time Poisson brackets, 284
Equal-time expectations, 335
Equations of motion, 19, 164, 241–243
Equations of motion (classical), 1
Equilibrium probability, 2
Ergodicity, 317, 341
Ergodicity threshold, 322
Escape trajectories, 121
Essential singularity, 201
Euler-MacLaurin formula, 247
Euler function, 93–94
Exponentially large time, 273, 324
Fermi energy, 277
Fermi statistic, 275
Fix point, 118
Fix point stability, 119, 217
Fixed point, 142, 162, 216
Fixed point Hamiltonian, 128, 130, 138, n. 20
Fotran program, 342
Forward Langevin equation, 327
Forward velocity, 336, n. 3
Fourier transform, 71
Free energy, 3–4, 121–123, 185, 222, 244
Free energy, field dependent, 186
Fugacity, 214, 296
Gap equation, 212
Gas–liquid transition, 215
Gaussian model, 27, 116, 126, 127, 243
Goldstone boson, 193
Goldstone-Salam-Weinberg theorem, 192, 206, n. 20

351

352 Index

Grand partition function, 214
Green functions, 235, 237, 243, 258
Ground state energy, 244
Hamilton-Jacobi equation, 235
Hamiltonian formalism, 284
Hard spheres, 316
Harmonic oscillator, 227–229, 243–247
Hartree method, 276
Heisenberg representation, 234
Heisenberg model, 23
Heisenberg principle, 249
Helium, 208, n. 34
High-temperature expansion, 211
Hilbert-Schimdt operator, 221, 233, n. 3
Holder functions, 254, n. 7
Imaginary time, 240, 249, 275, 301
Infinite dimensions, 36, 213, 213–214
Infinite volume limit, 2, 41–42, 123, 136, n. 2
Information, 6
Infrared divergences, 105, 169
Integral operator, 221, 230
Intersection of two sets, 138, n. 20
Intersections of random walks, 297
Irreversibility, 321
Ising model, 23, 46–48, 59–64, 130, 207, n. 28, 209, 222, 225, 334
Ito differential calculus, 336, n. 2
Jumping trajectory, 272
Kubo-Martin-Schwinger condition, 302, 307, 319
Large momentum region, 150, 154
Large-scale computations, 338
Lattice spacing, 70, 229
Legendre transform, 187
Lennard-Jones potential, 339
Levinson theorem, 270
Limiting cycle, 137, n. 17
Linear response, 12, 33, 303–307
Local operator, 167
Longitudinal susceptibility, 192, 195
Long rang forces, 205, n. 11
Lorentz group, 283
Low-temperature expansion, 46–48, 195–197
Markof chains, 340
Mass, 290

Mass gap, 291, n. 2
Mass insertion, 157
Massless theory, 169
Maxwell relations, 12
Mean field, 24–30, 183, 210, 213
Mermin-Wagner theorem, 188
Metric, 283
Microcanonical distribution, 3
Mixing, 317
Momentum space, 80
Monte Carlo renormalization group, 138, n. 26
Non-Gaussian fixed point, 135, 136, 140, 146
Noninteger dimensions, 163
Noninteracting fermions, 275
Nonrenormalizable theory, 167
Normalization factor, 230, 233, n. 11, 246
Number of bound states, 264
O (N) invariant theory, 108, 146, 147, 161, n. 7, 174, 187–201, 219
One line irreducible diagrams, 187
Order, 320
Oscillations in eigenvalue density, 265–267
Periodic boundary conditions, 222, 301
Periodic trajectories, 238
Perturbative expansion (asymptotic behavior), 149
Perturbative expansion nonanalytic terms, 160, 162, n. 21, 17
Phase shift, 268
Planck formula, 314
Poisson bracket, 307
Poisson distribution, 273
Poisson formula, 101, 313
Positivity condition, 287
Power-law decay, 201, 205, n. 6
Proper time, 298, n. 1
Pure states, 17
Quantum Hamiltonian, 234
Quantum probability distribution, 300
Random force, 326
Random numbers, 342
Random potential, 279
Random walk, 54–58, 293
Real gas, 213–215
Recursion equation, 113, 134
Renormalizable, 167
Renormalization conditions, 141, 170
Renormalization constants, 141, 142, 151, 156, 171

Renormalization group, 144, 171–173, 181, n. 25
Renormalization group in momentum space, 139, n. 34
Renormalized correlation functions, 150, 154, 156
Renormalized field, 141, 170
Renormalized coupling constant, 141, 159, 198, 216, 226–227
Renormalized coupling constant, Monte Carlo determination of, 161, n. 6
Renormalized perturbative expansion, 154–159
Resolvent of the Hamiltonian, 261
Response function, 303–304
Rigorous bounds, 87
Running coupling constant, 152, 161, n. 16, 173, 177, 200
Scaling corrections, 144
Scaling laws, 116, 121–126, 141, 150–154, 233, n. 7
Schrödinger equation, 231, 235, 330
Self-energy, 145, 155, 168
Self-repulsion of relativistic particles, 296
Square well potential, 249
Stochastic differential equation, 235
Subdiagrams, 155, 165
Superficial convergence, 84
Superposition principle, 236
Superrenormalizable theory, 167
Surface tension, 208, n. 41
Thomas-Fermi equations, 277
Time dependent correlations, 301
Time ordered product, 239, 242
Transfer matrix, 221–232, 288
Transfer matrix in many dimensions, 232
Transition probability, 340
Transversal susceptibility, 192
Truncation errors, 350, n. 3
Tunneling probability, 272
Two-body interaction, 214, 275
Typical trajectory, 331
Ultraviolet divergences, 72, 83, 144, 165
Upper complex plane, 306
Vacuum fluctuations, 296
Variational principle, 6
Viscosity, 325
Watson theorem, 87
Zero lattice spacing limit, 73–75, 178, 198, 229, 298